Technology and Global Change

Technology and Global Change describes how technology has shaped society and the environment over the last 200 years. Technology has led us from the farm to the factory to the internet. Technology's impacts are now global, and change continues to accelerate. Technology has eliminated many problems, but has added many others (ranging from urban smog to the ozone hole to global warming).

This book is the first to give a comprehensive description of the causes and impacts of technological change and how they relate to global environmental change. It organizes history into a sequence of technology clusters, each with its distinctive environmental "footprint". The result is a new, original explanation of change – illustrated with innumerable quantitative examples, data, and graphics – that makes this book required reading for all now looking to technology for environmental solutions: technologists, environmentalists, policy makers, and academics.

Written for specialists and nonspecialists alike, this book will be useful for researchers and professors, as a textbook for graduate students, for people engaged in long-term policy planning in industry (strategic planning departments) and government (R&D and technology ministries, environment ministries), for environmental activists (NGOs), and for the wider public interested in history, technology, or environmental issues.

ARNULF GRÜBLER is a Research Scholar at the International Institute for Applied Systems Analysis and is one of the world's leading scientists in the history of technological change and its impact on the environment. Afer taking a PhD at the Technical University of Vienna, Austria, he has been a lecturer at various institutions: International Centre for Theoretical Physics, Trieste, Italy; University of Vienna, Austria; Technical University of Graz, Austria; Mining University Leoben, Austria. He is a Lead Author of the Intergovernmental Panel on Climate Change (IPCC), and an Editorial Board Member of the *Journal of Industrial Ecology* and *Technological Forecasting and Social Change*. He is the author, co-author, or editor of five books, three special journal issues, and over 50 peer reviewed articles and book chapters and some 30 additional professional papers in the domains of technological change, energy systems, climate change and mitigation options, and resource economics. His published books include *The Rise and Fall of Infrastructures* (1990, Physica Verlag, Heidelberg), *Diffusion of Technologies and Social Behavior,* (1991, Springer Verlag, Berlin), *Environment and Development* (1994, Jugend und Volk Verlag, Vienna, in German), *Global Energy Perspectives to 2050 and Beyond* (1995, World Energy Council, London).

The International Institute for Applied Systems Analysis

is an interdisciplinary, nongovernmental research institution founded in 1972 by leading scientific organizations in 12 countries. Situated near Vienna, in the center of Europe, IIASA has been for more than two decades producing valuable scientific research on economic, technological, and environmental issues.

IIASA was one of the first international institutes to systematically study global issues of environment, technology, and development. IIASA's Governing Council states that the Institute's goal is: *to conduct international and interdisciplinary scientific studies to provide timely and relevant information and options, addressing critical issues of global environmental, economic, and social change, for the benefit of the public, the scientific community, and national and international institutions.* Research is organized around three central themes:

- Global Environmental Change;
- Global Economic and Technological Change;
- Systems Methods for the Analysis of Global Issues.

The Institute now has national member organizations in the following countries:

Austria
The Austrian Academy of Sciences

Bulgaria*
The Bulgarian Committee for IIASA

Finland
The Finnish Committee for IIASA

Germany**
The Association for the Advancement of IIASA

Hungary
The Hungarian Committee for Applied Systems Analysis

Japan
The Japan Committee for IIASA

Kazakhstan*
The Ministry of Science –
The Academy of Sciences

Netherlands
The Netherlands Organization for Scientific Research (NWO)

Norway
The Research Council of Norway

Poland
The Polish Academy of Sciences

Russian Federation
The Russian Academy of Sciences

Slovak Republic*
The Slovak Committee for IIASA

Sweden
The Swedish Council for Planning and Coordination of Research (FRN)

Ukraine*
The Ukrainian Academy of Sciences

United States of America
The American Academy of Arts and Sciences

*Associate member
**Affiliate

Technology and
Global Change

by

Arnulf Grübler

**International Institute for Applied Systems Analysis
Laxenburg, Austria**

CAMBRIDGE
UNIVERSITY PRESS

PUBLISHED BY THE PRESS SYNDICATE OF THE UNIVERSITY OF CAMBRIDGE
The Pitt Building, Trumpington Street, Cambridge, United Kingdom

CAMBRIDGE UNIVERSITY PRESS
The Edinburgh Building, Cambridge CB2 2RU, UK
40 West 20th Street, New York NY 10011–4211, USA
477 Williamstown Road, Port Melbourne, VIC 3207, Australia
Ruiz de Alarcón 13, 28014 Madrid, Spain
Dock House, The Waterfront, Cape Town 8001, South Africa

http://www.cambridge.org

First published 1998
First paperback edition 2003

A catalogue record for this book is available from the British Library

ISBN 0 521 59109 0 hardback
ISBN 0 521 54332 0 paperback

Meinen Eltern gewidmet

Contents

Acknowledgments

Writing a book is largely a lonely exercise and also a lengthy one.

First, therefore, I must thank my family and friends for their understanding as they watched this book steal much more of my time than originally planned. I dedicate this book to my parents; their love has endured more than five decades – longer than an entire technology cycle.

This book has benefited enormously from the encouragement and assistance given generously by a large number of friends and colleagues who have graciously lent their time and expertise and have contributed written material, reviews, and helpful comments. For their complementary perspectives, which are included in boxes throughout the book, I am particularly grateful to Stefan Anderberg, Jesse Ausubel, Dominique Foray, Jean-Paul Hettelingh, Nathan Keyfitz, Sten Nilsson, Max Posch, Hans-Holger Rogner, and David Victor.

Six anonymous reviewers helped to shape the book's structure, content, and exposition. Very useful and constructive comments were provided also by Dominique Foray, Helga Nowotny, Max Posch, Vernon Ruttan, and David Victor. David's especially helpful and productive comments were most valuable in helping me overcome the inevitable writer's fatigue at the very end of this project. I have tried to be responsive to all comments and criticism (the book is some 150 pages longer than originally planned) and hope that all who have provided valuable input will forgive my remaining shortcomings and occasional stubbornness in not incorporating every comment and suggestion.

Finally, I am deeply indebted to all those who made the production of this book possible. Foremost I wish to thank Alan McDonald, whose editorial and writing skills transformed the original draft – with all its endless sentences and grammatical assaults on the English language – into something readable. Special thanks go to members of the IIASA Publications Department for their professionalism, patience, and untiring efforts in reworking numerous drafts and revisions. Ewa Delpos, Jane S. Marks, and Lilo Roggenland have created this book out of my confused electronic bytes, and Eryl Maedel at IIASA and Matt Lloyd at Cambridge have always been understanding and helpful both in keeping me reasonably close to deadlines and in adjusting to inevitable delays (though how important is one year's delay when writing about 300 years of technology and global environmental change?). I would also like to thank Nebojša Nakićenović and my colleagues in the ECS project for stimulating discussions over the years and for tolerating, on occasion, my encroaching on project time to complete this book. Finally, I wish to thank IIASA's former Director, Peter de Jánosi, who helped to put this book project on a solid track, and the current IIASA Director, Gordon J. MacDonald, for support during the final phase.

Chapter 1

Introduction

1.1. Purpose

The primary purpose of this book is to interest the reader in technology. Interest leads to curiosity and further study. The secondary purpose is to deepen the reader's understanding of technology and of the dual role of technology as a source and remedy of global (environmental) change. At the end of the book's "technology journey" it is hoped that readers in general, and students, researchers, and practitioners in particular, can more fully incorporate technology issues in their reflection, conceptualization, analysis, modeling, and ultimately policy formulation when addressing global change.

My personal motivation for writing this book was dissatisfaction with the treatment of technology in studies, scenarios, models, and textbooks about global change. At worst, technology is entirely ignored, or treated as an "externality" that falls from heaven rather than evolving from *within* our societies and economies. At best, technology issues are included as an afterthought in a "pro forma" chapter or as an *ex post* model sensitivity analysis. Technology relates to all major drivers of global change such as population growth, economic development, and resource use. Technology is also central in monitoring environmental impacts and implementing response strategies. There are no textbooks on technology and global change, and technology's treatment in global change models is also rather poor. These are gaps this book hopes to start to fill.

But its principal objective remains generating interest and curiosity by hosting a guided tour through 300 years of technology. This should provoke new ideas and inspire some humility given technology's history of uncertainty, surprises, and persistently wrong forecasts. Such humility may be particularly useful considering the current fashion for creating 100 year scenarios of future global change.

1

1.2. Approach

1.2.1. What is global change?

The now omnipresent term "global change" is very recent, although global change research has already moved into the category of billion dollar mega-science (OECD, 1994).[1] The term traces its origins to the scientific planning for the 25th anniversary of the first International Geophysical Year. The first International Geophysical Year took place in 1957–1958 and stimulated, among other things, C. David Keeling's first measurements of atmospheric CO_2 concentrations at Mauna Loa, Hawaii. Preparatory work for the 25th anniversary celebration started in 1982 (for an excellent personal account see Malone, 1995). It was there (i.e., in the preparatory committee reports) that the term "global change" first surfaced. In 1986 the International Geosphere–Biosphere Program (IGBP) was formally created to further the understanding of the physical, chemical, and biological systems that regulate Earth's environment and of the role of human activity in changing that environment.

But the origins of the concept of "global change" are not only scientific. Perceptions of the planet as a complex, self-regulating system, operating as if it were a single complex organism (e.g., the "Gaia" hypothesis of Lovelock, 1979) were emerging. With the publication of models simulating "world dynamics" (Forrester, 1971) and "limits to growth" (Meadows *et al.*, 1972), popular culture incorporated the notion of Earth as a closed system, in which natural resources and the environment imposed severe limits on population and economic growth. The accompanying "icon" was provided by technology, i.e., by the first space missions' photographs of Earth as a small, blue shimmering planet, surrounded by the dark hostile emptiness of space.

Global change was originally firmly in the hands of the natural sciences: geology, physics, atmospheric chemistry, hydrology, soil science and plant biology, etc. The focus was on planetary processes and transformations. At the core of the research was the *environmental* change induced by anthropogenic activities in planetary processes and in the "grand cycles" of carbon, sulfur, etc. However, it was soon recognized that the "global environment" encompasses not only processes operating on a planetary scale,

---

[1]The International Group of Funding Agencies for Global Change Research (IGFA), an informal gathering of agencies in 25 countries, estimates that total global change research funding (institutional and project related) exceeded US$2 billion in 1990. However, funding for core global environmental change projects, particularly the International Geosphere–Biosphere Program (IGBP) and the World Climate Research Program (WCRP), is much smaller, about US$220 million in 1993 (OECD, 1994).

such as climate change, but also processes operating on shorter time and smaller spatial scales that assume planetary importance because of their pervasiveness. This category includes, for example, water pollution and acidic precipitation, which happen all over the world in different local environments. Moreover, the natural sciences had to acknowledge that any analysis of environmental changes must also address the anthropogenic causes that lead to these changes in the first place. Increasingly, therefore, all major social science disciplines are, to different degrees, now involved in global change research (cf. HDP, 1990).

"Global change" thus encompasses much. In this book we adopt a deliberately wide definition of global change as: *transformation processes that operate at a truly planetary scale* plus *processes that operate at smaller spatial scales* (local, regional, and continental) but *that are so ubiquitous and pervasive as to assume global importance.* The prime example of the first type of global change is global warming. A prime example of the latter is urbanization and urban air pollution.

We also do not limit our discussion to causes of environmental change. The reason is that technology's influences on the environment are sometimes direct and sometimes indirect. For example, the invention of chlorofluorocarbons (CFCs) that destroy the stratospheric ozone layer was a technological change with direct environmental impacts, as CFCs do not exist naturally, and without technology they could not be produced or released into the environment. But in most cases technology's influence is indirect. It affects the environment primarily by influencing the type, magnitude, and spatial location of human activities, as well as their constraints. Even in the absence of modern technologies, agriculture, for example, has transformed local environments for millennia. Modern biological, chemical, and mechanical innovations in the form of new crops, pesticides and machinery have made it possible to intensify agricultural production and provide more food for a rising population. In turn, population growth and the increased demand for agricultural products have required the expansion of croplands and ever larger environmental transformations. With trade, enabled by modern transport technologies, these transformations are no longer confined to areas of population growth, but occur in locations thousands of kilometers away. With a global agricultural system, localized changes have become global (i.e., ubiquitous and pervasive) phenomena.

Thus, "global change" as used in this book refers to both *direct* environmental changes and those that result *indirectly* from changes in the spatial patterns of human activities, from changes in production and output, and from changes in consumption patterns. The distinction can be illustrated by considering the human enterprise as a machine transforming inputs to outputs. Environmental inputs include nature's resources. Outputs are effluents such as degraded resources and pollutants from production and consumption. Technological change can influence the input/output relationship directly by changing the type and quantity of flows and thereby lessening or amplifying *direct* environmental impacts. For example, a new production process may use only half the original amount of water, but require an additional rare metal catalyzer that will need to be mined (generating wastes of its own). Alternatively, technological change acts indirectly through its influence on nonenvironmental inputs and outputs. A new production process that halves the labor or capital required to produce a good, allows us either to maintain existing production levels but halve the inputs, or to double production (and consumption) and maintain existing input levels. Such changes are at the heart of the vast increases in economic output since the Industrial Revolution. They result from the productivity increases enabled by continuous technological change. Increased human activity, the changing spatial patterns of human activity, and the associated environmental impacts are examples of *indirect* global change impacts of technology.

Of course, neat distinctions between direct and indirect impacts are more difficult to draw in practice than in theory. In most cases the two go hand in hand. But precise distinctions are less important than continued attentiveness to impacts *beyond* the direct environmental inputs and outputs of the human enterprise. From this perspective "global change" refers not only to environmental transformations, but to all changes that affect opportunities and constraints for human sustenance on this planet.

The indirect impacts of technology and the feedback effects these have on technology itself make it impossible to draw simplistic linear causality chains between technology and global change. Simplistic notions of "demand pull" or "supply push" are inappropriate in explaining technological change as we will see in Chapter 2. In fact, technology is now so pervasive that it affects, and is affected by, nearly every aspect of our societies. The appropriate mathematical metaphor is that of multiple dynamic feedbacks, and the appropriate conceptual model is that of coevolutionary processes that shape technology and are shaped by it at the same time. Such a coevolutionary perspective may frustrate those seeking simple linear cause–effect relationships to guide targeted technology planning and forecasting. It remains a huge intellectual challenge to incorporate the numerous interrelationships among technology, the economy, society, and the environment in theory, models, and policy. These interrelationships are at the core of inevitable "surprises" that characterize the history of technology, and no doubt its future too.

1.2.2. Technology and global change

Consistent with the broad definition of global change adopted in this book, no individual technological artifact, as important or fascinating as it might be in its own right, is so important that it can be seen as the single driver of "global change". Neither the steam engine in the 19th century, nor the automobile in the 20th, was solely responsible for the pervasive transformations that occurred with urbanization, the expansion of fossil fuel use (and carbon emissions), and the growth of a mass consumption society. The steam engine and automobile were linked to these developments and advanced them, but numerous additional changes – some technological, some not – contributed to a final outcome that classifies as "global change".

Thus, whole bundles of technologies, or what we later refer to as "technology clusters", are needed to explain the historical record of pervasive transformations within societies, affecting how goods are produced, where people live, what materials are created or grown or dug out of the ground, and what environmental impacts we suffer. Thus, throughout this book we adopt a macroscopic perspective, focusing on "grand" patterns and technology *clusters* rather than on individual technologies.

Our focus on global change also dictates that we will deal mostly with technological winners that have "made it". They have diffused pervasively, in fact so pervasively as to contribute to global change. Our emphasis is therefore on technology *diffusion*, i.e., the widespread adoption of technologies over time, in space, and between different social strata. Understanding diffusion is crucial. Only through diffusion do technologies exert any noticeable impact on output and productivity growth, on economic and social transformations, and on the environment. Without diffusion, a new technology may be a triumph of human ingenuity, but it will not be an agent of global change.

Technology diffusion does not follow a single uniform pattern. The process tends to last the longest in the region where a technology originates, which we call the innovation center. Regions where diffusion begins later see a quicker diffusion process as they "catch up" with the innovation center. The extent of diffusion within a region, the adoption level, tends to be highest in the innovation center. In the "catch-up" regions, diffusion times are shorter, but adoption levels are generally also lower. In the countries where railways and automobiles were first introduced, for example, they took nearly 100 years to reach maturity. Late adopters began several decades later, but diffusion took only a few decades instead of an entire century. The intensity of use of late adopters was, however, lower than in the innovation centers for both railways and automobiles. The timing of diffusion also sets the pace for pervasive technological change, i.e., the emergence of the sort of technology clusters that determine global change. Important technology clusters needed several decades to develop initially, and about half a century to reach maturity in the innovation centers and to diffuse at the international level. Altogether, the overall temporal envelope of any particular technology cluster spans up to a century, with its main growth period covering about five decades. Such time-scales are comparable to those characteristic of many global change processes.

As an initial illustration, *Table 1.1* summarizes the two dominant technology clusters of the 20th century and their principal direct and indirect global change impacts. These technology clusters are discussed in more detail in Chapters 5, 6, and 7, together with the technology clusters that preceded them and their global change implications.

Our focus on technological "winners", as mentioned previously, does not diminish the importance of looking at technological "losers". A critical dimension of the innovation and diffusion process is the uncertainty and "maze of ingenuity" (Pacey, 1976) that surrounds technological change. Out of this uncertainty new potential solutions emerge. A few of them ultimately catch on and enjoy widespread diffusion, while many more fall by the wayside. Such uncertainty and diversity, including all the technological losers, appear to be essential *prerequisites* for technological evolution rather than a hindrance or "inefficiency".

Our macroscopic perspective focusing on whole technology clusters has some drawbacks. It largely leaves out the specific contributions of individual actors, be they heroic inventors or institutions that promoted or obstructed particular technologies. However, these are the traditional focus of technology history and are well documented in the literature.

This book does not offer suggestions for quick policy "fixes" for steering technological change in any particular direction. The author is personally skeptical about technology "planning", forecasting, and selection by central authority. Recall the enthusiasm surrounding nuclear energy up through the 1970s. Nuclear growth potential was considered to be unlimited with future costs declining and electricity becoming "too cheap to meter". Billions of dollars were spent on the development of nuclear aircrafts and ships. Today the picture is very different, and few believe that current nuclear technologies will ever live up to earlier expectations. The experience with nuclear forecasting is typical, and because of the poor track record of such efforts we largely leave out issues of technology "planning", forecasting, and selection by central authority. In so doing, policy issues are largely left out, except in the concluding chapter where they are raised as questions rather than answers.

Table 1.1: Summary of the two important technology clusters of the 20th century and some of their global change impacts. Each cluster spans approximately a period of 50 years, with a 20 year period overlap to the preceding and successive clusters, respectively.

	Technology cluster of period	
	1850–1940	1920–2000
Agriculture (Chapter 5)	Mercantilistic agriculture	Industrialization of agriculture
Industry (Chapter 6)	Heavy engineering	Mass production
Services (Chapter 7)	–	Mass consumption
Dominant energy and transport systems	Coal, railways	Oil, roads
Diffusion geography "Center"	Benelux, England, France, Germany, USA	OECD countries
"Catching up"	Other OECD countries	Eastern Europe, Russia
Illustrative direct/indirect global change impacts	Yield increases: reduced land-use changes ("center")	Yield increases: reconversion of agricultural land ("center")
	Agricultural trade ("export" of impacts to ROW)	Reduced land-use changes (ROW)
	Urbanization ("center")	Urbanization (ROW)
	Urban air pollution, coal smog	Acid rain ("center"), urban smog (ROW)
		Ozone depletion
		Greenhouse gas emissions

Abbreviations: OECD, Organisation for Economic Co-operation and Development; ROW, rest of world.

1.3. Structural Overview

This book consists of three parts. The first defines the basic terminology, concepts, and models used to describe and analyze technology and technological change. The second part gives an overview of technological changes throughout history and how they have affected productivity, output, and the environment. Agriculture, industry, and services and leisure are discussed separately. The third and final part offers conclusions and a postscript illustrating analytical "next steps" that build on the ideas developed in this book. An Appendix is also given that provides the sources for selected data sets used throughout the book.

Throughout the text contrasting perspectives or extensions of selected issues are presented in the form of self-contained boxes. Most are written by others, not out of laziness on the part of the author, but in an attempt to offer alternative views and interpretations. At this point, I should disclose my basic ideological predisposition as one of cautious optimism. I believe that human ingenuity and its manifestation in technology can be better harnessed than in the past to further social and environmental objectives. I recognize there is uncertainty about these objectives that adds to technological uncertainty, but I hope that by the end of the book the reader will share my personal conviction that the goal of better harnessing technological change to our evolving social and environmental aims is very much worth the effort. Whether we as a society are currently prepared to make the necessary investments in research and experimentation, and whether existing institutions are appropriate for the tasks ahead, remain open questions.

1.3.1. Organizational principles

Discussion of global change can be structured in a variety of ways. From an environmental perspective it often makes most sense to use a structure based on different environmental media such as the atmosphere, hydrosphere, pedosphere, etc. Economic perspectives often use a sectoral structure focusing on industrial activities, transport, or the like. Such an environmental perspective was used, for example, in the second assessment report of the Intergovernmental Panel on Climate Change (IPCC, 1996a). It is a natural approach in view of the disciplinary divisions in science, but it has two drawbacks. First, there is always an inherent danger of compartmentalization, of viewing the world too much through a single "lens". Second, it increases the likelihood of missing important interdependencies or joint causes of global change that cut across different sectors and environmental media.

This book takes a different approach. The basic organizing principle is the concept of technology clusters, introduced and elaborated in the next chapter. The presentation also extensively incorporates the concept of economic structural change. The French economist Jean Fourastié showed in 1949 (based on the work of Colin Clark first published in 1940) that technological change leads to differential productivity increases across sectors, with particularly high increases in manufacturing. Increasing productivity enables rising incomes and consumer expenditures. These however are not

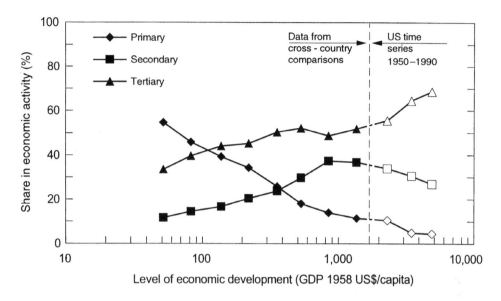

Figure 1.1: Changes in economic structure from primary to secondary and tertiary sectors (percent of economic activity) as a function of economic development (GDP per capita). Source: adapted from Kuznets (1971:111). Note that the last three observations that extend the original Kuznets data refer to the USA for the years 1950, 1970, and 1990.

necessarily spent only on products from the sectors with the highest productivity growth. Rather, they go increasingly to products in the "tertiary" (i.e., services) sector where productivity increases are smaller. This shift in expenditures also leads to shifts in employment among sectors. Fourastié suggested a simple model of economic structural change from activities dominated by the primary sector (resource-intensive activities like agriculture and mining), to secondary sector activities (industry, especially manufacturing), and finally to tertiary sector activities (services). Such shifts in economic activities and employment can be observed both in longitudinal analysis (i.e., evolution within one country over time) and in cross-sectoral analysis (i.e., comparisons across countries at different levels of economic development). *Figure 1.1* illustrates these shifts. They are used in this book to structure the discussion of historical changes in technology clusters and their relationship to environmental change.

The sequence of sectors also corresponds roughly to the *timing* of global change impacts. The first truly planetary transformations were land-use changes associated with agriculture. Agriculture has been transforming land use for millennia, and the environmental transformations associated with agriculture have been global phenomena ever since the 19th century. Industrial activities emerged from locally confined impacts to a source of global change only at the end of the 19th century. Their global change impacts have been especially significant during the second half of the 20th century. Finally, services and leisure activities could emerge as possibly the largest source of global change in the 21st century, given how dominant consumption has recently become relative to production in the most advanced industrialized countries.

The three sectors also illustrate in simplified form different *dynamics* of environmental impacts. Agriculture provides the first example of technology-driven resource savings measured in *absolute* terms. Specifically, yields in industrialized countries have risen to such levels as to allow reconversion of agricultural land to forests while maintaining, even increasing, output. In industry, productivity increases have been enormous, but in absolute terms they have translated at best into a stabilization of material use. Finally, current trends for services are in the direction of increasing resource use. We consume ever more material goods, generate more wastes, drive longer distances, and so on. Whatever technology improvements have been achieved for lowering resource inputs per unit activity, they have been largely offset and overtaken by continued demand growth and changes in consumer behavior.

1.3.2.   Main themes

For each of the three sectors we focus on either a prime resource input or a major environmental impact. For agriculture the principal focus is on land, with a secondary focus on water. These are the two environmental resources for which no easy substitutes can be found, as opposed to energy and other raw materials. Land and water are also the areas where agriculture has the greatest environmental impacts. For industry, the focus is on materials and energy and their main environmental impacts. These include resource depletion, waste disposal and, in the case of energy, atmospheric pollution.

While the agricultural and industrial sectors deal with production, the third sector, services and leisure activities, deals primarily with consumption. In this section our focus will be on transportation both as a main source of global environmental impacts and as the dominant source of changes in spatial patterns of human activities. The section discusses a somewhat novel source of global change: free time and leisure activities. In the case of production activities, productivity increases due to technology can lead toward improved environmental compatibility. In contrast, consumption activities are decentralized and driven by complex motivational structures that to economists often appear "irrational". For these reasons consumption activities do not lend themselves to quick technological "environmental fixes". The ultimate constraints are no longer natural and economic resources and technology, but human preferences and the ultimate constraint on human activities: time.

1.3.3. Summary of the subsequent chapters

Chapter 2. Technology: Concepts and Definitions

The chapter provides an overview of diverse conceptualizations and terminologies that have been introduced to describe technology and how it evolves. First, technology is defined as consisting of both hardware and software (the knowledge required to produce and use technological hardware). Second, the essential feature of technology – its dynamic nature – is outlined. Technologies change all the time individually, and in their aggregate, typically in a sequence of replacements of older by newer technologies. Finally, the chapter emphasizes the multitude of linkages and cross-enhancing interdependencies between technologies giving rise to successive technology "clusters", which are the focus of the subsequent historical analysis chapters. The most essential terminology distinguishes between invention (discovery), innovation (first commercial application) and diffusion (widespread replication and growth) of technologies. As a simple conceptual model the technology life cycle is introduced. In this model, new technologies evolve from a highly uncertain embryonic stage with frequent rejection of proposed solutions. In the case of acceptance, technology diffusion follows and technologies continue to be improved, widen their possible applications, and interact with other existing technologies and infrastructures. Ultimately, improvement potentials become exhausted, negative externalities apparent, and diffusion eventually saturates, providing an opportunity window for the introduction of alternative solutions. Technology diffusion is at the core of the historical technological changes of importance for global (environmental) change. This is why the main emphasis in this book is on technology diffusion, which also provides the central metric to measure technological change. Less emphasis is placed on the complex microphenomenon of technology selection. The main generic characteristics of technological change are presented and some generalized patterns of technology diffusion are outlined. The chapter concludes with a discussion of sources and mechanisms, i.e., the "who's and how's" of technological change.

Chapter 3. Technology: Models

The chapter gives an overview of the efforts to model technological change, which to date have been largely disappointing. Macroeconomic models that treat technology as a residual quantity are discussed first, both in their original classical growth accounting formulation as well as in their contemporary use in macroeconomic energy and environmental models. The chapter then presents sectoral models as well as models based on microeconomic foundations. The latter two types of models offer greater insights and explanations of the dynamics of technological change. These are characterized by features of path dependency, i.e., change in a persistent direction influenced by past decisions, technological uncertainty, diversity, learning, and interaction between economic agents.

Chapter 4. Technology: History

Technological changes since the onset of the Industrial Revolution are summarized. The concept of technology clusters, i.e., a set of interrelated technological, infrastructural, and organizational innovations driving output and productivity growth during particular periods of time is used to explain these changes. Four historical technology clusters are identified, with a prospective fifth, emerging one. The most salient characteristics of each cluster are discussed with illustrative examples. The chapter concludes with a discussion of quantitative and statistical approaches that corroborate the concept of technology clusters.

Chapter 5. Agriculture

An overview of agricultural output and productivity growth is outlined. Three broad historical periods are distinguished. In the first, agriculture improves primarily through biological innovations in the form of new crops and new agricultural practices. In the second, new transport technologies enable agricultural production and trade to expand to a continental and then a global scale. In the third, mechanization, synthetic factor inputs, and new crops, all developed through systematic R&D, push agricultural output and productivity to unprecedented scales. Throughout all three periods labor productivity rises, requiring ever fewer farmers to feed growing populations both at home and abroad. The reduced demand for farmers precedes a related migration from rural to urban areas, labeled urbanization. Progress in agricultural technologies and techniques also progressively decouples the expansion of arable land from population growth and food consumption growth. Initially, this decoupling simply slows down the expansion of agricultural land. Subsequently, international trade effectively transfers the expansion of agricultural land to other countries, limiting further expansion in the industrialized countries. Finally, agricultural productivity increases to such an extent that agricultural land in the industrialized countries can be reconverted to other uses. Thus technological change, combined with saturating demands for food, translates into *absolute* reductions in agricultural land requirements. Technology begins to spare nature. In contrast with its decreasing land requirements, the overall expansion of agricultural production has more problematic impacts on global water use and global nutrient and geochemical cycles. The chapter concludes with a discussion of urbanization and urban environmental impacts. These can be seen as an important indirect impact of the productivity increases in agriculture driven by technology.

Chapter 6. Industry

The chapter starts with a brief quantitative overview of global industrial expansion and the disparities that remain between centers of industrialization and those regions that are catching up. Overall expansion has been enormous. It has been possible only through successive replacements of manufacturing technologies, materials, and energy sources, and through continuing improvements in the organization of industrial production. These changes have yielded enormous productivity gains in labor, materials, and energy use per unit of production. Such productivity gains have sustained increasing levels of industrial output, increased work force incomes, and reduced working time. Productivity gains have also eased the demands on natural resources and reduced traditional environmental impacts such as indoor and urban air pollution. At the same time, however, new environmental concerns have emerged at the global level. Synthetic substances are depleting the ozone layer, and increased concentrations of greenhouse gases, mostly from fossil energy combustion, are causing global warming. Historically, environmental productivity gains have been outpaced by output growth. Only in the last two decades have gradually saturating demands in bulk materials combined with continued productivity increases resulted in near stabilization of materials and energy use in the most advanced industrial countries. The history of energy and carbon use illustrates the predominant pattern. Energy use per unit of economic output has declined by 1% per year, and carbon emissions per unit of energy use has declined by 0.3% per year. This is a combined carbon productivity increase of 1.3% per year. However, economic growth has averaged 3% per year. Thus carbon emissions increased in absolute terms. What is more promising is that until now environmental productivity gains have been the only unplanned side effects of overall technological productivity gains. That these gains follow classical technology learning curve patterns suggests there is a large potential for future environmental productivity gains once they become an explicit objective.

Chapter 7. Services

For the service sector the most important impacts of technological change are changes in how individuals use their time – their "time budgets" – and changes in consumer expenditures. Longer life expectancies, shorter working hours, and vastly rising incomes have changed time budgets and expenditure patterns in ways that have significant environmental impacts. A principal example is increased personal mobility – a consumer demand that appears far from satiated. Increased demands for ever more personal mobility have been largely met by motorized vehicles. Thus emissions from transportation, along with a whole variety of other environmental impacts, have grown substantially. Fortunately, projecting future transportation growth from historical innovation diffusion patterns indicates lower environmental impacts than are suggested by traditional linear extrapolations, assuming business-as-usual. Yet, the growth of the service economy and the consumer society is such that these could soon rival agriculture and industry as major sources of global change. Thus individual lifestyle decisions, particularly decisions about *which* artifacts are used and *how*, become ever more important in determining the type and scale of environmental impacts. One important example described in more detail is that of food. With rising incomes food demands become increasingly saturated. In the industrialized countries, further agricultural productivity increases from biological and mechanical innovations can then be translated into actual absolute reductions in agricultural land use, even while production and exports continue to increase.

Chapter 8. Conclusion

The final chapter summarizes the technology–environment paradox – technology as both source and remedy of environmental change – and mentions technology's additional critical role as an instrument for observing and monitoring environmental change. Examples are presented of how the technology–environment paradox has been resolved, and has reemerged, throughout history. The critical questions are, first, which aspect of technology – as source or remedy of environmental change – currently has the upper hand and, second, how to tilt the scales toward the latter? To answer the first question, the chapter summarizes the balance of evidence from agriculture, industry, and the service sector. Answering the second question requires better models of technological change than we have today. The chapter reviews the major insights from the previous chapters that should be incorporated in improved models and lays out the major challenges that remain. The chapter concludes with a discussion of open issues that remain for a deeper understanding of the interactions between technology and global (environmental) change. Technology's most important historical role has been to liberate humanity from environmental constraints. That job is not complete, and the immediate challenge is to include the billions of people who have so far been excluded from the benefits of technology. The next challenge is to wisely use the power of technology to "liberate" the environment from human interference.

Chapter 9. Postscript: From Data Muddles to Models

The postscript briefly reviews useful theoretical formulations and empirical data that are available for building improved models of technological change. Elements of a stylized model are outlined, emphasizing uncertainty, mechanisms of continual technological improvement, and their influence on technology diffusion and substitution. Uncertainty introduces stochasticity in model formulations. Technological improvement through R&D and learning by doing introduces nonconvexities due to increasing returns. A number of models with these essential features are presented. The chapter concludes with a simplified model that integrates uncertainty, R&D, and technological learning as sources of technological change. The model demonstrates the feasibility of dealing simultaneously with stochasticity and nonconvexity arising from uncertainty and increasing returns from R&D and learning by doing. The postscript concludes with the optimistic outlook that modeling approaches do exist that can improve the traditional treatment of technological change as an "externality" to the economy and society at large.

Chapter 10. Appendix

The Appendix briefly presents data sources and descriptions for representative data sets presented in the preceding chapters that may be useful in coursework and modeling of technological change. After presenting data sources, a description, and formats, instructions are given on how to obtain the data sets in electronic form through internet access.

Part I

What is Technology?

Chapter 2

Technology: Concepts and Definitions

Synopsis

The chapter provides an overview of diverse conceptualizations and terminologies that have been introduced to describe technology and how it evolves. First, technology is defined as consisting of both hardware and software (the knowledge required to produce and use technological hardware). Second, the essential feature of technology – its dynamic nature – is outlined. Technologies change all the time individually, and in their aggregate, typically in a sequence of replacements of older by newer technologies. Finally, the chapter emphasizes the multitude of linkages and cross-enhancing interdependencies between technologies giving rise to successive technology "clusters", which are the focus of the subsequent historical analysis chapters. The most essential terminology distinguishes between invention (discovery), innovation (first commercial application) and diffusion (widespread replication and growth) of technologies. As a simple conceptual model the technology life cycle is introduced. In this model, new technologies evolve from a highly uncertain embryonic stage with frequent rejection of proposed solutions. In the case of acceptance, technology diffusion follows and technologies continue to be improved, widen their possible applications, and interact with other existing technologies and infrastructures. Ultimately, improvement potentials become exhausted, negative externalities apparent, and diffusion eventually saturates, providing an opportunity window for the introduction of alternative solutions. Technology diffusion is at the core of the historical technological changes of importance for global (environmental) change. This is why the main emphasis in this book is on technology diffusion, which also provides the central metric to measure technological change. Less emphasis is placed on the complex microphenomenon of technology selection. The main generic characteristics of technological change are presented and some generalized patterns of technology diffusion are outlined. The chapter concludes with a discussion of sources and mechanisms, i.e., the "who's and how's" of technological change.

2.1.   From Artifacts to Megamachines

What is technology?[1] In the narrowest sense, technology consists of manufactured objects like tools (axes, arrowheads, and their modern equivalents)
and containers (pots, water reservoirs, buildings). Their purpose is either to
enhance human capabilities (e.g., with a hammer you can apply a stronger
force to an object) or to enable humans to perform tasks they could not
perform otherwise (with a pot you can transport larger amounts of water;
with your hands you cannot). Engineers call such objects "hardware". Anthropologists speak of "artifacts".

But technology does not end there. Artifacts have to be produced. They
have to be invented, designed, and manufactured. This requires a larger
system including hardware (such as machinery or a manufacturing plant),
factor inputs (labor, energy, raw materials, capital), and finally "software"
(know-how, human knowledge and skills). The latter, for which the French
use the term *technique*, represents the disembodied nature of technology, its
knowledge base. Thus, technology includes both *what* things are made and
how things are made.

Finally, knowledge, or *technique*, is required not only for the production
of artifacts, but also for their use. Knowledge is needed to drive a car or use
a bank account. Knowledge is needed both at the level of the individual,
in complex organizations, and at the level of society. A typewriter, without
a user who knows how to type, let alone how to read, is simply a useless,
heavy piece of equipment.

Technological hardware varies in size and complexity, as does the "software" required to produce and use hardware. The two are interrelated and
require both tangible and intangible settings in the form of spatial structures and social organizations. Institutions, including governments, firms,
and markets, and social norms and attitudes, are especially important in
determining how systems for producing and using artifacts emerge and function. They determine how particular artifacts and combinations of artifacts
originate, which ones are rejected or which ones become successful, and, if
successful, how quickly they are incorporated in the economy and the society.
The latter step is referred to as technology diffusion.

For Lewis Mumford (1966:11) the rise of civilization around 4000 B.C.
is not the result "of mechanical innovations, but of a radically new type of

---

[1] From the Greek τεχνε (*techne*, art, the practical capability to create something) and
λογοσ (*logos*, word, human reason). Thus, τεχνολογια (*technologia*) is the science and
systematic treatment of (practical) arts. In a most general definition technology is a system
of means to particular ends that employs both technical artifacts and (social) information
(know-how).

social organization: ... Neither the wheeled wagon, the plow, the potter's wheel, nor the military chariot could of themselves have accomplished the transformations that took place in the great valleys of Egypt, Mesopotamia, and India, and eventually passed, in ripples and waves, to other parts of the planet". To describe the organization of human beings jointly with artifacts in an "archetypal machine composed of human parts", Mumford introduced the notion of a "mega-machine", with cities as a primary example.

Some may consider such semantics as philosophical overkill and irrelevant for a book on technology and global change. Others might find in them confirmation of a general uneasiness that technology is something large, opaque, and pervasive, which constrains rather than enhances our choices. Nevertheless it is important to present at the outset the broad continuum of conceptualizations of technology. It emphasizes that technology cannot be separated from the economic and social context out of which it evolves, and which is responsible for its production and its use. In turn, the social and economic context is shaped by the technologies that are produced and used. And through technology humans have acquired powerful capabilities to transform their natural environments locally, regionally, and, more recently, globally.

The circular nature of the feedback loops affecting technological development cannot be stressed too much. All the numerous technology studies of the 20th century share one conclusion: it is simply wrong to conceptualize technological evolution according to a simple linear model, no matter how appealing the simplification. Technological evolution is neither simple nor linear. Its four most important distinctive characteristics are instead that it is *uncertain, dynamic, systemic,* and *cumulative.*

Uncertainty is a basic fact of life, and technology is no exception. The first source of technological uncertainty derives from the fortunate fact that there always exists a variety of solutions to perform a particular task. It is always uncertain which might be "best", taking into account technical criteria, economic criteria, and social criteria. Uncertainty prevails at all stages of technological evolution, from initial design choices, through success or failure in the marketplace, to eventual environmental impacts and spin-off effects. The technological and management literature labels such uncertainty a "snake pit" problem. It is like trying to pick a particular snake out of a pit of hundreds that all look alike. Others use the biblical quote "many are called, but few are chosen". Technological uncertainty continues to be a notorious embarrassment in efforts to "forecast" technological change. But there is also nothing to be gained by a strategy of "waiting until the sky clears". It will not clear, uncertainty will persist, and the correct strategy is experimentation with technological variety. This may seem an "inefficient" strategy for progress. To the extent that it is, it is one of the many areas in which writers have drawn useful analogies between technology and biology.

Second, technology is dynamic; it keeps changing all the time. Change includes a continuous introduction of new varieties, or "species", and continuous subsequent improvements and modifications. The varying pace of these combined changes is a constant source of excitement (and overoptimism) on the one hand, and frustration (or pessimism) on the other. As a rule, material components of technology change much faster and more easily than either its nonmaterial components or society at large. The main factors governing technology dynamics are, first, the continuous replacement of capital stock as it ages and economies expand and, second and most important, new inventions.

Third, technological evolution is systemic. It cannot be treated as a discrete, isolated event that concerns only one artifact. A new technology needs not only to be invented and designed, but it needs to be produced. This requires a whole host of other technologies. And it requires infrastructures. A telephone needs a telephone network; a car needs both a road network and a gasoline distribution system, and each of these consists of whole "bundles" of individual technologies. This interdependence of technologies causes enormous difficulties in implementing large-scale changes. But it is also what causes technological changes to have such pervasive and extensive impacts once they are implemented. From historical research we know particular periods of economic development correspond with clusters of interrelated developments in artifacts, techniques, institutions, and forms of social organization. These mutually interdependent and cross-enhancing "sociotechnical systems of production and use" (Kline, 1985:2–4) cannot be analyzed in terms of single technologies, but must be considered in terms of the mutual interactions among all concurrent technological, institutional, and social change.

Fourth and finally, technological change is cumulative. Changes build on previous experience and knowledge. Only in rare cases is knowledge lost and not reproducible. A new artifact, like a new species, is seldom designed from "scratch". (The beginnings of the space program are a notable exception.) Hence, technological knowledge[2] and the stock of technologies in use grow continuously.

The following chapters emphasize the dynamic, systemic, and cumulative nature of technological change. In describing the history of technological

---

[2]One question is how much of the growth in information represents growth in *usable* knowledge? Rescher (1996) argues unconventionally that (usable) scientific knowledge only grows with the logarithm of the brute volume of scientific information.

change we discuss technological diffusion (i.e., technology's dynamic nature) largely in terms of technological "clusters" or "families", thus also highlighting technology's systemic and cumulative characteristics. We relate these to pervasive transformations in the economy, the spatial division of production, and also to environmental impacts. We have the benefit of hindsight, which conceals to a large extent the considerable uncertainties prevailing at the beginning of each technology cluster.

No claim to originality is made in adopting the notion of technology clusters as the organizing principle here. In 1934 Lewis Mumford characterized four phases of sociotechnical development according to dominant materials and energy sources used from preindustrial times to the 20th century (Mumford, 1934). Mumford's clusters set a useful historical stage, and we will build on them later as we extend the history of technology up through the last 200 years.

2.1.1. Terminology

The Austrian economist Joseph A. Schumpeter distinguished three important phases in technology development: invention, innovation, and diffusion.

Invention is the first demonstration of the principal, physical feasibility of a proposed new solution. An invention is usually related to some empirical or scientific discovery, frequently measured through patent applications and statistics. However, an invention by itself often offers no hints about possible applications despite the technological romanticism surrounding the inventor's human ingenuity. Even where applications are apparent, as in the recent frenzy surrounding the discovery of high temperature superconductivity, an invention by itself has no economic or social significance whatsoever.

Innovation is defined succinctly by Mensch (1979:123) as the point when a "newly discovered material or a newly developed technique is being put into regular production for the first time, or when an organized market for the new product is first created". A distinction is frequently made between process and product innovations. The former refers to new methods of production, for example, the Bessemer process of raw steel production. The latter refers to directly usable technological hardware, for instance, consumer products such as video recorders and compact disc players.

Numerous attempts have been made to discriminate between innovations that might be labeled "radical" or "basic" and others considered of lesser importance. But such distinctions are *ex post* rationalizations. At the moment of innovation proper it is nearly impossible to guess the ultimate or potential significance of an innovation (cf. Rosenberg, 1996). This inherent uncertainty (or inefficiency) is reflected in the fact that only a small percentage of innovations eventually "make it". The success rate is comparable to that of biological mutations. It is an essential feature of the evolutionary character of technological change, and we will return to it later when discussing technology selection.

Diffusion is the widespread replication of a technology and its assimilation in a socioeconomic setting. Diffusion is the final, and sometimes painful, test of whether an innovation can create a niche of its own or successfully supplant existing practices and artifacts. Technology assumes significance only through its application (innovation) and subsequent widespread replication (diffusion). Otherwise it remains either knowledge that is never applied, i.e., an invention without subsequent innovation, or an isolated technological curiosity, i.e., an innovation without subsequent diffusion.

One can elaborate on this basic framework of distinguishing between invention, innovation, and diffusion, by identifying additional intermediary steps and important feedbacks. Different methods of knowledge generation can be distinguished. For example, research efforts are classified into basic and applied research. Distinctions can also be made between research, development, and demonstration (RD&D). Distinctions can be made between radical and incremental innovations. The latter label is given to continuous improvements that extend applications, lower costs, and transfer new technologies into different sociocultural settings. Such continuous improvements are especially important as new technologies, like all innovations, are initially rather crude, deficient, and imperfect. Therefore considerable effort (research, development, marketing, etc.) is required to sustain pervasive diffusion.

Anyone who has driven a Model T Ford will appreciate that the artifact that we call a car today is markedly different from, and definitively easier to drive, than a similar artifact produced at the beginning of the century. Or compare the first brand of instant coffee to the hundreds of varieties that now cater to different tastes in such diverse places as Austria, Brazil, France, Saudi Arabia, and the USA.

In short, nothing could be more misleading than a simple linear model of knowledge and technology generation. To be successful, innovations must be continuously experimented with, and continuously modified and improved. Suppliers and users must work together; information from the marketing department must be fed back to the research lab in order to suggest new promising avenues for both applied and basic research. The appropriate metaphor or model is therefore that of *networks,* operating to generate innovations and to modify and tailor them in the course of diffusion.

2.1.2. Invention and innovation: Chronology and lags

Table 2.1 gives an abridged chronology of the development of railways, a particularly important technological innovation of the 19th century. The chronology is a good example of a long evolutionary line of developments with important precursor technologies and infrastructures. For example, the innovation represented by Stevenson's steam locomotive plant and the first 20 km Stockton & Darlington railway line in 1825 cannot be understood independent of earlier important developments in stationary steam engines and mine railways. *Table 2.1* also illustrates the considerable time lags that can take place in technological developments. For example, 55 years passed between invention and innovation dates of railways.

Although the timing of particular historical events is indeed important, most dimensions of technological development are continuous rather than discrete. They are either rooted in precursor technologies or rely on a confluence of various streams of developments, like the marriage of a new mobile power source (the steam locomotive) to an entirely new infrastructure system (rails). It is particularly the confluence, complementarity, and synergy between various streams of developments that characterize technological evolution. As a simple illustration consider a new product for which applications need to be found, production processes need to be established, materials must be chosen, and so forth. These activities require time and effort, and unless all aspects are addressed successfully, the new innovation may never appear on the market.

Table 2.2 shows a similar chronology for Neoprene, a synthetic rubber used, for example, in diving suits. In this case, more than two decades elapsed between invention and innovation. *Figure 2.1* indicates that, in general, decades are indeed the appropriate unit for measuring invention–innovation lags.

Figure 2.1 also reveals substantial variability. Of the 140 major innovations analyzed by Rosegger, 20 have lags over five decades, but nine have lags of less than a year. *Figure 2.1* includes innovations ranging from the electric railway, the jet aircraft, the telephone, and the transistor, to DDT, dynamite, margarine, and insulin. There is no clear decrease over time of the invention–innovation lags shown in *Figure 2.1*. Any advantage of modern organized R&D at the corporate level must therefore lie with other kinds of innovations rather than those traditionally considered in samples, such as that of *Figure 2.1*, of "basic" or "major" innovations. [Other examples are given in Mensch (1979:124–128) and van Duijn (1983:176–179). For a critical discussion, particularly of the Mensch sample, see Freeman *et al.* (1982) and Kleinknecht (1987).]

Table 2.1: A chronology of invention, innovation, and diffusion of railways.

Year	Event
1769	Watt patents low-pressure steam machine (invention)
1770	Cugnot develops steam-gun vehicle
1790	Read develops steam-powered road vehicle
1800	Watt's patent expires
1804	Evans constructs road steam locomotive
1813	Hadley develops locomotive to ride on rails
1814	Stephenson begins building locomotives
1820	About 40 private horse railways are operated between coal mines and the rivers Tyne and Wear in Northern England (Marshall, 1938)
1824	Stephenson builds first locomotive plant (innovation)
1825	Stephenson opens 20 km Stockton & Darlington line (beginning of diffusion)
1830	Opening of the Manchester–Liverpool railway, national railway network extends over 157 km
1845	UK railway network extends over 3,931 km; 0.2% of coal reaching London arrives by rail
1875	UK railway network extends over 23,365 km, transporting 490 million passengers and 200 million tons[a] of goods; 65% of London's coal arrives by rail
1900	UK railway network extends over 30,079 km
1900–1925	Railways achieve absolute dominance in UK transport market, transporting between 70% and 80% of all passenger- and ton-kilometers of the country; freight traffic reaches all-time peak with 570 million tons (including Ireland) in 1913; passenger traffic reaches its all-time high with 1.5 billion passengers in 1920
1928	UK railway network reaches maximum size with 32,846 km (end of diffusion and beginning of saturation and decline)

[a]Throughout this book ton is defined as metric ton, i.e., equal to 1,000 kg.
Source: Based on Marchetti (1980), and Grübler (1990a:90–122).

A few other illustrations of time lags include the example of nuclear energy in the USA; Fermi's Chicago reactor demonstrated the feasibility of a controlled nuclear fission reaction (invention) in 1942. It was not until 1957, 15 years later to the day after Fermi's demonstration, that the Shipping Port reactor went into operation (innovation).[3] It took over 30 additional years for nuclear reactors to account for 20% of US electricity generation. The prospects for further diffusion are highly uncertain.

---

[3]The sad military equivalent would be the first nuclear test bomb explosions and the first application in warfare, i.e., Hiroshima in 1945.

Table 2.2: Events in Neoprene development.

Year	Event
1906	Julius A. Nieuwland observed the acetylene reaction in alkali medium and worked for more than 10 years on the problem of higher yield of the reaction (invention)
1921	Nieuwland demonstrates that his material, "divinylacetylene", a polymer, can be produced through a catalytic reaction
1925	E.K. Bolton of Du Pont listens to a lecture of Nieuwland at the American Chemical Society; Du Pont assumes the further development of this type of rubber material
1932	E.I. Du Pont de Nemours and Company introduces Neoprene, a synthetic rubber, onto the market as a new, commercial product (innovation)

Source: Mensch (1979) based on Jewkes *et al.* (1969).

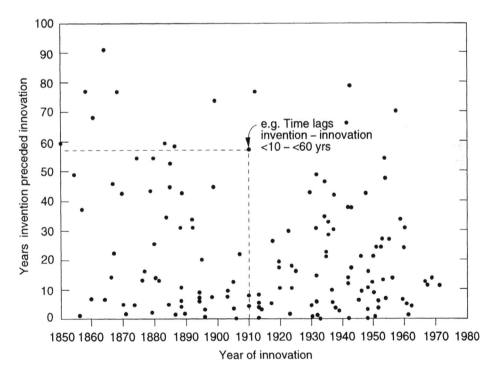

Figure 2.1: Time lag (in years) between invention and innovation of a sample of 140 major innovations introduced in the period 1850–1970. Source: Rosegger (1996:175).[1]

Postage stamps were first introduced in England in 1840 (innovation), but it took close to 50 years for a sample of 37 independent European, North American, and South American countries to follow suit (Pemberton, 1936). Compulsory school attendance in the USA was first introduced in 1847. It took until 1927 for the final state to follow suit.

These examples illustrate that changes in technologies and social techniques are not one-time discrete events. Technologies and techniques are neither developed nor changed instantaneously. Technology development is characterized by considerable time lags between development, first implementation, and widespread replication; all requiring considerable effort. Technology is not free. It is the result of deliberate research and development in university, government and private laboratories and by creative individuals. It requires cooperation between suppliers and users of new knowledge, between suppliers and users of technologies, and between proponents and opponents of particular technological solutions. Freeman (1994) provides an excellent review of recent research[4] identifying important linkages that exist between demand and supply, between users and providers of technology, between private and public R&D, and between knowledge and competencies internal to firms and those outside them. All of these shape the patterns and timing of invention and innovation.

2.1.3. The wider context of technology

In this section we present some general overall tendencies of technological evolution in the course of history. Counterexamples exist, and we admit that the discussion is not entirely free of our own analytical and personal biases. Nonetheless it provides a wider context of technological evolution that will be useful for the reader forming his/her own opinion of respective "progress"[5] or "regress" in the subsequent discussion.

Four general tendencies are identified:

- Increasing scale (cf. *Figure 2.2*), output, and productivity.
- Increasing variety and complexity.
- Increasing division of labor, both functionally and spatially.
- Increasing interdependence, interrelatedness and "network externalities".

These four tendencies should be seen not only as consequences of technological development, but also as resulting from technological "expectations"

---

[4] For a concise perspective from industry cf. Frosch (1984:56–81).

[5] For a critical appraisal of the value-laden concept of technical "progress", see Marx and Mazlish (1996).

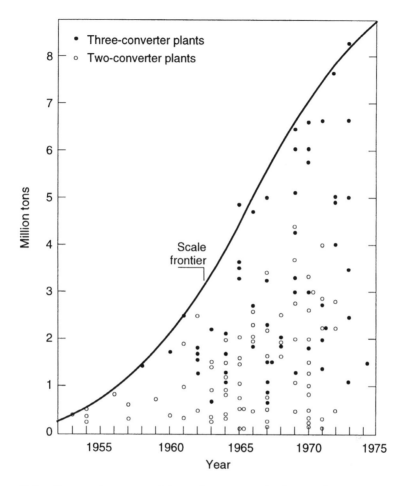

Figure 2.2: Increasing capacities of new steel plants (basic oxygen furnaces), in million tons. Source: adapted from Rosegger (1996:101).

(Rosenberg, 1982) that explicitly or implicitly shape the visions, missions, and expectations of those involved in the "technology business". We will return to this point in Section 2.3 when discussing entrepreneurship as a source of technological change.

Increasing Scale, Output, and Productivity

Increasing output, productivity, or efficiency is both a prime motivation and an effect when creating a new artifact. Increases can be quantitative or qualitative. A new production process can increase output either by scaling up existing production, or by reducing costs and thereby stimulating demand.

Economies of scale have been a pervasive phenomenon in increasing industrial output and lowering production costs. [Economies of scale exist when
production costs increase less proportionally than the size of a production
unit or a plant. Thus the costs of a 4-million-ton steel plant will be lower
than the costs of two separate 2-million-ton plants. It is useful to distinguish
(technology driven) economies of scale from (price driven) *economies of size*.
In the latter, changing relative prices can lead to a different use of factor
inputs, e.g., land versus labor in a farm, with technologies and production
techniques otherwise unchanged. For instance, substitution of capital for
labor in farming can lead to increasing farm size even in the absence of
economies of scale proper.] *Figure 2.2* illustrates the extent to which the
"scale frontier" has been pushed in oxygen steelmaking.

Other sources of output growth include growth in productivity and efficiency that enable to overcome resource limitations or to lower costs (and
prices). Historically, growth in productivity and efficiency (lower input requirements per unit output) in most cases has led to increases in output
rather than maintaining existing output levels and reducing inputs.

Improvements in economies of scale, productivity and efficiency do not
come "automatically". They require engineering effort and experimentation.
Such efforts and experimentation are an important source of technological
learning and subsequent performance improvements.

A good example of an improvement that cut costs and stimulated demand comes from Henry Ford. With the assembly line he introduced standardized mass production to an industry characterized by small-scale production of customized items. That, after all, was how the automobile's
predecessor, the horse carriage, had been produced. Reducing complex operations to a sequence of well-defined routinized jobs also enables better
quality control and more focused learning and improvements in work routines. These, in turn, lead to further cost reductions.

Together with new materials (steel sheets), new forms of management and production organization (e.g., Taylorist time metering and
optimization),[6] the Fordist assembly line reduced the selling price of a
Model T Ford from US$850 in 1908 to US$290 in 1926 (Abernathy, 1978).
This was possible despite increased wages to compensate for the increased
work pressure that accompanied stepped-up output. The Model T production was standardized to such an extent that Henry Ford's quote that

---

[6]Frederick Winslow Taylor (1856–1915) developed a system of scientific management,
primarily aimed at increasing labor productivity. The exact analysis and timing of production and work patterns, improvements in machinery, organizational changes, as well as
financial incentives (bonuses) are characteristic elements of "Taylorism".

consumers "can have any color, provided it is black" became proverbial. To-day even a "Fordist" assembly plant is run to provide substantial varieties of car models, colors, additional equipment, engines, and the like. New forms of production organization have also increased output, variety, and quality further. Volvo in Sweden, for example, pioneered a system combining as-sembly line operation with small assembly work teams. The result combines high output and productivity with more diverse and varied job responsi-bilities, thereby raising work satisfaction, lowering absenteeism, and raising productivity.

Output increases are not confined to industrial production. They also apply to new products and services. In industrialized countries, items such as the telephone, radio, television, home video recorder, and microwave oven are now standard equipment in most households. These expand people's communication and entertainment options, both quantitatively and quali-tatively. Enlarging consumer choices at reasonable costs creates precisely the demand to sustain increases in output. There is no mass production without mass consumption. Mass consumption, in turn, may have powerful environmental consequences – but that is a topic for a later chapter.

Finally, output increases qualitatively. Even if the number of cars or computers produced were constant, increases in performance, features, and designs would all increase output. Volumes and prices do not capture the full story of output growth. The comfort, safety, and reliability of today's cars relative to their ancestors are as different as a Pentium PC from a 286 model, dubbed "advanced technology" at the moment of introduction. Both old and new "run", but they "run" very differently. This presents serious problems in macroeconomic growth accounting, to which we will return when we turn to modeling issues. In emphasizing qualitative improvements we recognize it is not always easy to distinguish between quantity and quality. When consumers switched from black-and-white to color TV sets, for example, the black-and-white sets were often not scrapped. Instead they were moved to the basement or a secondary residence. Therefore, as a result of qualitative changes, the total number of TV sets in use increased also.

In addition to increasing output, technological change can also reduce in-puts. Producing the same with less means a rise in productivity (efficiency), and historical productivity gains in terms of input reduction per unit of out-put have indeed been impressive. Industrial labor productivity (discussed in more detail in Part II) has increased by a factor of 200 or more since the middle of the 18th century. What took two weeks of work at 12 hours per day 200 years ago, is now produced in one hour. The energy requirements for producing a ton of iron or steel have dropped by a factor of more than 10 in the last 100 years.

Productivity gains are thus a central mechanism for improving the efficiency with which natural resources are used, and thereby reducing environmental impacts. But input reduction and output expansion often go hand in hand, and increases in productivity do not always lead "automatically" to resource conservation. Where productivity gains overcome resource constraints on further growth, output and its environmental impacts can expand. Technology is thus a double-edged sword in cutting the Malthusian resource limitation knot. Productivity increases have helped historically to overcome resource constraints so successfully as to expand output to unprecedented scales. Output has risen to such an extent as to face yet new limitations. Some are familiar input constraints on land, materials, and energy. But some are less familiar, such as limits on environmental capacity to absorb production and consumption wastes from ever larger output volumes.

Increasing Variety and Complexity

Another driver – and consequence – of technological change is increasing variety and complexity. Modern industrial systems produce not only a greater volume, but also an ever increasing variety of products. To the extent that variety multiplies a product's markets, it can generate cost reductions and profits. Thus economists speak of "economies of scope", in addition to economies of scale discussed previously.

The great variety of cars, computers, and travel packages to the remotest parts of the planet prove that mass production and standardization need not mean standardized products. There needs to be a functioning market that responds to consumer tastes for variety, as evidenced by the limited variety of consumer products in the former USSR. And much product variety may be classified as "pseudo-innovation", providing superficial variations in design or color, serving competitive and advertising strategies of firms. Consider the differences in the results of using alternative detergents in comparison with, for example, the marketing and advertisement effort devoted to different brands. Variety is exploding. The average number of items on sale in a typical large US supermarket has increased from 2,000 in 1950 to 18,000 items in the 1990s (Ausubel, 1990). The number of new items introduced into US grocery stores in 1993 alone totaled 17,000 (Wernick *et al.*, 1996). Of course, not all were successful. Westinghouse Electric Co. produces over 50,000 different steam turbine blade shapes, and the IBM Selectric typewriter, consisting of 2,700 parts, could be made in 55,000 different models (Ayres, 1988).

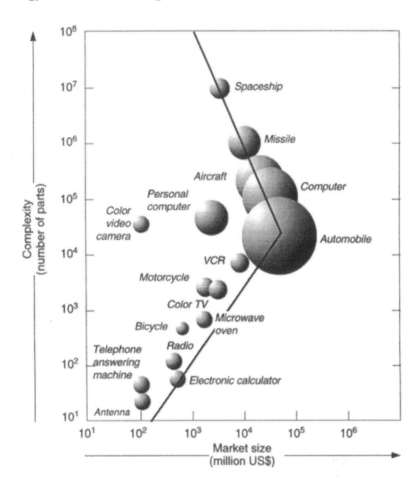

Figure 2.3: Market size (million dollars) and complexity (number of parts per item) of major durables produced in the USA. Source: Ayres (1988:28) based on Nagayama and Funk (1985).

Complexity is also increasing. Early hand tools like hammers, tongs, and shears typically involved two or three parts. A late 19th century hand drill accommodating various drill diameters involved 20 parts. A modern electric drill, including the motor, may have up to 100 parts. Vehicles are even more complex. The 1885 Rover safety bicycle consisted of approximately 500 parts, a modern car involves as many as 30,000 component parts, and a Boeing 747 roughly 3.5 million (all data from Ayres, 1988). The apogee (and nightmare) of mechanical complexity is the space shuttle with 10 million parts (see *Figure 2.3*).

Increased complexity means increased risk of production errors and consequent failures. A car with 50,000 components, and a failure rate of 1 per 1,000,[7] means 50 defective components per car. Inspection and quality control systems eliminate many defects, and design safety margins reduce the consequences of those that slip through. Ayres (1988:29) estimates that a single large US car manufacturer provides three billion opportunities for human assembly line error per day. Even with quality control and inspections reducing undetected errors to one in a million, the result would be 3,000 serious undetected production flaws per day, or about one in every three cars. Consumer surveys repeatedly report several manufacturing defects per car, although most are minor. Design safety margins, multiple inspections, and quality controls can be successful in reducing defects and their consequences. But eventually they are limited either by extreme complexity (as in the Challenger space shuttle), or in the case of an aircraft or nuclear power plant, by catastrophic consequences of failure (cf. Perrow, 1984). Multiple safety and backup systems are the usual response strategies, but they come at considerable additional cost.

Production risks due to complexity are only one part of the story. The other is risk due to human error when using the technology. Such risks are perhaps orders of magnitude larger than those from design and manufacturing defects, and they too increase with complexity. The history of large industrial accidents (e.g., Seveso and Chernobyl) reinforces this estimate. Technology, in the form of monitoring, automatic safety shutdown, and detailed safety procedures and protocols, can help reduce risks, but can never eliminate them entirely. Recent trends toward miniaturization (nanotechnology) and biotechnology promise reduced complexity. But biotechnology is still in its early stages, and may yet prove dauntingly complex. Living organisms like humans are, after all, several orders of magnitude more complex than even the most complex technological artifacts.

Increasing Division of Labor

Increasing complexity, sophistication, and skill requirements in both producing and using technologies require specialization. Metal tools, pottery and textiles have long been produced by specialists: craftsmen and craftswomen. Services have also long been provided by specialists: doctors, astronomers, accountants, writers, etc. In economics this specialization is called *division of labor*, enabled by increases in market size as described by Adam Smith in 1776 in his *Wealth of Nations*.

---

[7]The photocopier manufacturer Xerox heralded the success of a substantial reduction on its parts reject rate from 8 to 1.3 per 1,000 (Ayres, 1988:26).

Specialization and division of labor are pervasive phenomena of all societies beyond the neolithic period, so much so that numerous family names like Smith and Miller derive from an ancestor's trade. Historically, the trigger to specialization was a sufficiently large market size, and the spatial concentration of demand, specifically in the form of cities. In the industrial age, output growth and large-scale trade, via modern transport and communication systems, have much the same effect.

Since the transformation to an industrial society, the number of specialized professions has grown substantially. The yellow pages of any larger city, such as Vienna, contain more than 5,000 specialized trades, businesses, and services. Each subdivides into many further professional specializations.

Cities also provide the earliest examples of spatial division of labor. All those listings in the Vienna yellow pages presuppose the existence of a market where the supply of specialized job opportunities and the demand for specialized trades can meet. A book dealer, specializing in antique books of astronomy and geography, may find enough customers in a large city like Paris or New York, but certainly not in a village in the Tyrolean Alps. But spatial division of labor also results from differences in resource endowments and climatic conditions. Copper is mined where deposits are found, and tropical fruit cannot be grown in temperate climates. Much spatial division of labor results from economics. Production moves to where total costs are lowest. All costs need to be considered. An industrial plant can only be located where highly skilled labor is available. Transportation costs and the size of markets can be critical. In many specialized activities "intangible" factors such as proximity and close interaction with clients are important. This explains the existence of "high-tech" zones with high spatial concentrations of specialized firms in the computer and aerospace industries. Taken together, all these factors make location decisions highly complex and worthy of study by geographers, regional scientists, economists, and sociologists. Location decisions also entail a great deal of irreversibility because of the high sunk costs that result in terms of buildings, infrastructure, and personnel recruitment.

Spatial division of labor occurs at all levels: local, regional, national, and international. Many street names in European cities preserve the concentration of specialized trades that once resided there: goldsmiths, butchers, tailors, and traders. "Rustbelts" bear witness to the concentration of the coal, iron, and steel industries in regions of North America and Europe that "rusted away" when these industries declined. But perhaps increasing spatial division of labor is best illustrated by the increase in international trade (see *Table 2.3*).

Table 2.3: Index of growth in volume of world trade (1913=100).

ca. 1700	1
1800	2
1850	10
1900	57
1950	117
1970	520
1990	1,380
World trade (total exports f.o.b.) in 1990	US$3,397 billion
Distribution (%)	
Foodstuff	8.7
Raw materials	5.2
Energy	10.3
Chemicals	8.8
Machinery	35.7
Other manufactured goods	31.3

Abbreviation: f.o.b., free on board.
Source: Rostow (1978:669), Kennedy (1987:414), and IMF (1996:111). For a critical discussion of data sources of these historical estimates see Rostow (1978:663–669).

Total world trade in 1990 was around US$3,400 billion, or 13% of world GDP.[8] Trade is dominated by manufactured goods (75%, including chemicals) and by exports from industrialized countries (72%), mostly among themselves (57% of all world trade). Conversely, the share of primary resources including energy is less than 25% and the share of developing countries is also less than 25%. This asymmetry reflects the much smaller economic output in developing economies, plus low prices for raw materials relative to manufactured goods, thus the unfavorable "terms of trade" experienced by the developing world.

Increasing Interdependence and Interrelatedness

The final and fourth category of features that both drive technological evolution and are a consequence of such evolution covers technological interdependence and interrelatedness. Although difficult to describe and to model, the basic idea is that technologies increasingly depend on one another for both production and use. Consider the personal computer. It is built of hardware that needs to be produced and assembled. To run it, you need software. Switching it on requires an electricity network, with power plants, fuel supply infrastructures, primary energy extraction, and more. Network

---

[8]US$ in this book refers to constant 1990 money and prices, unless otherwise stated.

surfing requires more hardware (a modem), software, a telephone line, a local telephone network, and the internet itself. To ecologists the notion that "everything depends on everything else" might be familiar. However, to students of technology and policymakers, interdependence and interrelatedness create formidable challenges. It is impossible to manage change through attention to just a few "key" technologies.

In fact, because of technological interrelatedness, it may even be easier to manage change where few technologies and related infrastructures exist, such as in many developing countries. Consider, for instance, the example of cellular or satellite telephones that can be put in place everywhere, compared to a conventional telephone network system. This is the essence of the argument that latecomers to development may have genuine advantages too in terms that they can "leap-frog" (Goldemberg, 1991) older technology systems altogether. Conversely, countries "locked-in" to large existing technology systems face difficulties to move rapidly to newer systems. A historical example (England) and model for such entrenchment in old technology systems was first given by the economist Marvin Frankel in 1955 (Frankel, 1955).

As a contemporary example, consider the introduction of "zero-emission" vehicles, already mandated in California. They are not a technological novelty. Applicable inventions and innovations have existed since the turn of the century. Thus the difficulty lies not in producing electric cars, but in solving the chronic problem of power supply and storage. Without significant progress in batteries, for instance, the speed and range of electric cars is severely limited and costs are high. And a new infrastructure is also required for charging or exchanging discharged batteries.

Technologies depend increasingly on infrastructures of transport, energy, and communication. The service these provide is much larger than the usually modest costs charged to users. We notice them most, however, when we miss them most – when they fail. Thus infrastructures and related technologies are important examples of what economists call "network externalities". Consider your telephone: even with all costs paid, it would be useless if only you owned a phone. Rather, the utility of your phone increases with the number of participants in the telephone network and the more people and services you can access, e.g., to enquire about a flight departure, to order a pizza, or to chat with family and friends. Because costs are shared among all participants of the network, but each participant has the full benefits (utility) of being able to communicate throughout the network, the real value of the service remains "exogenous" to the price paid by an individual. This presents serious issues when new infrastructure networks need to be put in place. The high initial costs are incurred when benefits are still comparatively low;

if no one is prepared to incur the initial set-up costs, future benefits cannot arise. Distributive issues are also raised because those who incur the initial high costs are not the same people who reap the ultimate full benefits.

Thus like the air we breathe, for which we pay nothing, but without which we could not exist, infrastructures create important "externalities". These can be ignored in the microeconomic calculus, but they cannot be ignored by those studying or aspiring to direct technological change.

With the terminology and these four central tendencies of technological change in place, we can now turn to the most exciting feature of technology: technological dynamics or the mechanisms and patterns of technological change over time.

2.2.   Technological Change

Some 10,000 years ago humans survived as nomadic hunters and gatherers. This required considerable sophisticated (technical) knowledge. (If you doubt this, try making a living today by hunting and gathering.) However, the first revolution in technology – the development of agriculture – changed the nomadic lifestyle dramatically. The development of markets and of money (institutional and organizational innovations or "technologies" in a larger sense) set people free from the need to be self-sufficient, enabling them to benefit from division of labor and specialization. Markets and agriculture (more precisely agricultural surplus production) were fundamental drivers for the emergence of cities.

Since that time, many further technological revolutions in fields such as materials, construction, navigation, and military technology have dramatically influenced the course of history. The past 300 years – the "age of technology" – have witnessed more momentous technological changes than any previous period in human history. Anthropologists, historians, and philosophers were quick to take an interest in technology and its role in shaping societies and cultures. Surprisingly, economists only came later to the study of technological change (Rosegger, 1996). Observing the Industrial Revolution from its midst, classical writers in economics from Adam Smith to Karl Marx could hardly fail to see the importance to economic growth of technological change, of new products and new production processes. But technological change – the "industrial arts" – was not seen as an integral element of the economic process. Even Karl Marx, who argued that transformations in the material structure of production determined changes in social relations, and who wrote extensively on technology, said relatively little about the sources of such changes (Rosegger, 1996).

Two economists deserve special credit for pioneering our thinking on technology: Thorstein Veblen and Joseph A. Schumpeter. Veblen (1904, 1921, 1953), perhaps best known for his *Theory of the Leisure Class* (first published in 1899), was the first to focus on the *interactions* between humans and their artifacts in an institutional context. He considered technology not as an exogenous force on entrepreneurs, engineers, or workers, but rather part of material and social relationships. Technology was developed and shaped by social actors, while at the same time shaping social values and behavior. Such a "circular" model of interactions was revolutionary at a time when technology was viewed as the exclusive domain of inventors, engineers, and "heroic" entrepreneurs (a kind of naive, romantic fascination adhered to even by the early Schumpeter). Such a unified view of technology contains a revolutionary message today, when many social scientists are trapped in a futile polarization between extreme positions of technology shaping society, or in turn society shaping technology.[9]

More widely acknowledged are the contributions of the Austrian economist Joseph A. Schumpeter (1883–1950),[10] who started his successful scientific career in Austria, passed through failed stages as an entrepreneur, served a short, unsuccessful interlude as Austrian finance minister, and completed his career at Harvard University. Schumpeter's *Theory of Economic Development*, published in 1911 and translated into English in 1934, is a landmark in considering the sources of technological change as endogenous to the economy. His later publications, in particular the monumental *Business Cycles* (1939) and the still eminently readable *Capitalism, Socialism and Democracy* (1942), deepened and extended the treatment of technology in his earlier work.

For Schumpeter the essence of technological change is "new combinations", particularly those that represent a discontinuity, i.e., new combinations that cannot be achieved by gradual modifications of existing artifacts, practices, and techniques. This Schumpeterian notion of technical change is referred to as "radical" technical (as opposed to incremental) change below.

> ... to produce other things or the same things by a different method, means to combine these materials and forces differently. In so far as the "new combination" may in time grow out of the old by continuous adjustment in small steps, there is certainly change, possibly growth, but neither a phenomenon nor development in our sense. In so far as this is not the case, and new combinations appear discontinuously, then the phenomenon characterizing development emerges. ... [the latter] ... is that kind of

---

[9]These extreme positions are referred to as "technological determinism" (e.g., Gille, 1978) versus the "social construction" of technology (e.g., Smith and Marx, 1994).

[10]For an excellent biography on the life and work of Schumpeter, see Swedberg (1991).

change arising from *within* the system which so displaces its equilibrium
point that the new one cannot be reached from the old one by infinitesimal
steps. Add successively as many mail coaches as you please, you will
never get a railroad thereby. [Joseph A. Schumpeter, *Theory of Economic
Development*, 1934:64–66]

For Schumpeter the essence of technological change is "changes in techniques and productive organization", i.e., changes in technological hardware
and software. As the above quote emphasizes, such changes are inherently
"nonlinear". They entail both quantitative and qualitative characteristics
that cannot be produced by simply adding linearly "more of the same" to
existing technologies and practices.

Schumpeter also draws an important distinction between changes that
emerge from an accumulation of small gradual changes (referred to as incremental improvements in the next section) and those that represent radical
"new combinations". He gives five examples (1934:66), listed as follows:

1. The introduction of a new good or product, or of a new quality of a good
   or product.
2. The introduction of new methods of production, not tested yet by experience in the relevant branch of manufacturing. New production methods
   may be based on a new scientific discovery, or on a new way of handling
   a commodity commercially.
3. The opening of a new market, either one that did not exist before or one
   that has previously not been entered.
4. Obtaining (Schumpeter uses the rather inappropriate term "conquest
   of markets") new sources of raw materials or semimanufactured goods.
   The new source may already exist, or it may have been newly created.
5. New forms of organization, e.g., the establishment or the break-up of a
   monopoly.

It cannot be stressed enough that any technological change, whether
incremental or radical, arises from *within* the economic system as a result
of newly perceived opportunities, incentives, deliberate research and development efforts, experimentation, marketing efforts, and entrepreneurship.
Technological change does not fall like "manna from heaven". Schumpeter
also emphasizes the nonequilibrium nature of new combinations. Technological change is not simply "more of the same"; it radically changes the relations
between economic inputs and outputs, and it changes the constraints under
which these can evolve.

As we will see in the next section most macroeconomic models still
largely ignore these two fundamental features of technological change, that is:
(i) evolution from *within* (i.e., technological change should not be exogenous

to the model); and (ii) the inherently dynamic and nonequilibrium nature of technological change, which static equilibrium models fail to capture. With this up-front pessimism about the treatment of technological change in much of economic modeling, let us return to Schumpeter's own words:

> ... Capitalism, is by nature a form or method of economic change and not only never is but never can be stationary. And this evolutionary character of the capitalistic process is not merely due to the fact that economic life goes on in a social and natural environment which changes and by its changes alters the data of economic action; this fact is important and these changes (wars, revolutions and so on) often condition industrial change, but they are never its prime movers. Nor is its evolutionary character due to a quasi automatic increase in population and capital or the vagaries of monetary systems of which exactly the same thing holds true.
>
> *The fundamental impulse that acts and keeps the capitalistic engine in motion comes from the new consumers' goods, the new methods of production or transportation, the new markets, the new forms of industrial organization that capitalist enterprise creates* [italics added].
>
> ... The history of the productive apparatus of a typical farm, from the beginnings of the rationalization of crop rotation, plowing and fattening to the mechanized thing of today – linking up with elevators and railroads – is a history of revolutions. So is the history of the productive apparatus of the iron and steel industry from the charcoal furnace to our own type of furnace, or the history of the apparatus of power production from the overshot water wheel to the modern power plant, or the history of transportation from the mail coach to the airplane. The opening of new markets, foreign or domestic, and the organizational development from the craft shop and factory to such concerns as US Steel illustrate the same process of industrial mutation – if I may use this biological term – that incessantly revolutionizes the economic structure from within, incessantly destroying the old one, incessantly creating a new one. This process of Creative Destruction is the essential fact about capitalism. [Joseph A. Schumpeter, *Capitalism, Socialism and Democracy*, 1942:82–83]

After setting the scene about the importance and essence of technological change, we can now introduce the finer conceptual and terminological detail in the following section, which presents a taxonomy of technological change.

2.2.1. A taxonomy of technological change[11]

Incremental Improvements

Occurring more or less continuously across all industry or service activities, incremental improvements resulting from scientific research and development, engineering, and learning effects improve the efficiency of all factors

---

[11] This section is based on Freeman and Perez (1988) and Freeman (1989).

of production. Although the combined effect of incremental improvements is extremely important, no single improvement by itself will have a dramatic effect. The accumulation of small incremental innovations in long-term overall productivity growth is extremely important, but the steps of individual improvements are difficult to document in detail. As a rule they can be documented through resulting aggregate productivity increases. Typical examples include reduced labor, materials, or energy requirements. The associated model is the "learning" or "experience" curve – with accumulated experience, humans learn to make things better, faster, and with fewer defects (see Section 2.3). Economists call this "learning by doing" (Arrow, 1962) and "learning by using".

The extent and rate of such learning effects vary according to the kind of learning involved. Most importantly they are not "autonomous". They should not be represented as an exogenous time-trend function, as is frequently the case in models trying to capture technological change. Learning depends on the actual accumulation of experience. Without "doing" there is no "learning".

Radical "New Combinations"

Radical "new combinations" are discrete and discontinuous events. In recent decades they have usually been the result of deliberate research and development efforts in industry, government labs, or universities. They may make quantum leaps in productivity possible and overcome resource limitations. Or they may enable the development of entirely new materials and products. Although they depart radically from existing engineering practice and technologies, they nevertheless often tie in with existing industrial structures. They therefore require no radical changes in overall industrial organization, although they do necessitate changes at the level of plants or even industrial sectors. The introduction of the Bessemer process, offering the possibility of low-cost, mass production of high-quality steel in the 19th century, the introduction of nylon, or the contraceptive pill both in the 20th century, are illustrative examples. Despite their importance for individual industrial sectors or submarkets, their aggregate economic impact remains comparatively small and localized, unless a whole cluster of radical "new combinations" is linked together to give rise to entirely new industries or services.

Changes in Technology Systems

Under this heading we refer to far-reaching changes in technology, affecting several branches of industry or occurring across several sectors of the economy. Such changes combine both radical and incremental innovations with organizational and managerial changes.

Technological change in one part of the economy triggers corresponding changes both upstream and downstream in related branches. A good example is the introduction of industrial electric motors (cf. Devine, 1982). Before their introduction, factories would have used a central steam engine with power distribution via transmission belts. Electric motors provided a new versatile decentralized source of motive power. They changed, first, the entire organization of the shop floor. Second, they required changes upstream in the production and distribution of electricity. Without such substantial changes in organization, both on the shop floor and in upstream electricity supply, the electric motor's impact on productivity would have remained localized and limited.

Devine (1982) estimates that the impact of the electric motor was multiplied by a factor of three through such organizational changes. The overall energy efficiency of a steam engine, coupled with mechanical power distribution, according to Devine's estimates is between 3% and 8%. If only the steam engine is replaced by self-generated electricity, the overall energy efficiency remains at 3–6%. However, combining utility-generated electricity and decentralized unit drives raises overall energy efficiency to 10–12%, or by a factor of three at the lower end of the range. These estimates report 1920s efficiencies. Current overall energy efficiencies for industrial drive systems are on the order of 25–28% (Nakićenović *et al.*, 1990), twice as large as 70 years ago.

Clusters and Families

Some changes in technology systems are so far-reaching that they impact upon the entire economy and nearly every aspect of daily life. Such changes involve whole clusters of radical and incremental improvements and may incorporate several new technology systems. The development of the automotive industry, for example, was contingent on developments in materials (high quality steel sheets), in the chemical industry (oil refining), in production and supply infrastructures (oil exploration, pipelines, and gasoline stations), in public infrastructures (roads), and a host of other technological and organizational innovations. The growth of the industry was based on a new way of organizing production, i.e., Fordist mass production combined with Taylorist scientific management principles. These yielded significant real-term cost reductions, making the car affordable to a wider social strata. This changed settlement patterns, consumption habits of the population, leisure activities, etc. And the automobile is just one among many consumer durables now considered standard in industrialized countries.

Clusters of interdependent radical innovations and technology systems give rise to whole families of hardware and software innovations with associated new institutional and organizational settings. Together they multiply the effects of each other on the economy and society. Thus their collective effect is more than the sum of their individual contributions. It would be impossible to calculate overall impacts even if detailed data on individual components were to exist. Qualitative descriptions are more appropriate. In the literature such clusters have been analyzed under the headings of "general natural trajectories" (Nelson and Winter, 1977) and "technoeconomic paradigms" (Freeman and Perez, 1988). Such clusters drive particular periods of economic growth, and will provide the central organizing concept for this book's analysis of technology and global change.

A Schumpeterian (1935, 1939) perspective on long-term economic growth and technological change sees overall development coming in spurts, driven by the diffusion of clusters of interrelated innovations and interlaced by periods of crisis and intensive structural change.[12] The existence of a succession of a number of such clusters over time does not mean that there is a quasi-linear development path, e.g., from textiles to basic metal industries to mass-produced consumer durables as alluded to in Rostow's (1960) stage theory of economic growth. Instead, such clusters are time-specific phenomena. The success of any one (in terms of economic growth) and the drawbacks (in terms of environmental impacts) cannot be repeated quasi-mechanistically at later periods in history or in different socioeconomic settings.

We adopt the concept of technology clusters and families to distinguish broadly between various historical periods characterized by different driving forces and patterns of technological change and their impacts. Our interest in global change issues together with technological interrelatedness and interdependence explains why we have adopted a taxonomy and perspective

---

[12]Such discontinuous paths of economic development have been corroborated by empirical studies ever since the seminal contributions of Nikolai Kondratiev (1926) and Joseph A. Schumpeter (1939). They received revived interest in the periods of economic crisis in the 1970s and 1980s (see e.g., van Duijn, 1983; Freeman, 1983; and Vasko, 1987). Beyond the empirical corroboration of important historical discontinuities, however, the interpretation and theoretical explanation of such long waves of economic and social development remains fragmented and open to further research. In particular, debate continues, first, on whether we are dealing with a recurring or cyclical phenomenon endogenous to the economy, and, second, on what causes the long waves that have been identified. For an excellent collection of classical, seminal papers of long wave theory including critical writings, see Freeman (1996).

with a deliberately large boundary. There are disadvantages to such an approach; we cannot dwell on the detail of individual artifacts and techniques. Instead, we must analyze them as systems and address their characteristics, and the scale and quality of their global change impacts, as a whole that is more than just the sum of its parts. In Chapter 4 we present briefly empirical evidence on the existence and timing of technology clusters, and identify appropriate indicator technologies that can be used as *pars pro toto* for their respective technology clusters and families. We focus on four major technology clusters since the beginning of the Industrial Revolution and identify a possible fifth cluster that in the next millennium could transform our entire technological and material base.

2.2.2. A taxonomy of global change: Impacts of technological change

With respect to (direct and indirect) global change impacts we group technological changes into four categories: (i) those that augment resources; (ii) those that diversify products and production; (iii) those that enlarge markets (output); and finally (iv) those that enhance productivity.

Technological Changes that Augment Resources

The tremendous historical expansion of industrial production has consumed enormous amount of natural resources in the form of raw materials and fuels. Technological changes that augment the resource base have therefore been essential. These include technologies that facilitate the discovery of new resource deposits and that improve the accessibility and recoverability of existing resources; technologies that represent new resource inputs altogether; and finally technologies that substitute for existing material and fuel inputs. Technologies that increase efficiency (i.e., enable to produce more with less inputs) can also be considered to augment resources, but we will discuss them separately under the general heading of productivity.

The onset of industrialization in 18th century England is usually associated with the emergence of coal as a major new industrial fuel. Although coal had been used in the brewing industry and to evaporate salt brines since the 13th century, its use remained limited because of restricted access to coal resources and limited applications. Coal was basically used in the same way as the fuelwood it was supposed to replace. Mining concentrated on comparatively shallow deposits, and coal could only be transported from mines located near riverways and the seashore. Hence the use of the term "sea coal" well into the 19th century. Two important technological innovations changed this situation. First was Abraham Darby's discovery of the coking process through which pig iron could be produced using coal instead of increasingly scarce and expensive charcoal. Second, the invention of stationary steam engines (Newcomen-Savary) allowed water to be pumped from greater depths than had been possible previously with mechanical pumps driven by horses. This increased physical access to deeper coal resources. These two technological innovations in turn paved the way for numerous subsequent innovations. The coking process eventually gave rise to an entirely new coal-based chemical industry that included city gas and synthetic versions of dyes like indigo. James Watt improved the thermal efficiency of the Newcomen stationary steam engine. It subsequently was used in mines not only for lifting water but also as a power source for mechanization, thus lowering mining costs and improving the economic accessibility of coal resources. Most importantly it became a mobile power source for railways. This further improved access to coal deposits and drastically lowered transport costs. With railway transport coal finally became just coal, and was no longer "sea coal".

Petroleum is another example of a new resource that both replaced other materials/fuels in existing uses and opened up new uses. Petroleum, in the form of kerosene, was initially used as a substitute illuminant for dwindling supplies of whale oil.[13] With advances in petroleum refining and the emergence of the internal combustion engine petroleum became a major transport fuel and petrochemical feedstock. That led to its use as a substitute for a variety of raw material inputs to industry (synthetic fibers, rubber, plastics, etc.). That the petroleum industry has grown to its current dominant position, despite recurrent fears of immediate resource exhaustion ever since the early 1920s, is a powerful illustration of the impact of technological change on augmenting resources through improved exploration, discovery, and access to increasingly remote and difficult environments.

Finally, entirely new resources have been made available through technological change. While copper and iron ores have been exploited since antiquity, it was only the introduction of aluminum that made bauxite a major resource for metal supplies. Similarly, nuclear technologies turned uranium into a new energy resource.

Technological Changes that Diversify Products and Production

This is the most familiar impact of technological change. Just compare the numbers and kinds of products and technological "gadgets" in nearly every

---

[13]For a concise account of how the industry drove whales nearly to extinction, see Ponting (1991:186–191).

household in the industrialized world today to the situation some 100 years ago. Electric lights, refrigerators, telephones, radio, TV, video, computers, automobiles, air travel, antibiotics, and vaccines were all either completely unknown or just curiosities with no social or economic relevance. Technological change has also opened up new production options. With steel, for example, production can now draw upon a variety of input materials (e.g., virgin iron ore or recycled steel scrap), energy sources, reductants, etc. to better match available inputs to production requirements, to increase product differentiation (e.g., speciality steels), and to increase quality.

Continuous change in product specifications makes it difficult to measure quality improvements outside "high tech" products such as aircraft or computers for which well-defined performance characteristics exist. Quality measurement problems are particularly relevant for consumer products. Therefore most analyses of technological change impacts on consumer product quality focus simply on falling real prices. A notable exception is a careful study by Payson (1994) analyzing a range of consumer products and their specifications from Sears Roebuck catalogues between 1928 and 1993. *Figure 2.4* reproduces his key findings for five different consumer products. (Note the semilogarithmic scale of *Figure 2.4*.)

Payson's analysis shows significant quality improvements even in consumer products with a low technology content such as sofas and shoes. Typically product quality improves at 2–3% per year. For higher technology products, such as gas ranges (ovens) and air conditioners, quality improvements range from 7% to 9% per year (Payson, 1994:119). These quality improvements are on top of price reductions (reflecting falling production costs) that have enabled mass diffusion of such products into nearly every household in industrialized countries. These quality improvements are generally not considered in macroeconomic statistics, which therefore tend to significantly underestimate the true impact of technological change [cf. also Nordhaus (1997) on this point and for an interesting case study on the costs of light].

Increased diversity as a result of technological change is continually counterbalanced by another tendency of technological change: standardization. Product and process innovations increase diversity, but the push to reduce costs increases standardization. The balance may well change in the near future in the age of new information technologies. These create the possibility of breaking the dominant paradigm of industrial mass production of standardized products. The sort of customized, one-of-a-kind products that are characteristic of preindustrial, handicraft production may reappear in industrial production. Current increasing product differentiation in aircraft, automobiles, and even textiles reinforces such a scenario.

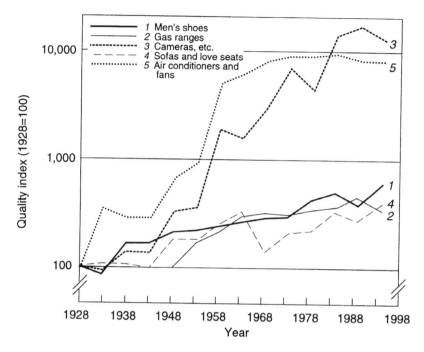

Figure 2.4: Evolution of the quality index (semilogarithmic scale) of five consumer products offered in the Sears Roebuck catalogues, 1928–1993. Source: Payson (1994:118).

Technological Changes that Enlarge Markets (Output)

Technological change has directly enlarged markets through successive transport revolutions from the canals, steam railways and ships of the 19th century to the road vehicles and aircraft of today. Higher transport speeds and falling costs have reduced the "economic" distance between production and raw material supplies on the one hand, and between production and markets on the other. These effects have enabled unprecedented increases in spatial division of labor through trade and market growth. Both permit increased economies of scale and have been important drivers in sustaining ever increasing output (and consumption) volumes.

Technological change also enlarges markets indirectly through improved productivity. Productivity improvements reduce production costs. Falling costs enable price reductions and expand the customer base and thus the market. The first automobiles and fax machines were expensive gadgets for a few wealthy individuals and institutions. With falling prices, the market for both products grew as they came within the financial reach of ordinary consumers. Mass consumption enables mass production, increasing economies of scale, further price reductions, and yet bigger markets. This positive feedback mechanism (here somewhat oversimplified) has driven the expansion of industrial production in domains as diverse as textiles, porcelain, cars, consumer durables, instant soups, electricity, and many more.

Technological Changes that Enhance Productivity

Productivity improvements are the key impact of technological change. Doing more with less is the central objective applying to all factors of production: land, labor, energy, and raw materials. Only with a long-term historical view can we grasp the scale of productivity increases due to continuous technological change over the last 200 years. The sources of these productivity increases are diverse and defy any simplifying summary. At this point, the key conclusion is simply that without such increases the spectacular historical expansion of human numbers, production and consumption could never have been sustained. It could not have been sustained in terms of resource availability, in terms of environmental impacts, or in terms of the economics of production and consumption.

In offering this simple taxonomy of technological changes we recognize the groupings are not clear cut. The impacts of technological change are frequently interdependent and overlap the categories defined above. We noted the relationship between productivity increases and expansions of the resource base and markets. It is similarly difficult to separate the direct impacts of productivity increases from their indirect impacts on mass consumption through increased wages and reduced working time. All are integral parts of the interwoven impacts of technological change that are relevant for global change, even if the impacts are too frequently subsumed under output growth and increasing environmental burdens.

2.2.3. Technological dynamics and interaction

The fact that the essential feature of technology is *change* causes an epistemological problem. In trying to describe a particular technology such as the railway or car, we have to face the problem that the object of our investigation keeps changing. Initially a new technology is imperfect, expensive, and limited in its applications. It must first prove itself in niche market applications where performance rather than cost is the overriding criterion. If successful, subsequent improvements and cost reductions can lead to wider applications. This evolution is the essence of the technology life cycle model described below. It is important to remember that the technology being analyzed in any particular case is only defined with the benefit of hindsight.

It is almost impossible to anticipate a new product's future applications or the new "combinations" that may become part of its life cycle.

To date no comprehensive method has been developed to describe and classify the myriads of technological artifacts and techniques. At the sectoral level, attempts have been made (e.g., Foray and Grübler, 1990) to use morphological analysis techniques, first, to describe the total evolutionary space of possible combinations capable of performing a specific task, and, second, to map the historical "branching" of the evolutionary tree of actual combinations. Such an analysis illuminates the functions that particular technological "combinations" can provide, and which combinations remain "locked out". It thus helps identify feasible, unexplored alternatives that may emerge later as possible "surprises" and competitors. However, such analyses are extremely data-intensive and therefore remain localized and very specific.

It is somewhat easier to classify technology dynamics than it is to classify technologies. As a first step, we simply consider the evolution of a particular artifact or technique with an "introspective" perspective, e.g., looking at its design features, performance, price, scale, and various productivity measures. This is the principal perspective of technology life cycle models. Second, we consider how a particular technology interacts with its environment: what are the factors determining its growth or failure; how does it perform in a particular market; and how does it complement or compete with other artifacts and techniques? This is the perspective of technology diffusion and substitution models. It is only through diffusion that inventive and innovative potentials are translated into actual changes in social practice, artifacts, and infrastructures. Diffusion phenomena are therefore at the heart of all changes in society and its material structures.

Technology Life Cycles

The world of technology is full of biological metaphors: for example, evolution, mutation, selection, and growth. Some are more appropriate than others. The clearest metaphor is between biological and technological growth or life cycles, and it is one that is widely used in the technological, management, and marketing literature.[14] The appeal of the life cycle model lies primarily in its considerable success as, first, an empirically descriptive tool and, second, as a heuristic device capturing the essential changing nature of technologies, products, markets, and industries. The essence of the technology life cycle model (like that of other growth models in biology) is that growth is nonlinear, and especially not unlimited. Typically growth in biology and

---

[14]For an excellent (and also critical) survey, see Ayres (1987).

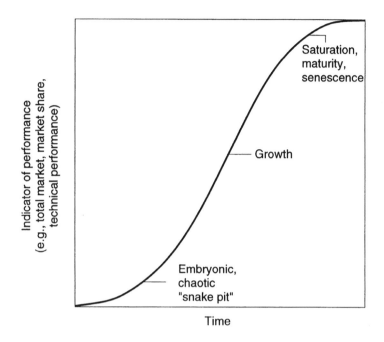

Figure 2.5: A stylized technology life cycle model.

of technologies alike proceeds along an S-shaped pattern: slow growth at the beginning, followed by accelerating growth that ultimately slows down leading to saturation. However, the S-curve or life cycle model is not an explanatory one. It does not explain *why* things evolve as they do.

The technology life cycle model (see *Figure 2.5*) classifies the phases of technology development into three phases: childhood, adolescence, and maturity. Subsequently, decline or senescence (and ultimate death) may follow. These correspond to a technology's introduction, growth, saturation, and eventual decline. Typically a technology's life cycle is described by indicators such as output volumes, market share, product characteristics (performance), sources of technological change, and the structure of industry. Most important with the last three of these is whether a life cycle phase is characterized by diversity or standardization. Associated with each of the three phases of the life cycle is a "stylized" pattern[15] as described below.

Introduction/childhood. The first phase is characterized by low production volumes and market shares and is the period with the greatest technological

---

[15]These patterns are "stylized" in that they represent a simplified summary of a large number of product and industry studies. In many individual cases deviations from these "stylized" patterns can occur.

and market diversity. Many possible technological designs are explored, development focuses on product innovations, and numerous firms try to gain a footing in the market. Emphasis is on demonstration of technical viability, and costs are of secondary importance. Learning effects and technology improvements derive primarily from experimentation and R&D. Overall the market is highly volatile and uncertain, characterized by a large number of "drop-outs", both of design alternatives and firms.

Growth/adolescence. Initial diversity gives way to increasing standardization as technical viability is established and efforts begin to be made to improve production economics. Increasing certainty of technological viability and applicability, reduced risks to innovators, and falling costs and prices lead to rapid market growth. Product innovations improve a technology's design features and enlarge its field of application. Process innovations improve production economics, and significant learning effects for both producers and users additionally reduce costs. Such innovations and learning effects provide positive feedbacks that further stimulate market growth. Eventually, however, the competitive environment becomes increasingly concentrated. This concentration applies first of all to firms and industry structure. Either because smaller firms go broke, or are absorbed in mergers and acquisitions, the number of producers declines rapidly. The history of the automobile industry is a case in point (*Figure 2.6*), although hardly an extreme example. For instance, there are fewer than five large commercial aircraft and aircraft engine manufacturers worldwide. Of course, *product variety* continues to be large, and is even increasing, as ever more specialized applications are searched (and found) for technologies and products.

Although the number of radically different designs diminishes in favor of a few demonstrated alternatives, these continue to be modified and adapted for increasingly diverse and remote applications. Whereas design changes in the early phases are characterized by a rapid succession of new models with increasing performance and productivity, later phases are characterized by incremental design changes. The passenger aircraft industry is a good example. Aircraft productivity, in terms of passenger-kilometers per hour, increased between the 1930s and 1970s through a rapid succession of different designs from the classic DC-3 of the 1930s to the Boeing 747 "jumbo" jet of the 1970s (*Figure 2.7*).

These rapid design changes allowed improvements to be made not only in aircraft productivity but also in fuel economy and crew productivity. Since 1970, however, improvements have been incremental. The B-747 has been "stretched" by increasing its length, stretching the double deck, and so forth. Incremental improvements can be impressive; a modern B-747 (400 series)

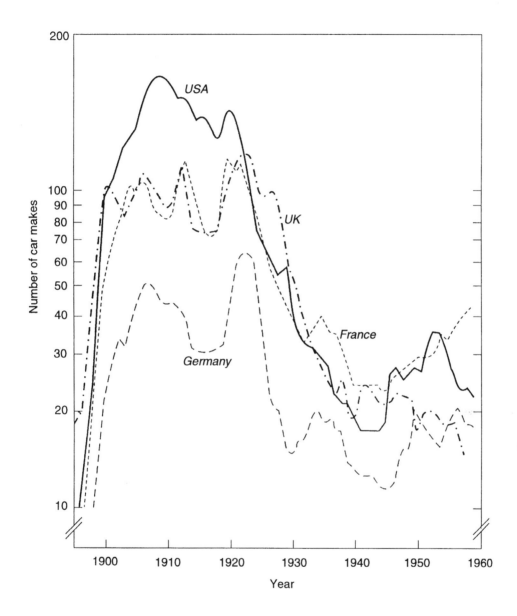

Figure 2.6: Number of car makes, 1895–1960 (on semilogarithmic scale), showing the increasing market concentration characteristic of a maturing industry. Note persistent differences between countries even under a similar overall trend of substantial reductions in car makes competing on the market. Source: Rosegger and Baird (1987:96).[2]

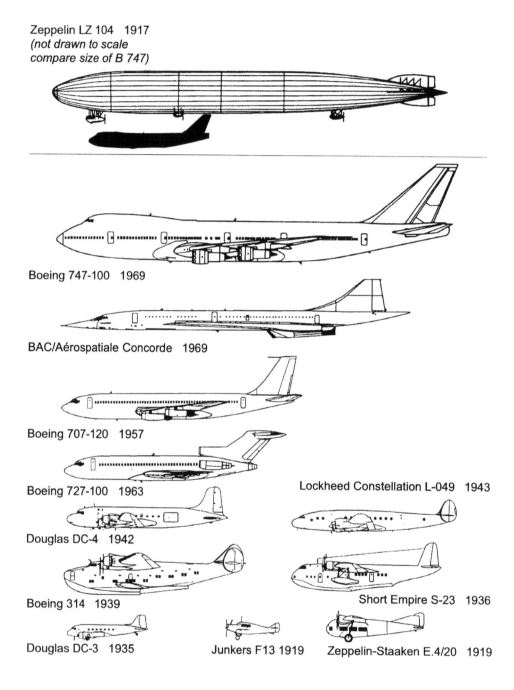

Figure 2.7: Size of selected commercial passenger aircraft. Note the comparatively modest size of today's commercially successful Boeing 747 jumbo jet relative to that of the unsuccessful Zeppelin from the beginning of the century. Source: Hugill (1993:256).[3]

consumes one-quarter less fuel than its 100 series counterpart of 1969 (Borderon, 1990:33). But the incremental nature of improvements reflects the increasing maturity of current aircraft technology, even if subsystems may continue to change radically (e.g., the new fly-by-wire system introduced in the Airbus 320/340 series).

Saturation/maturity. Growth rates slow down as markets become saturated and improvements face diminishing returns. Competition is based almost entirely on cost reduction rather than design improvement, and the market is concentrated in the hands of a few suppliers. The labor and skill intensity of production becomes increasingly "internalized" in machinery and mechanization. Large plants operate with almost no labor.

The management literature is full of examples of industries "taken by surprise" by market saturation and the slow down of market growth (e.g., Porter, 1983, 1990). Marketing departments typically continue to forecast a recovery in growth "just around the corner", and there are considerable lags in adjusting investment and expansion plans. As a result, the industry faces considerable overcapacity and intensified competition and market volatility. Common responses are to concentrate production to squeeze out the last marginal cost improvements from scale economies, or to outsource production altogether. This is one of the core areas of current concerns about job losses due to "globalization", but it should be related to increasing market saturation and industry maturity phenomena, rather than globalization per se. On the product side, design innovations focus on packaging and appearance rather than intrinsic features and qualities. The technology or product finally turns into a mass-produced commodity increasingly subject to regulation and an increasing awareness of its disbenefits. Disbenefits, such as environmental impacts, are generally either not anticipated in the earlier phases of a technology's life cycle or considered of secondary importance. Many problems also emerge nonlinearly with increasing application densities, and these in particular constitute genuine "surprises" (Brooks, 1986) to industry, consumers, and governments. The classic example is the automobile, which increases congestion and pollution as the number of them on the road grows. Thus, even small additional growth can suddenly generate important "externalities" that limit the usefulness of further growth.

We next turn to the mechanics of diffusion that underlie the progression through the three life cycle stages. As an initial illustration let us turn back the clock nearly 1,000 years and return to monastic life in 11th-century Burgundy.

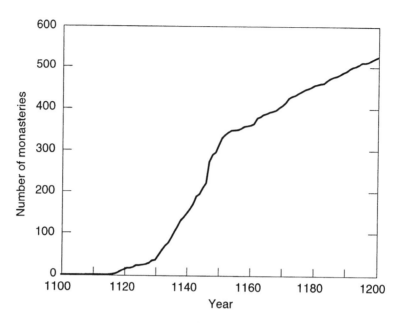

Figure 2.8: Diffusion of Cistercian monasteries in Europe: the first 100 years. Data source: Janauschek (1877).

A Medieval Prelude

In 1098 movement for the reform of Benedictine rule led St. Robert to found the abbey of Citeaux (Cistercium). Citeaux would become the mother house of some 740 Cistercian monasteries, about 80% of which were founded in the first 100 years of the Cistercian movement. Nearly half were founded between 1125 and 1155, and many traced their roots to the Clairvaux abbey founded as an offshoot of Citeaux in 1115 by the tireless St. Bernhard. The nonlinear, S-shaped time path of the spread of Cistercian rule (*Figure 2.8*) resembles the diffusion patterns we will observe later for technologies. In terms of the terminology introduced previously (Section 2.1.1), we might say that St. Robert invented Cistercian rule, St. Bernhard innovated, and diffusion followed. This basic pattern of temporal diffusion is essentially invariant across centuries, cultures, and artifacts: slow growth at the beginning, followed by accelerating and then decelerating growth, culminating in saturation. Sometimes a symmetrical decline follows.

Diffusion is a spatial as well as a temporal phenomenon. The topology of the Cistercian network reveals a hierarchy of centers of creation and structured channels of spread. *Figure 2.9* illustrates some example pathways in the spatial spread of two Cistercian "subfamilies", named after their respective mother houses as lines of Clairvaux and of Morimond.

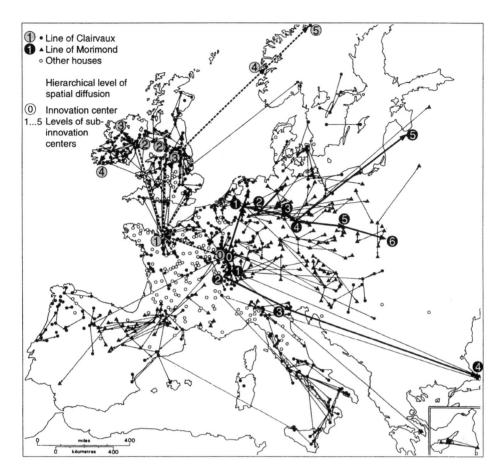

Figure 2.9: Spatial diffusion of Cistercian settlements (lines of Clairvaux and Morimond). Note in particular the hierarchical topology of spatial diffusion, from innovation centers to subcenters, and from the respective centers out to the hinterlands as illustrated for selected traits of the two houses. Adoption densities (settlements in this case) are highest in the innovation centers, and lowest in the hinterlands. Bottom right hand box shows diffusion to Cyprus. Source: adapted from Donkin (1978:28–29).

The patterns bear witness to the existence of networks, and today there is a growing literature on the role social and spatial networks play in the diffusion process (cf. Kamann and Nijkamp, 1991). *Figure 2.9* also shows significant differences in the spatial density of settlements. The origin of the innovation, Burgundy, was home to all four mother houses and had the highest spatial concentration of settlements. From there daughter houses were founded (regional "subinnovation centers" in the terminology of spatial diffusion), from whence Cistercians further spread to their respective hinterlands (the "neighborhood effect" in spatial diffusion) to found other subregional centers, which in turn led to further settlements.[16] The density of settlements decreases the further one moves away from the original center and from each subsequent regional and subregional center. The result is persistent regional differences and disparities.

The importance both of social networks and of diversity is exemplified by differentiation into different Cistercian "subfamilies". Each was named after its respective mother house, and each followed its own pattern of settlements, regional specialization, and implementation of Cistercian rule. Some additions to Cistercian rule were not genuine new settlements, but were "takeovers". For example, Savigny, with all its daughter houses, submitted to Clairvaux rule in 1147 and subsequently became the mother house of all Cistercian settlements on the British Isles. Despite differentiation and regional specialization, close communication existed between all the monasteries, creating an important channel for the spread of 13th- and 14th-century innovations like the water mill, new agricultural practices, and Gothic cathedral architecture.

The Cistercian movement had significant social, economic, and environmental impacts. It was particularly instrumental for the introduction of new agricultural practices and manufacture of textiles. Moreover, Cistercian rule commanded location of settlements in remote areas. This made Cistercian monasteries important local nodes for the internal colonization of land in Europe, and for early deforestation as well (see Part II).

Technological Diffusion and Substitution

Technological growth is the central feature of the technology life cycle, and is measured either in terms of growing volumes (e.g., tons of steel, number of cars) or growing market shares. Such growth cannot be analyzed by focusing narrowly on an artifact or product itself, but can be understood only by examining how a technology interacts with its environment, including other technologies. This interaction is the essence of technological diffusion and substitution. As illustrated in our medieval prelude, diffusion phenomena are not linked to the spread and growth of technological artifacts alone, but are

---

[16] Spatial diffusion proceeds in a kind of patchwork and hierarchical manner. Originating from innovation centers diffusion proceeds first to the areas in close proximity to the center (the center's neighborhood, or its "hinterland"). At the same time, the innovation is "exported" to other, more remote places (regional subinnovation centers) and spreads from there to the respective hinterlands as well as to further remote (third or even higher level hierarchical) subcenters of innovation diffusion. The classical work of spatial innovation diffusion remains the seminal book of Torsten Hägerstrand (1967).

a much wider social phenomenon (see Rogers, 1962, 1983). The most general definition of diffusion is: an innovation (idea, practice, artifact) spreads via different communication channels in time and space, among members of a social system. A primer on diffusion, as well as some elementary mathematics describing diffusion and growth, is given in *Box 2.1*.

Some instances exist of what might be called "pure" diffusion where an idea, practice, or artifact represents such a radical departure from existing solutions that it creates its own niche for diffusion. More frequently, however, a new solution does not evolve in a vacuum but interacts with existing practices and technologies. This is referred to as technological substitution,[17] with the new solution either competing one-on-one with an existing alternative or competing with several different technologies simultaneously. These interactions are usually best understood by examining relative (i.e., market) shares of competing alternatives, rather than absolute volumes.

Figure 2.10 illustrates the growth of the US canal network in the 19th century, along with other important transport infrastructures. The empirical data are approximated by a symmetrical growth curve (a three parameter logistic in this case).[18] The estimated asymptote (saturation or maturity level) of the diffusion processes of canals is approximately 4,000 miles and in good agreement with the actual maximum of 4,053 miles (6,400 and 6,485 km, respectively) reached in 1851 (shown as 100% diffusion level in *Figure 2.10*). The standard measure of diffusion speed is the time a process takes to grow from 10% to 90% of its ultimate saturation level (see *Box 2.1*). In the case of symmetrical growth this also equals the time required to grow from 1% to 50% of the saturation level.

In *Figure 2.10* the diffusion rate for canals, Δt, equals 31 years, and the entire diffusion cycle spans about 60 years. Thus, it took more than half a century to develop the canal network in the USA, with most canals (80%) constructed within a period of 30 years. The year of maximum growth (t_0) was 1835. After reaching its saturation level, the canal network declined rapidly due to vicious competition from railways.

---

[17]A distinction can be made with respect to the concept of "substitution" as used in economic theory. There substitution describes a case when a particular product is produced through a different combination of factor inputs, without necessarily entailing changes in technologies, processes, or techniques. Consider, for instance, an industrial boiler that can burn oil or natural gas. If prices change, oil may be substituted for gas or vice versa without requiring a new boiler or changes in industrial processes. In most cases, substitution between various factor inputs also entails changes in technologies and techniques. Thus, substitution in an economic sense, i.e., from scarce to more abundant raw materials as inputs to production, is generally impossible without technological change.

[18]For statistical measures of fit and parameter uncertainty of this and subsequent examples, see Grübler (1990a).

Box 2.1: Innovation Diffusion and Technological Substitution

The patterns and pace of the spread of innovations – in the form of new ideas and artifacts (diffusion) and the way these interact with existing ones (substitution) – are, as a rule, nonlinear. No innovation spreads instantaneously, if it spreads at all. Instead, the temporal pattern of diffusion is usually S-shaped: slow growth at the beginning, followed by accelerating and then decelerating growth, ultimately leading to saturation. The adage "Only the sky is the limit" certainly does not hold true for technologies.

As a simple and representative S-shaped diffusion/substitution curve, the logistic curve has been widely used. (Note though that the model is entirely *descriptive*, it shows how a diffusion/substitution process looks, but does not explain *why* it behaves as it does. Various causality mechanisms from learning theory to capital vintage, or turnover, models have been suggested explaining the empirically observed S-shaped diffusion/substitution patterns. In the diffusion literature, parameters of the logistic curve – like its growth rate – are linked to other explicatory economic or sociological variables such as profitability, compatibility with social norms, or even systemic variables, like complexity and size of the system being analyzed.)

The logistic curve is given by the following equation:

$$y = \frac{K}{1 + e^{-b(t - t_0)}}$$

where K denotes the upper limit (asymptote), t_0 denotes the inflection point at K/2, where growth rates reach their maximum, and b denotes the diffusion rate (the steepness of the S-curve). The diffusion rate is frequently also denoted by Δt, the time a process takes to grow from 10% to 90% of its ultimate potential K. It is related to the growth rate b by:

$$\Delta t = \frac{1}{b} \log 81 = \frac{1}{b} 4.39444915 \ldots$$

Δt also denotes the time to grow from 1% to 50% of K. Hence the entire diffusion life cycle spans $2 \times \Delta t$.

The logistic curve can be rewritten with a linear right hand side, frequently used when plotting relative market shares F = y/K:

$$\log \frac{y}{K - y} = b(t - t_0)$$

Here the interaction between the growth y achieved (or market share F), versus the growth $K - y$ (market shares $1 - F$) remaining to be achieved, yields a straight line when plotted on a logarithmic scale. This linearization, subsequently referred to as *logit transform*, highlights in particular the often turbulent early and late phases of the diffusion process. Note though that in this linearization zero or exactly 100% market share (K = 1) cannot be shown.

The following graph (from the classic 1971 paper by Fisher and Pry) illustrates the life cycle in the diffusion of 17 technological innovations, measuring their relative market shares F. For simplification, the symmetrical declining shares of the older technologies being substituted are not shown. Examples of technological substitution studied by Fisher and Pry include the replacement of natural by synthetic fibers, and the replacement of traditional steel making processes by the basic oxygen process.

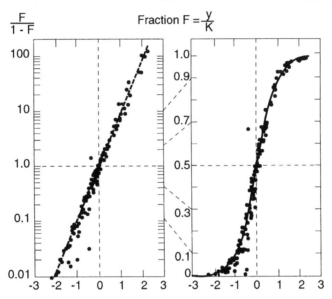

Statistical uncertainties of parameter estimation of logistic curves are discussed by Debecker and Modis (1994). Corresponding uncertainties and measures of goodness of fit of numerous examples are given in Grübler (1990a). As a rule, however, the human eye is an excellent guide for judging whether a particular technological diffusion or substitution path follows an S-shaped, e.g., logistic, pattern. Hence, for the sake of brevity, no curve-fitting statistics will be reported here.

Diffusion or substitution processes can also show deviant behavior from simple logistic patterns. In almost all cases this is due to the fact that a new technology, initially replacing an old technology along a logistic substitution pattern, becomes challenged by yet a newer technology, and is substituted in turn.

In the *logit transform* this shows as follows: a technology initially follows a linear diffusion/substitution pattern, that with a curvature passes through a peak significantly below the maximum possible (K = 1, i.e., 100%), in order to decline again along a linear (i.e., logistic) path. This is due to the fact that it is being substituted by yet a newer technological solution. Therefore it is quite misleading to analyze particular technologies in isolation, e.g., in the form of binary (one-to-one) substitution models. Only a holistic analysis can allow conclusions to be made on the particular shape of the diffusion/substitution trajectory technologies follow.

A generalized model for multiple competing technologies was first proposed by Marchetti and Nakićenović (1979), and some illustrative examples are given in the subsequent chapters (cf. e.g. *Figure 2.12*).

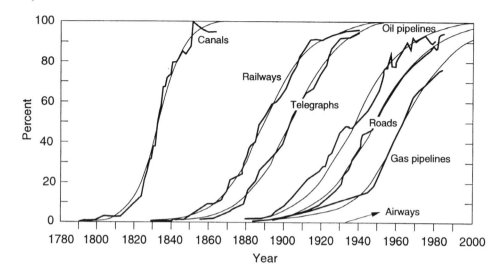

Figure 2.10: Growth of US transport infrastructures as a percentage of their maximum network size, empirical data (bold jagged lines) and model approximation (thin smooth lines). Source: Grübler and Nakićenović (1991). For the data of this graphic see the Appendix.

Figure 2.10 illustrates that subsequent transportation infrastructures, e.g., railways and roads, followed a similar pattern. In the figure the different sizes of individual networks have been renormalized to emphasize their similar diffusion patterns. The absolute saturation size of the railway network is an order of magnitude greater than that of canals. For the road networks, the saturation size is two orders of magnitude greater. Not surprisingly, their diffusion rates are slower. Δt equals 55 years for railways and 64 years for roads, compared with 31 years for canals. It is also interesting to note the regular spacing in *Figure 2.10* – about half a century between the three major historic transport infrastructures – and to note the close relationship between different infrastructures. Railways and the telegraph evolved together, as did road networks and oil pipelines necessary to transport the oil fueling the road vehicles. These examples illustrate the importance of technological interdependence and cross-enhancement, and the necessity of analyzing the diffusion of technologies in the larger context of technology "families" and "clusters".

Figure 2.11 illustrates a particularly striking case of technological substitution: the replacement of horses and carriages by cars. The figure shows the numbers of (urban) riding horses and cars in the USA and the practical disappearance of the horse as a transport technology within less than

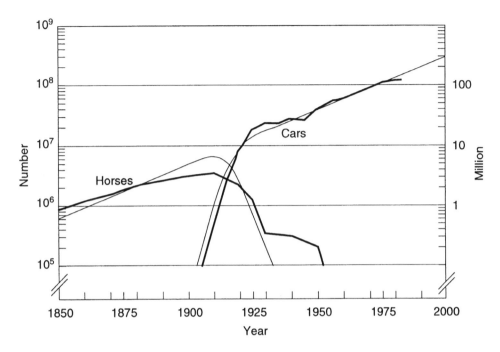

Figure 2.11: Number of (urban) draft animals (horses) and automobiles in the USA, empirical data (bold jagged lines) and estimates (thin smooth lines) from a logistic model of technological substitution. Source: Nakićenović (1986:321).

three decades. Δt equaled approximately 12 years. [The Nakićenović (1986) estimate refers to nonfarm horses only, peaking at over three million in 1910. Farm horses (many of them also used for transport purposes) totaled over 20 million in that year.] The substitution was undoubtedly fast enough to traumatize oat growers and blacksmiths, but it also created new job opportunities in gasoline stations, in the oil industry, in auto repair shops, and elsewhere.

The substitution of an old technology by a new technology shown in *Figure 2.11* is a simple example of the general case of technological change in which there are several competing technologies. *Figure 2.12* shows the introduction of the first generation of emission controls in the US automobile fleet followed later by the technology of catalytic converters. Note that the diffusion rates (Δt) in *Figure 2.12* are about 12 years, the same as that in *Figure 2.11*. This suggests that the replacement dynamics of road vehicle technologies have not changed very much. The most likely explanation is that the lifetime of road vehicles has remained relatively constant: the working lives of horses and cars are both about 10 to 12 years.

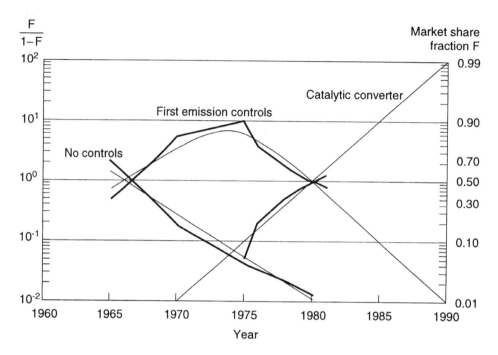

Figure 2.12: Diffusion of cars with first emission controls and catalytic converters and displacement of cars without emission controls in the USA, in fractional shares (F) of total car fleet, empirical data (bold jagged lines), and estimates (thin smooth lines) from a logistic substitution model. Source: Nakićenović (1986:332). For the data of this graphic see the Appendix.

The example of the automobile illustrates yet another dynamic feature of technological evolution: growth beyond the initial field of application. The car industry grew initially by replacing horses. That stage of its growth was completed in the 1930s. Subsequently new markets developed: long-distance travel in competition with the railways and short-distance commuting that enabled, and responded to, changing settlement patterns characterized by suburbanization. The result is approximately 135 million automobiles registered in the USA, roughly 0.6 cars per capita. As mentioned above, however, other countries will not necessarily follow an identical path. The high density of cars in the USA results from specific initial conditions including high individual mobility, even before the automobile, and from a long sustained period of diffusion that created precisely the lifestyles, spatial division of labor, and settlement patterns of an "automobile society". In short, it is yet another example of "path dependency".

Some "Stylized" Facts on Diffusion

The above brief description reiterates the main result derived from thousands of diffusion studies: no innovation spreads instantaneously. Rather, diffusion follows a very consistent pattern of slow growth at the beginning, acceleration of growth via positive feedback mechanisms, and finally saturation. Of course timing and regularity of such processes vary. But the important lesson to retain is that diffusion in most cases of any economic or social significance takes several decades. (For a comparative cross-national study of technology diffusion in industry, see Nasbeth and Ray, 1974; and Ray, 1989.) For large-scale and long-lived infrastructures it may take up to 100 years (Grübler, 1990a).

Diffusion is also a spatial phenomenon. It spreads from focused innovation centers, through a hierarchy of subcenters, to the "periphery" of diffusion (cf. Hägerstrand, 1967). *Figure 2.13* illustrates the spatial diffusion of railway networks in Europe. The construction of railway networks in England spanned approximately 100 years, while it took only half as long in Scandinavia. Railway networks were also more extensive in the countries leading the introduction of this technology (i.e., England and the USA) than in countries that followed later (*Figure 2.14*).

By 1930 the core countries in railway development (England, the rest of Europe, and the USA) had constructed 60% of the world's 1.3 million km of railways. The global railway network has not increased since then because of the introduction of newer transportation systems. These systems follow patterns that are similar to those of the railways. Automobile diffusion at the global level corroborates the accelerated diffusion rate (learning of late adopters) and their lower adoption densities (Grübler, 1990a). Thus, uneven adoption levels are likely to persist, particularly as new transport systems are developed in response to concerns over environmental impacts and changing societal needs. In the case of the automobile, we might expect alternatives to the internal combustion engine to become available within the next few decades, a development that would lead to considerably lower future energy demands than currently assumed (Grübler *et al.*, 1993b; see also Chapter 7 below).

Figure 2.15 summarizes the following main "stylized" facts representative of both theoretical and empirical diffusion research:

- No innovation spreads instantaneously. Diffusion typically follows an S-shaped temporal pattern. The basic pattern is invariant, although the regularity and timing of diffusion processes vary greatly.

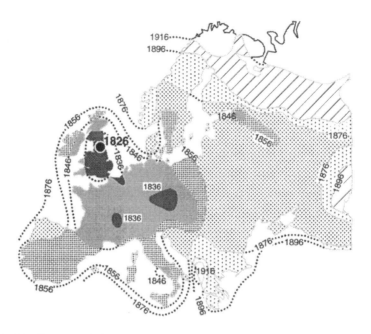

Figure 2.13: Spatial diffusion of railways in Europe, in 10 year isolines of areas covered by railway networks. Source: adapted from Godlund (1952:34).

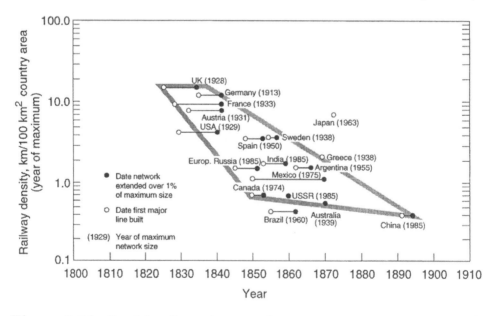

Figure 2.14: Spatial railway densities (in km railway lines per 100 km² country area) as a function of the introduction date of railways. Source: Grübler (1990a:98).

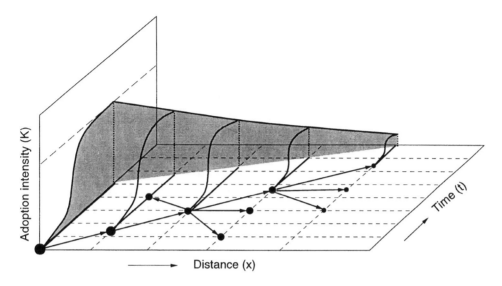

Figure 2.15: A conceptual representation of the diffusion process in time and space. Source: adapted from Morill (1968).

- Diffusion is both a temporal and spatial phenomenon. Originating in innovation centers, a particular idea, practice, or artifact spreads within a core area and then, via a hierarchy of subcenters, to the periphery.
- Although starting later, the periphery profits from the experience gained by the core and generally has faster adoption rates. Quicker adoption, however, results in a final lower adoption intensity than in core areas.
- Because of such differences, application densities and the timing of diffusion are not uniform in space, among the population of potential adopters, or across different social strata. In particular, there is little theoretical or empirical evidence to assume that adoption intensities of early diffusion starters are any guide to the adoption levels of late followers.

What governs the pace of technological diffusion? At the microlevel of the individual consumer or firm a number of factors have been identified (see e.g., Rogers, 1983):

- *The perceived relative advantage of a new artifact or technique.* This has been the focal point of diffusion studies in economics. Key variables include profitability and the required size of investments. Other things being equal, the higher the perceived profitability and the lower the required investments, the faster diffusion proceeds.

- *Compatibility.* Sociological and anthropological studies have identified compatibility with social values and with existing practices and techniques as important determinants of diffusion rates. In economics "network externalities", i.e., requirements for additional infrastructures or the existence of standards facilitating interchanges, have also been identified as important variables. For example, the diffusion of electric appliances in areas without an electricity grid is unlikely. In the early days of video recorders the existence of three different major cassette standards reduced the possibilities of sharing or renting cassettes, thereby slowing diffusion.

- *Complexity.* By complexity we refer to the learning and knowledge requirements for producing and using new artifacts and techniques. Anthropological, technological, and economic diffusion studies invariably identify complexity as an important variable. However, because quantitative measures for complexity are difficult to develop, its influence is usually described in qualitative terms.

- *Testability, observability, and appropriability.* Diffusion proceeds faster if a new artifact or technique can be tried out, if experience and information from peers is available, and if an innovation is easy to obtain. Starting with the French sociologist Gabriel Tarde in 1890, a number of research streams (e.g., Bandura, 1977) have analyzed diffusion processes primarily as learning and social imitation phenomena. In the words of a Chinese proverb, "If you want to become a good farmer, look at your neighbor". While mass media like television or the press are effective in spreading general information about an innovation, actual adoption decisions appear to be made based on interpersonal communication with peers and neighbors. It may be reassuring that today's PC users are not very different from Chinese farmers of 1,000 years ago. The fundamental lesson is that interactions within small social networks are important, take considerable time, and should not be "shortcut" through top-down centralized marketing efforts. Economic studies also emphasize the importance of informal information networks and close cooperation between buyers and suppliers, i.e., good appropriability conditions.

The macrolevel factors governing the rate of technology diffusion include, first, the size of the system involved (bigger systems entail longer diffusion time) and, second, whether the process is one of technological substitution or pure diffusion. Substitution involves replacing existing techniques or artifacts, while pure diffusion entails creating an entirely new social, economic, and spatial context, which obviously takes a longer time to achieve, or can even block diffusion in the first place. These are the macrolevel equivalents of the complexity and compatibility variables discussed above.

Although the driving forces and factors determining the speed and extent of diffusion are varied and change over time, at the macrolevel the transition paths have a very ordered structure. Diversity and complexity at the microlevel result in overall orderly transition paths, and according to recent theoretical findings (see e.g., the discussion and simulation models of Dosi *et al.*, 1986; Silverberg *et al.*, 1988; and Silverberg, 1991), such diversity appears to be even a prerequisite for diffusion.

Finally, it is important to recognize the pervasiveness of uncertainty and imperfect information in all decisions concerning technology diffusion. These factors affect the assessment of existing artifacts and practices, and more particularly, of new alternatives. Any adoption decision involves personal "technological forecasts" and varying degrees of risk aversion. Individuals, firms, and organizations cannot be modeled as economic "robots" with perfect foresight and economic "rationality". This is particularly true for the early diffusion phase of a technology, where decisions are especially complex and uncertain.

2.2.4. Technology selection: Abundance of nonstarters, uncertainty, and opposition

Any realistic history of social and technological innovation would consist mostly of "nonstarters", i.e., examples of innovations that failed to diffuse altogether. The existence of a possible solution (innovation) is therefore by no means a guarantee for subsequent diffusion. *Figure 2.16* shows an amusing failure suggested in 1828 by Henry R. Palmer – a monorail railway using sails. By then Stephenson had built his first railway line, and the dependence of Palmer's innovation on the vagaries of the winds would seem to have made for long odds. Nonetheless, it is fair to assume that the race was still far from settled at that time, and the ultimate success of the steam railways would have been very difficult to predict.

A good example of both the uncertainty in the early phases of technology development and the abundance of nonstarters is the problem of preventing dangerous smoke sparks from steam railways. Smoke sparks from wood-burning steam locomotives in the USA represented a serious fire hazard. Over 1,000 patents for "smoke-spark arresters" were registered in the 19th century (some illustrated in *Figure 2.17*) in a futile search for a solution. Ultimately none of these was successful, and the problem was solved not by an incremental "add-on" technology, but by the replacement of steam by diesel and electric power.

Figure 2.16: A failed innovation: monorail using sails, as proposed by Henry R. Palmer in 1828. Source: Marshall (1938:171).[4]

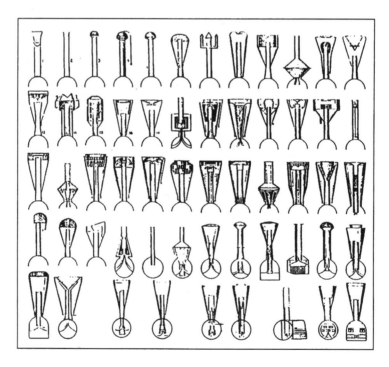

Figure 2.17: Technological variety in response to an environmental hazard. A few examples of the more than 1,000 patented "smoke-spark arresters" for wood-burning steam locomotives in the USA. Source: Basalla (1988:136).[5]

This large variety of possible alternatives illustrates the extent of the diversity and experimentation that precede successful diffusion. "Many are called, but few are chosen". Often the period of experimentation is lengthy. The current standard railway gauge in Europe emerged only after considerable time. (Spain and Russia continue to use different gauges, creating the inconvenience of changing trains at the border or changing the train bogeys.) Even in the USA, a single country, standardization of different railway gauges took several decades, as each company was reluctant to make the costly investments to retrofit their railway lines. In the case of road traffic the decision to drive on the left or the right side was also not straightforward. There were even instances where both standards prevailed at the same time.[19]

Standards are essential for technological systems to function smoothly. We can define standards simply as a set of technical specifications that assure intra- and interoperability of technologies (see *Box 2.2*). Intraoperability refers to technologies functioning within their specific infrastructures (e.g., a locomotive that can run on standard gauge railway lines). Interoperability refers to standards enabling the "exchange" between otherwise distinct technologies (e.g., standard dimensions for containers that can be loaded from a ship onto a truck, or the now ubiquitous data file transfer protocols for exchanging data between computers with different operating systems and file structures).

Optimality is of secondary importance, as any standard is better than none at all. Indeed, the issue of "bad" technology choices has received considerable attention recently, stimulated by the work of Brian Arthur (1983, 1988) and Paul David (1985). The two most prominent examples cited are the choice of the internal combustion car at the turn of the century over steam and electric alternatives (Arthur, 1988) and the choice of the QWERTY keyboard standard for typewriters (David, 1985). Arthur and David argue that both choices were inferior to the alternatives available at the time, and are therefore examples of suboptimal choice "by historical accident". They have been challenged both by economists defending the neo-classical dogma (e.g., Liebowitz and Margolis, 1990, 1995) and by historians (e.g., Kirsch, 1996). Although the steam car won a number of early automobile races, the internal combustion engine offered a much higher power to weight ratio (especially important considering the bad roads at the time) and no requirement for frequent water refilling. It also had a much larger range

---

[19]Between 1918 and 1938, the western part of Austria drove on the right side of the road, and the eastern part drove on the left. Italy in the 1920s was even more complicated; in major cities where tramways drove on the left (reflecting their origin in England), cars also drove on the left. In the countryside, cars drove on the right.

Box 2.2: Technology Standards

Technology standards are a set of codified technology characteristics that enable:

- *Interchangeability* (e.g., electricity plugs of different devices all fit into the wall sockets.
- *Product information* (e.g., producers and consumers alike can rely on standard-ized product qualities).
- *Interoperability* (e.g., a train can operate throughout the entire railway network if gauges are standardized).
- *Regulation* (e.g., through establishing environmental standards).

Standards can emerge spontaneously (*de facto* standardization), or can be the inten-tional outcome of a formal process of cooperation between companies (e.g., between different equipment manufacturers of compact discs) or of administrative procedures (*de jure* standardization).

The first wave of standardization originated at the end of the 18th century and aimed toward industrial rationalization. A typical example of this would be standardized metal construction parts that could be used for building a whole range of structures, from bridges to the Eiffel tower. The main economic rationale of such standards is the exploitation of economies of scale.

A second (and in some ways parallel) wave of standardization originated from the increasing complexity of products and the increasing size of markets. This created information asymmetry problems between sellers and buyers of products. Quality standards help to evaluate product quality without requiring costly inspection and test procedures. (For instance, at a gas station, the consumer needs to be sure that "unleaded" is indeed unleaded gasoline.) The main economic function of this type of standard is the reduction of transaction costs.

The third category of standards enables exploitation of so-called *network externali-ties*, where the (economic and user) value of a network (from railways to information technologies such as the telephone system) increases with its size. This requires interoperability and interconnectivity (interface or compatibility standards) among initially independent and incompatible networks that can co-exist sometimes for ex-tended periods of time. For instance, it took nearly 50 years before the different gauges of private railway companies became standardized in the USA enabling a train to run from the east to the west coast. Spain and the former USSR continue to use a different (wider) railway gauge from the rest of Europe (and as illustrated below, a diversity of electric plugs standards still persists).

The last standardization movement is more recent: the use of standards as regulatory instruments to increase social welfare such as health, or environmental quality. Mini-mum quality standards or levels fix the maximum allowable levels (e.g., of emissions, noise, or of pollution and toxics in water and food). Obviously these standards change over time, influenced by increasing knowledge of negative effects and the availability of new technologies to monitor and measure ever more dilute concentrations.

Dominique Foray
Centre National de la Recherche Scientifique, Paris, France

than the electric car, an advantage that continues today. The QWERTY keyboard design is argued to be ergonometrically inferior to alternative layouts (e.g., the Dvorak design). In the age of mechanical typewriters, however, the resulting reduction in typing speed and less frequent hammer blocking may have been desirable features of the QWERTY layout.

The QWERTY keyboard is a good example of the extent to which we are often "locked-in" to particular configurations, artifacts, technological systems and standards (Arthur, 1988).[20] A particular solution that may have been best at an earlier time, but now faces superior alternatives, can often only be dislodged with great difficulty and at high costs. Not only do technologies change; so do social, environmental, and technological priorities and requirements. Given such changes, the existence of a large stock of technologies and infrastructures strongly influenced by past decisions creates formidable challenges, and can even become an obstacle for the introduction of newer systems and of economic growth (cf. the classic paper by Marvin Frankel, 1955). However, this is no real news. Societal concerns have been, and continue to be, important forces shaping technology systems. In turn, dominant technological systems are difficult to change within a short period of time.

Such challenges are not insurmountable, and indeed technologies eventually become adapted to changing social preferences. The bicycle is an example of such an adaptation and of the extent to which social fashion drives initial technological designs. Today's bicycle, with front and rear wheels of equal size, is derived from the safety bicycle design that emerged at the end of the 19th century. Its design is radically different from earlier bicycles, particularly the famous Penny-farthing (*Figure 2.18*).

Why were the Penny-farthing and (the name tells all) *Boneshaker* designs successful in the 19th century, whereas the safety bicycle only emerged at the end of the century? The answer lies in the changing expectations that people projected onto the technology. The Penny-farthing's main appeal was to "young men of means and nerve" (Pinch and Bijker, 1987:34). Such an athletic image conveyed by customers and producers alike neglected women

---

[20]Technological "lock-in" is often referred to as "path dependency". We prefer to use the term "lock-in" to describe a particular historical choice that becomes almost irreversible, standards being the most apparent example. We will use "path dependency" for describing apparent stabilities in macropatterns of technological change resulting from the accumulation of many decisions moving in a persistent direction. These are not the result of a discrete historical event or "accident". They result from persistent "signals" driving technological change in one particular direction and thereby creating irreversibilities, or at least substantial inertia. We return later to the issue of path dependency when we address theories of induced technical change.

Figure 2.18: A typical Penny-farthing bicycle (Bayliss-Thomson Ordinary of 1878), a design for "young men of means and nerve". The safety bicycle design (resembling the bicycle of today) evolved much later. Its rather bumpy development history was apparently strongly influenced by the social construction of "what a bicycle had to be". Quotations from Pinch and Bijker (1987:28–34). Photograph courtesy of the Science Museum London/Science & Society Picture Library.

with their cumbersome 19th-century dress code. It took many unsuccessful design innovations, several confluent technology developments (Dunlop pneumatic tires and the rear chain drive), and 20 years before the alternative design and social image of the bicycle that we know today stabilized: a bicycle as a safe and comfortable transport device, that anybody could ride. This "social constructivist" perspective emphasizes feedbacks between consumers and designers, between actual and potential users, and among different social groups promoting or resisting particular technological configurations and designs.

Such interactions usually pass unnoticed. They become most apparent in instances of violent opposition to technological change. Such opposition is a recurrent historical phenomenon – from the Luddites, to resistance against railway construction (*Figure 2.19*), and modern-day concerns over job losses and NIMBY (Not In My Back Yard) resistance. The Luddites were organized bands of English handicraftsmen who sought to destroy the textile machinery that was displacing them. They were named after their imaginary leader, King Ludd. The movement started in 1811 in Nottingham and spread quickly. It was halted by severe repression, culminating in a mass trial at York in 1813, with many hangings and deportations. The pattern was to be repeated later in 1830 in the resistance of the Captain Swing movement to new agricultural machinery (see *Figure 2.20*). [The best overview of resistance to technology continues to be Stern (1937:39–66).] Interestingly, the opposition to mechanical threshing machines in rural England in the 1830s also follows the classic diffusion pattern (*Figure 2.20*). The diffusion rate of about two weeks shows the effectiveness of social networks even in the absence of modern transport and communication technologies.

Opposition to technological change is a source of uncertainty, but it can also serve as an effective selection mechanism that either eliminates socially unsustainable solutions or prompts technological designs to be responsive to societal concerns. As such, opposition illustrates best the complexity of the forces driving technological change. The interplay among social groups shapes the context in which technologies evolve and can trigger an exploration for new alternatives when existing technological combinations no longer appear sustainable.

2.3. Sources of Technological Change

There are three principal sources of technological change: (i) new knowledge; (ii) improved application of knowledge, i.e., learning; and (iii) entrepreneurship and organization. All three represent "disembodied" aspects of technology regulated through social "techniques", including institutions such as universities and R&D laboratories, media such as scientific and applied journals, and incentive systems such as patent protection. New developments in these disembodied (software) aspects of technology need to occur before embodied (hardware) technological change can take place, although embodied technological change can then lead to further advances in knowledge. New scientific knowledge leads to new technologies, but science also depends on technologies for measurements, experiments, and disseminating new knowledge. Thus, there is no simple one-way street between science and technology, or between technology (instruments, new observation technologies) and science, as convincingly argued by Adams (1995:32–33).

Galileo's discovery of Jupiter's moons and his challenge of the Aristotelian dogma of the sun revolving around Earth were made possible by a new technological artifact from the Netherlands: the telescope. In turn, new

Figure 2.19: Resistance to US railways: January 1838. Source: Grübler (1990a:105), courtesy of Metro-North Commuter Railroad, New York.

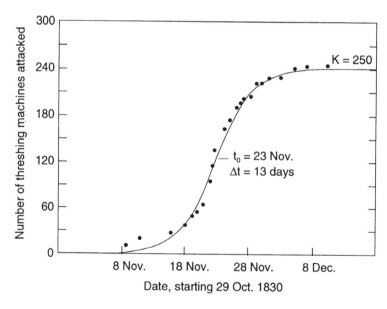

Figure 2.20: Resistance to technology as a diffusion process: number of threshing machines attacked during the Captain Swing movement in 1830. Data source: Hobsbawn and Rudé (1968:Appendix III:1–24).

scientific theories (astronomical in this case) directed and guided the further development of observational and measurement technologies. The spectrum of signals analyzed by astronomers today far extends beyond the human eye.

Knowledge takes many forms and comes from many sources. Over the past 300 years, science has emerged as the principal "technology" for generating new knowledge. Distinctions are commonly made between basic and applied science and research, and between public knowledge, proprietary knowledge, and truly private or tacit knowledge. Public knowledge is what anyone can acquire, e.g., by reading *Nature* or *Science*, or other information in the open literature. Proprietary knowledge is protected by patents and access is limited through licensing arrangements. Private or tacit knowledge includes special "tricks" in manufacturing that are largely unrecorded, known only to experienced workers and passed on largely through "hands-on" experience. There is a correlation between the institutional source of knowledge and its *appropriability*. Scientific knowledge is largely a public good, and much applied knowledge is either proprietary or tacit.

The primary institutions of science – universities, learned societies, and academies of science – date from the Age of Enlightenment, but professional and industrial R&D is a relatively recent phenomenon (Rosenberg, 1991). The first R&D labs were created for elementary tasks such as measurement and quality control. Typical first applications were measuring the metal content in ores and measuring the quality of metallurgical products. Another early application was research on possible uses for "by-products" of petroleum refining in the production of illuminating oil (Rosenberg, 1991). These early "by-products" are now principal products of oil refining: motor fuels, petrochemical feedstocks, and lubricants. Only at a much later stage did industrial R&D labs move into process and product innovation.

The distinction between basic and applied science and the development of many technologies from scientific results suggests a linear model of technological change. This model is a more detailed stage representation of the life cycle typology *invention, innovation*, and *diffusion* discussed previously. The stages of this model are as follows:

- Basic research produces new scientific knowledge (discoveries).
- Applied research leads to proposed applications (patents).
- Further applied research and development refines this knowledge sufficiently to justify substantial investments in new technology (development).
- Investments are made in new production facilities, equipment, and specific products (innovation).

- Experience leads to improvements and adaptation in early applications (early commercialization).
- Widespread commercialization leads to new levels of technical standards, economic performance, and productivity (diffusion).

To these stages we could add three more:

- Experience, learning, and feedbacks from customers lead to further technological and economic improvements and to wider fields of application.
- Pervasive diffusion leads to macroeconomic, social, and environmental impacts.
- Such impacts lead to scientific research and new information on causes of and possible solutions to adverse impacts.

This takes us back to square one, and the whole sequence starts again. Following these steps in the order just presented represents a science or technology "push" view of technological change. Were we to follow essentially the same steps but in the reverse order, we would have a "demand pull" view of technological change. Both are extreme perspectives. The first views technology development as driven exclusively by opportunities; the second views it as driven exclusively by needs.

Both linear models have been largely dismissed in the literature (see e.g., Mowery and Rosenberg, 1979, or the review article by Freemann, 1994) in favor of models with multiple feedbacks and various factors driving different phases of a technology's life cycle. In early phases science/technology push factors may dominate, whereas in later phases demand pull factors may be more important (see e.g., the work of Walsh, 1984; and Fleck, 1988).

There are certainly examples of a linear development sequence where "science discovers and technology applies", e.g., nuclear energy and the transistor and semiconductor. But counterexamples also abound. The first steam engines were built without much understanding of thermodynamics, which was developed only much later. The Wright brothers flew propelled, heavier-than-air machines, even while some physicists still proclaimed this to be impossible (Rosegger, 1996:4). Aviation developed in the 1920s and 1930s without the knowledge and technology to fly in difficult weather conditions or at night. Radar, today considered essential for aircraft navigation, was not developed until World War II.[21] Such examples emphasize the inadequacy

---

[21] The eminent sociologist of invention S. Colum Gilfillan (1935) listed 25 different technological means to overcome the limitations that fog and similar bad weather conditions represented for aviation (NRC, 1937). None of these eventually contributed toward the solution that was provided by radar. But Gilfillan was right in predicting "quite confidently" that the problem of fog would soon be overcome, and he was justified in exploring scenarios of industry development that assumed no danger from fog.

of models in which technological change proceeds linearly with "neat" divisions between science and technology. This does not create problems for scientists doing basic research within industry (AT&T scientists have, for example, been pioneers in atmospheric chemistry and the discovery of cosmic background radiation). But it can embarrass modelers who treat knowledge generation and improvements in a technology's application as exogenous to the economic system.

2.3.1. Who performs and who pays for knowledge generation (R&D)?

Tables 2.4 and *2.5* present a statistical overview of the R&D enterprise in the USA, the country with the largest R&D expenditures. Some US$160 billion were spent on R&D in 1993, about 2.5% of the gross domestic product (GDP). This is similar to the percentage of GDP spent on R&D in most of the advanced industrial economies. About two-thirds of R&D expenditures are devoted to (expensive) development work, 25% to applied research, and about 15% to basic research. By far the largest part of this research effort (70% or US$112 billion) is performed by industry, simply because it is industry that typically does development work, and development dominates R&D expenditures. Overall, industry provides slightly more than half of the total R&D funding in the USA. The role of government and other nonindustry institutions in R&D is also very important. It is justified first by the fact that much of new knowledge produced by research, especially basic research, is a public good. Nonindustrial R&D is also justified by the potentially very long lead times between the generation of new knowledge and its possible applications and the fact that new knowledge may never produce any direct economic "spin-offs". For these reasons firms are likely to underinvest in R&D that would be beneficial to society. Public expenditure in research is justified because society must consider the long-term future more than firms and value the noneconomic social and cultural spin-offs and new knowledge simply for its own sake.

Quantitative statistics, such as R&D expenditures or R&D personnel, only measure the inputs to knowledge generation. Outputs are even harder to quantify in the aggregate. Where attempts have been made to measure the R&D output of corporations, in terms of new products, improved production methods, etc., the results indicate significant economic returns to R&D. Frosch (1996:27) for instance, reports (internal) rates of return from 38% to 70% for the R&D operations of companies such as General Electric or General Motors, respectively.

Table 2.4: R&D activities in the USA in 1993, by institutional sector.

Sector	Basic research Mill. US$	%	Applied research Mill. US$	%	Development Mill. US$	%
Federal government	2,900	11.1	4,900	12.3	8,800	9.3
Industry	4,700	17.9	26,500	66.8	81,100	85.5
Universities and colleges	16,350	62.4	6,360	16.0	3,140	3.3
Nonprofit institutions	2,270	8.6	1,920	4.9	1,810	1.9
Total	26,220	100.0	39,680	100.0	94,850	100.0

Source: National Science Board (1993).

Table 2.5: R&D funders and performers in the USA in 1993 (in million dollars).

R&D performer	Source of funds Federal gov.	Non-fed. gov.	Industry	Uni. & colleges	Nonprofit inst.	Total	%
Federal gov.	16,600					16,600	10.3
Industry	31,000		81,300			112,300	70.0
Universities and colleges	16,700	1,850	1,500	4,150	1,650	25,850	16.0
Nonprofit institutions	3,700		750		1,550	6,000	3.7
Total	68,000	1,850	83,550	4,150	3,200	160,750	
%	42.3	1.1	52.0	2.6	2.0		100.0

Abbreviations: gov., government; Uni., University; inst., institutions.
Source: National Science Board (1993).

Still, in as far as the main output of R&D is new knowledge, or rather *new combinations* of knowledge, that can subsequently be applied in production (where economic returns accrue), it is indeed a formidable challenge to try to measure R&D "output" directly. Unlike measuring the capital intensity, or the energy intensity of an economic sector or industry, it is extremely difficult to measure "knowledge intensity" (Smith, 1995). Patent statistics suffer two weaknesses. Not all new knowledge is patented, and not all patented information is used. Nevertheless patent research has identified patterns of inventive activities (e.g., Pavitt, 1984) that provide useful insights into important sectoral and industry differences in knowledge generation and innovation.

Tables 2.4 and *2.5* indicate that R&D extends well beyond government-sponsored basic research and should therefore not be treated as "external" to economic activities. On the contrary, knowledge generation and technological development are an integral part of economic activity and constitute the single most important "input" to growth in a modern economy. Such an endogenous view of knowledge generation becomes even more important when analyzing improvements in technological applications as reflected in "learning curves".

2.3.2. Learning

The performance and productivity of technologies typically increase substantially as organizations and individuals gain experience with them. Such improvements reflect organizational and individual learning. Learning can originate from many sources. It can originate from "outside" an organization – an example is a company that, in order to facilitate its own introduction of a new process technology, hires a production engineer from a competitor that has already done so. Or learning can originate from the "inside" through R&D and investments in new technologies. Learning can come through improving "know-how", i.e., learning how to "make things better" with the "things" (artifacts, designs, practices, jobs, etc.) basically unaltered. Or learning can come through improving design features and economies of scale, i.e., reducing costs by building and using larger and larger units. There is, however, one strict precondition for learning. It requires effort and the actual accumulation of experience. It does not come as a free good.

Technological learning phenomena – long studied in human psychology – were first described for the aircraft industry by Wright (1936), who reported that unit labor costs in air-frame manufacturing declined significantly with accumulated experience. Technological learning has since been analyzed for manufacturing and service activities ranging from aircraft, ships, refined petroleum products, petrochemicals, steam and gas turbines, even broiler chicken. Applications of learning models have ranged from success rates of new surgical procedures to productivity in kibbutz farming and nuclear plant operation reliability (Argote and Epple, 1990). In economics, "learning by doing" and "learning by using" have been highlighted since the early 1960s (see e.g., Arrow, 1962; and Rosenberg, 1982). Detailed studies of learning track the many different sources and mechanisms (for a succinct discussion of "who learns what?", see Cantley and Sahal, 1980). Here we focus on the productivity gains from learning, and these can be very large indeed. During the first year of production of World War II Liberty ships, for example, the average number of labor hours required to produce a ship decreased by 45%, and the average time decreased by 75%. There are also cases, however, where no learning is evident, and we briefly discuss the reasons for such learning failures.

Learning phenomena are described in the form of "learning" or "experience" curves, where typically the unit costs of production decrease at a decreasing rate. Unit costs decrease along an exponential decay function. Because learning depends on the actual accumulation of experience and not just on the passage of time, learning or experience curves are generally described in the form of a power function where unit costs depend on cumulative experience, usually measured as cumulative output:

$$y = ax^{-b},$$

where y is the unit labor requirement or cost of the xth unit, a is the labor requirement or cost associated with the first unit, and b is a parameter measuring the extent of learning, i.e., the unit labor or cost reductions for each doubling of cumulative output. The resulting exponential decay function is frequently plotted with logarithmically scaled axes so it becomes a straight line (see *Figure 2.21*). Because each successive doubling takes longer, such straight line plots should not be misunderstood to mean "linear" progress that can be maintained indefinitely. Over time, cost reductions become smaller and smaller as each doubling requires more production volume, and the potential for cost reductions becomes increasingly exhausted as the technology matures.

Figure 2.21 plots the costs per kW as a function of total cumulative installed capacity for several electricity generation technologies. The figure shows how costs drop as experience accumulates. The learning curve patterns shown in *Figure 2.21* illustrate several general features characteristic of technological learning.

First, the learning rates, at about a 20% reduction in specific investment costs for each doubling of cumulative output, are quite similar across the three technologies of wind, gas turbines, and PV cells. This is true despite the initial costs of PV cells being ten times higher than the costs of gas and wind turbines. The learning rates are also similar between countries as shown by the PV costs in the USA and Japan.

Second, when costs are plotted as a function of accumulated experience rather than time, it is easier to draw useful analogies. For example, *Figure 2.21* shows that the dynamics of cost reductions for windmills in the USA in the 1980s are quite similar to those for gas turbines in the early 1960s.

Finally, note the two distinct phases of cost reductions in the case of gas turbines. There is an early rapid phase associated with R&D and technical demonstration (in the innovation phase), followed by distinctly slower cost reductions during commercialization (the diffusion phase). This illustrates

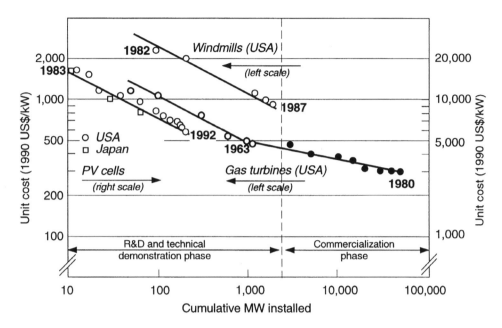

Figure 2.21: Technology learning curves: unit cost (US\$ per kW) versus cumulative experience, measured by output (installed MW) for photovoltaics (right hand side scale), wind and gas turbines (left hand side scale). Note in particular the similar slope of the learning curves of the three technologies and that photovoltaics start off at costs ten times higher than the two turbine examples. Source: IIASA–WEC (1995:29).

important differences in the sources of technological learning in different phases of a technology's life cycle. As a rule, cost reductions are most substantial in early phases where R&D and design improvements yield the largest return on investments, even though benefits may not accrue directly to investors. Later entrants have the benefit of "external" learning from the improvements achieved by the "internal" learning financed by early innovators. New technological knowledge is costly to produce, but cheap to imitate. To limit external learning, or "free-riding", and to protect R&D performers, regulatory measures, particularly the patent system, have been created. Such protection is far from perfect, however. Information, learning, and experience can leak out through staff turnover, key R&D personnel being hired elsewhere, or through straightforward espionage. However, such "leakage" may be socially desirable – leading to fast diffusion of new knowledge – even if it may not be desirable for the individual firm.

The rate of learning and experience can vary enormously among different sectors and technologies. *Figure 2.22* illustrates the range of learning rates

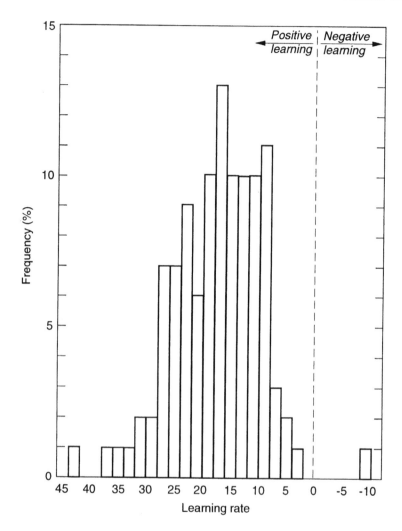

Figure 2.22: Distribution of learning rates (unit costs reduction, in %, for each doubling of cumulative output) for a sample of 108 technologies synthesized from 22 field studies. Source: adapted from Argote and Epple (1990:921).

(cost reductions per doubling of cumulative output, i.e., the parameter b in the previous equation) from a sample of 108 different technologies and products. Learning rates range from a high of 45% to only a few percent. There are also examples of negative learning, or "organizational forgetting", where costs increase rather than decrease.

In addition to learning via R&D and actual experience (investments), significant learning takes place through large-scale production. We divide large-scale production learning into three classes (see also Cantley and Sahal, 1980). These three classes are listed as follows:

1. Learning by upscaling production units (e.g., the examples of steel converters and steam turbines given previously).
2. Learning through consecutive repetition or mass production (e.g., the Model T Ford).
3. Learning through both increasing scale and consecutive repetition, referred to here as "continuous operation", i.e., the mass production of standardized commodities in plants of increasing size. The best examples are base chemicals such as ethylene or PVC (polyvinylchloride), where cost reductions have been particularly spectacular (Clair, 1983).

Such large-scale production learning usually begins at the individual plant level, but later spills over to other plants (for which this represents a source of external learning) and eventually spreads to an entire industry.

A statistical analysis of learning rates across many technologies and products (Christiansson, 1995) confirms the value of the above taxonomy and concludes that learning rates are typically twice as high for "continuous operation" as for either upscaling or mass production alone. (The mean learning rate for continuous operation in the Christiansson sample was 22%, compared to 13% for upscaling and 17% for mass production.)

A learning rate of 20% is a representative mean value advanced in the literature (Argote and Epple, 1990). Twenty percent is also the mode of the distribution function shown in *Figure 2.22*.

The example of negative learning shown in *Figure 2.22* deserves some elaboration. The example comes from the Lockheed L-1011 Tristar aircraft production. Production started in 1972 and reached 41 units in 1974. It subsequently dropped to 6 units in 1977, and increased again thereafter. The drastic reduction in output led to large-scale layoffs. When production increased again, new personnel were hired, and the experience gained initially was lost with the staff turnover. As a result, production cost reductions could not be maintained, and the planes built in the early 1980s were in real terms (after inflation) more expensive than in the early 1970s.

Thus, stop-and-go operations in R&D, and "hire and fire" strategies in production, seem to be detrimental to technological learning. Continuity in effort, in accumulation of experience, and the maintenance of human know-how seem essential for technological learning. The converse of "learning by doing" is "forgetting by not doing". This holds for R&D and production alike. Massive technology crash programs that are abandoned after a few years (e.g., the multi-billion dollar US synthetic fuel program), "stop-and-go" production schedules (e.g., of the Lockheed Tristar), or frequent design

changes at considerable cost (e.g., in nuclear reactors to improve safety features) all illustrate that learning and cost reductions are not always related
to scale of effort. It also depends on how efforts are organized and on the
continuity and commitment of the effort. Technological "forgetting", or creating conditions not conducive to learning, can sometimes be as powerful as
"learning".

2.3.3. Entrepreneurship and organization

We have discussed R&D and learning as important sources of technological
change. None of these activities can take place without dedicated human
effort, and it is therefore important to conclude this chapter by mentioning
the human and organizational factors in technological change. These factors were particularly stressed by Joseph A. Schumpeter. He believed that
the organizational entity bringing about new technological "combinations"
is the firm, and that innovative activities usually do not arise out of existing
firms. "It is not the owners of stagecoaches who build railways" (Schumpeter,
1911:66).[22] The creation of such firms and the promotion of particular new
"combinations" was the domain of Schumpeter's "entrepreneur". Schumpeter's emphasis on the entrepreneur as the bearer of change seems to have
been unduly influenced by the writings of Nietzche and prominent capitalistic
entrepreneurs such as Vanderbilt, Carnegie, Edison, and Rockefeller. Schumpeter later acknowledged the importance of large organizations in performing
R&D. A development engineer in the R&D department of a large electrical
firm would equally qualify as a Schumpeterian "entrepreneur" (Freeman,
1994), as would a manager keen to introduce a new production process, or a
marketing salesperson (a "change agent" in the terminology of the diffusion
literature, see Rogers, 1983) promoting a new product.

As an example, the now ubiquitous yellow "post-it" notes were originally conceived by a 3M company employee who sang in a choir and was
annoyed that the paper slips used to mark the hymns kept slipping away.
The technological ingredients that were combined in post-it notes already
existed; the innovation consisted of creating the new combination. The prototype, however, fell flat. Major office supply distributors thought it was
silly; market surveys were negative. The product, which is now a US$100
million plus business for 3M, eventually succeeded because 3M's secretarial
staff liked to use the specimens available within the company. The eventual breakthrough came with a mailing of product samples to *Fortune 500*

---

[22]A more contemporary quote in the same spirit is attributed to C.F. Kettering, the
founder, and patron saint, of the GM research labs: "Never put a new technology in an
old Division" (as observed by an anonymous reviewer of this manuscript).

CEO executive secretaries under the letterhead of the 3M executive secretary (Peters, 1986). While post-it notes may not classify as a major technological innovation, they are certainly a major "entrepreneurial" innovation – realized, promoted, and brought to success by individuals within a large corporation.

Such individualistic conceptions of technological change may appear naive in the age of large multinational corporations, institutionalized R&D and "big science" (de Solla-Price, 1963). But they point to the importance of organizational and institutional factors in the promotion of, or opposition to, technological change. Organizations and institutions represent social "techniques" to organize and to regulate individual human actions.

For instance, large corporations do not usually entrust the development and commercialization of new innovations to departments responsible for the existing, dominant technology. For promotion of rapid development, organizational "offsprings", such as "skunkworks", largely liberated from bureaucratic routines and tedious accountancy, have become an accepted organizational strategy. The US Army asked Lockheed in 1943 to design a new fighter aircraft, stipulating that the prototype must be delivered within 180 days. Lockheed entrusted the task to Clarence L. "Kelly" Johnson, who drew together a small team of designers, engineers and shop mechanics. They were located in temporary quarters in California near a foul-smelling industrial site, hence the name "skunkworks". [Another, or perhaps complimentary, explanation for the word comes from a popular comic strip (Lil Abner), where two brothers produce mysterious elixirs in their "skunkworks".] Johnson had 14 management rules that assured considerable informality, autonomy, and flexibility. The prototype fighter was ready in just 137 days. It was the first US jet fighter aircraft. Later technological marvels of Johnson's skunkworks were the U2 spy plane and the famous SR71 "blackbird" aircraft, which has held the speed record for air-breathing aircraft since 1962. For an autobiography of Kelly Johnson (1910–1990), see Johnson and Smith (1985).

Of course innovations continue to be created by individuals and small firms, even if the latter – if successful – do not necessarily stay "small" for long. Much has been written on the impact of firm size on innovations and their diffusion. The conclusion is that "bigger" is not necessarily "better". Internal organization within large firms is as important to innovation and diffusion as is the role of small enterprises.

New actors appear increasingly on the scene. Government-sponsored agricultural research institutions and dissemination efforts have been instrumental in introducing new crops and farming practices in the USA. Networks of institutions rather than "monolithic" R&D organizations have emerged.

The Consultative Group for International Agricultural Research (CGIAR), for example, is a network of 17 agricultural research institutions. It conducts primary research on crops and exchange of genetic resources, and also plays a major role in the diffusion of new, high-yield strains to farmers, particularly in the tropics. Environmental NGOs play an increasing role not only in opposing certain technologies, but also in actively promoting more environmentally compatible innovations. Greenpeace Germany, for example, commissioned a small company (Freon) in the former German Democratic Republic to design a refrigerator without ozone-depleting CFCs. The successful design forced all major refrigerator companies to quickly offer CFC-free models also (much to the detriment of the small, innovative company).

Thus, the portfolio of change agents is larger than ever, and their motivations, incentives, risk perceptions, and views of the future are ever more diverse. The notion of a single representative "agent" of technological change is outdated, although it continues to be used in much of the mainstream modeling of technological change, as discussed in the following chapter.

Finally, it is important to dismiss the notion of "lonely heroes" as innovators and agents of technological change. People communicate with each other, exchange ideas and information, and thereby create joint "technological expectations" (Rosenberg, 1982). These influence the visions, missions, and expectations of all those involved in research and development, marketing, etc. Because everybody expects things will develop in a particular direction, research and development focus on that direction. The model of the self-fulfilling prophecy is entirely appropriate here. It has been shown that joint expectations in the microchips business, expressed in shared technological forecasts,[23] helped establish targets, drive research, and achieve results in line with the motivating expectations (Mackenzie, 1991; Benzoni, 1992). Motivating expectations also encompass consumers. Those in the market for personal computers, for example, time their purchases based on shared expectations that prices will inevitably drop and that the next generation of models will be more powerful and their performance will be better than their forebears.

---

[23]Gordon E. Moore, Director of Fairchild Semiconductors (and one of the co-founders of Intel Corporation), postulated in 1964 that, based on trends since 1959, the number of transistors per integrated circuit would double every year or so (Benzoni, 1992:25). By mid-1995 the number of transistors per chip had reached about 100 million, basically on track with "Moore's law".

Copyright acknowledgments

Chapter 3

Technology: Models

Synopsis

The chapter gives an overview of the efforts to model technological change, which to date have been largely disappointing. Macroeconomic models that treat technology as a residual quantity are discussed first, both in their original classical growth accounting formulation as well as in their contemporary use in macroeconomic energy and environmental models. The chapter then presents sectoral models as well as models based on microeconomic foundations. The latter two types of models offer greater insights and explanations of the dynamics of technological change. These are characterized by features of path dependency, i.e., change in a persistent direction influenced by past decisions, technological uncertainty, diversity, learning, and interaction between economic agents.

3.1. Models of Technological Change

We start with a disappointing confession. There is no single model, or class of models, that captures all the aspects of technological change outlined in Chapter 2 in even a rudimentary integrated fashion. It is not from lack of trying. Hundreds of models have been developed with various levels of aggregation, theoretical underpinnings, and empirical corroboration. We cannot do justice to them all here. They are considered in two categories: at a macrolevel and a sectoral/microlevel. Illustrative examples for each are presented, along with their (few) strengths and (numerous) deficiencies and omissions. How to improve this situation (at least partially) will be discussed in the Postscript chapter. Ideas and applications developed by the author and his colleagues at IIASA will be presented, together with illustrations primarily from the energy sector. These novel attempts integrate the three main clusters of drivers of technological change: uncertainty, R&D, and increasing returns (learning) into a model formulation of endogenous change, drawing on new mathematical tools that enable stochasticity and nonconvex model behavior to be dealt with efficiently.

91

The essential criticism advanced here (and elsewhere) is that traditional models adopting a macroscopic view of technology either treat technological change as exogenous or represent it in the form of a "black box", thereby allowing no insights into its internal drivers and dynamics. Conversely, models adopting a microscopic view are rich in detail, but are unable to capture important macroscopic transformations and feedbacks. We start the discussion with macro(economic) models and continue with micro(economic) models. We focus on economics because it is the research discipline that has done the most quantitative modeling on technology. It is the only discipline where both the driving forces of technological change (such as price changes) and impacts (such as productivity changes) have been the focus of research. We do not cover phenomenological or descriptive modeling traditions such as those prevailing in management science or sociology.

3.1.1. Macroeconomic perspectives

Modern economics is the science of the allocation of scarce resources. Since its inception, the focus of macroeconomic modeling has therefore been on how (scarce) factor inputs are allocated (e.g., capital, labor, and land) and how they contribute to economic output growth. The focus on output growth constitutes an immediate important limitation. To focus on output, e.g., as measured by gross domestic product (GDP)[1] and its growth means to focus on *flows*. *Stock* variables are neglected. Key stock variables include knowledge (especially technological knowledge), the level of capital stock ("How rich are we?"), and the stock of natural and environmental resources ("Are we really as rich as we thought we were if our resources and environment are becoming quickly depleted and degraded?").

The essence of macroeconomic modeling is to explain output and output growth as functions of available inputs. In the eyes of the classical writers, such as Adam Smith and David Ricardo, land was the principal constraining input. Modern macroeconomic growth models focus more on capital and labor inputs.

---

[1] GDP measures the value of all goods and services produced by the factors of production located in an economy that are subject to economic transactions, i.e., exchange via money. Extending the geographical boundary to include factors of production located outside a country that nonetheless contribute to the country's economic growth (e.g., income from patent fees from abroad or the transfer of profits from companies owned abroad) results in the gross national product (GNP). Note that *neither* measure takes into account inputs/outputs for which there are no market transactions, no matter how essential and important they might be. Factors and activities that are excluded range from household work and voluntary social activities to environmental amenities and degradation.

Technology as a "Residual"

Let us turn now to the first attempts to quantify technology's contribution to economic output growth. In his classic 1957 paper on technical change and the aggregate production function, Robert M. Solow pioneered a macroeconomic conception of technological change that continues to this day.[2] To model the relationship between economic output and inputs of labor and capital he used a Cobb-Douglas production function:

$$Y = kL^{\alpha}C^{(1-\alpha)}$$

where Y represents economic output, L labor, and C capital; k and α are positive constants. α is less than one and represents the "elasticity of substitution" between capital and labor.

Such a production function and its variants [e.g., constant elasticity of substitution (CES) functions] continue to be *en vogue* in both macroeconomic and economy–environmental modeling. Frequently the production function is expanded ("nested") by including additional factor inputs like energy or agricultural land to complement the traditional inputs of capital and labor.

We should first note briefly that we harbor reservations about important features of this (and similar) production function. First is its "(practically unavoidable) assumption of constant returns to scale" (Solow, 1957:317). Constant returns to scale mean that an additional unit of capital or labor will produce exactly the same amount of economic growth as the units of capital and labor already used in the economy. Second is the assumption that both capital and labor are independent of technology, i.e., the fact that the production function model treats the explaining variables of output growth as well as the residual (the unexplained output growth) as independent from each other (see Abramovitz, 1993). We return below to this, but at this point, of principal interest is Solow's quantitative evidence.

In Solow's production function model and its modern successors, economic growth is explained by endogenous developments, such as capital formation, and exogenous developments. Exogenous developments include

---

[2]Credit for the earliest model calculations apparently belongs to the Dutchman Jan Tinbergen, later awarded the Nobel prize (Tinbergen, 1942). [His paper was published in German in the *Weltwirtschaftliches Archiv*, incidentally at a time (1942) the Netherlands were occupied by the Nazis.] Tinbergen added an exponential term for "technical developments" to a Cobb-Douglas production function and computed the average value of this trend component (as a measure of "efficiency") for four countries. Solow's 1957 model was very close to Tinbergen's original formulation, although Solow was apparently unaware of Tinbergen's article. For a concise history of the "residual" in economics, see Griliches (1996:1324–1330).

Table 3.1: Growth of (actual) US national income (% per year) and contribution by source of growth (%), total and per person employed, 1929–1982.

	Actual national income			
	Total		Per person employed	
	Whole economy	Non-residential business	Whole economy	Non-residential business
Contribution to growth rate (%)				
Labor input (excl. education)	32	20	−12	−25
Education per worker	14	19	27	34
Capital	19	14	20	13
Advances in knowledge	28	39	55	68
Improved resource allocation	8	11	16	18
Economies of scale	9	12	18	22
Changes in legal and human environment	−1	22	−3	−4
Land	0	0	−3	−3
Irregular factors	−3	5	−7	−8
Other determinants	−5	8	−10	−13
All sources	100	100	100	100
Growth rate (%/yr)	2.9	2.8	1.5	1.6

Source: Denison (1985:30).

both labor availability (the domain of demographics) and technological development, generally subsumed under "advances in knowledge". Assessing the relative contributions of different factors to per capita economic growth in the USA between 1909 and 1949, Solow concluded, "Gross [economic] output per man-hour doubled over the [1909–1949] interval, with 87.5% of the increase attributable to technical change and the remaining 12.5% to increased use of capital" (Solow, 1957:320). In other words, only 12.5% of economic output growth per person-hour is explained by the model endogenously, while the majority (87%) represents an unexplained "residual" exogenous to the model. The numbers speak for themselves.

Since then, numerous "growth accountants" have extended Solow's production function methodology by trying to quantify various components of the unexplained "residual" at ever finer detail. The results confirm Solow's conclusion that technological change is the single largest contributor to per capita economic growth. Ironically, despite empirical and methodological sophistication, growth accounts and related neoclassical growth models perpetuate technology's treatment as an "externality" and unexplained "residual".

Table 3.2: Contribution of factor inputs to average annual growth (%/yr) of US national income, 1929–1982.

Sources of growth	%/yr
Level of employment	1.12
Reductions in average hours worked	−0.51
Improved efficiency per hour worked	+0.24
Age–sex composition of work force	−0.08
Education	0.40
Other (unallocated) factors	0.17
Total labor	1.34
Capital	0.56
Other factors (incl. advances in knowledge and technology)	1.02
Total national income	2.92

Source: Denison (1985:111).

Table 3.1 shows the results of work by Edward Denison (1962, 1985) for the USA.[3] Denison builds on Solow's approach and is ingenious in sifting out important factors, particularly in the area of labor inputs. Nevertheless, despite the increased level of detail in Denison's analysis, the "residual" (which he calls "advances in knowledge") remains large indeed. For the economy as a whole it accounts for 55% of national income growth per person employed (the comparable measure to Solow's). For nonresidential business it is 68%. Even for total income growth, advances in knowledge remain the second largest contributor after increases in employment.

Denison's analysis also illustrates the economic impacts of population growth. There are effectively two opposing views on population growth. One can be traced back to Thomas Malthus in 1798 (1986) and considers population growth primarily a burden: "more mouths to feed". The other (e.g., Simon and Khan, 1984) views population growth as a resource: "more brains to think and more hands to work". *Table 3.2* summarizes Denison's estimates of the contribution of labor to US national income growth from 1929 to 1982. Increased employment is estimated to have contributed 1.12 percentage points of a total of 2.9% per year average of US national income growth (see *Table 3.2*). Among the labor input characteristics studied by Denison, this is by far the largest contributor. The second largest

---

[3]The best data source for long-term data series at the international level of the factors accounting for long-term economic growth within a framework similar to that used by Denison are the exceptionally rich and comprehensive publications of Angus Maddison (e.g., 1995).

contributor, which had a negative impact on growth (see the negative entries in the labor input category in the "Per person employed" columns in *Table 3.1*), was reductions in working time (–0.51% per year), although this was partly offset by improved efficiency (and intensity of work) per remaining hours worked (0.24% per year). Finally, improvements in the quality of the labor force through education contributed a further 0.4 percentage points to growth.

Thus, within this production function framework, population growth has a definite positive impact on economic growth provided enough jobs can be found. And the beneficial impact of population growth on economic growth is further enhanced by better education. Thus, both opposing views on population growth are partially right. If enough jobs can be generated and the quality of labor improved through education, then the impact of population growth on economic growth is positive. Without jobs and education, however, population growth can indeed be viewed simply as "more mouths to feed". There exists, therefore, no economic "law" that the influence of population growth goes invariably in a particular direction.

The contribution of different components of the labor input to economic growth highlights an important weakness in this type of growth accounting. It goes beyond possible important measurement and aggregation errors. The key weakness is that the underlying causes of economic growth remain unexplained. Moreover, even if explained, the results would be misleading due to the assumed independence of various growth components. In fact, Moses Abramovitz (1993:218) states that the residual (i.e., that part of economic growth that cannot be explained by more labor and more capital employed in production) as a "measure of ignorance" grossly understates our ignorance.

Consider labor and technology. The spectacular decline in working hours, and the resulting increase in leisure time discussed in Part II, is an important welfare achievement that counts only as a negative factor, reducing production growth within this framework. The analysis also does not reflect how these reductions in working time came about. They are the consequence of productivity increases resulting from numerous technological and managerial changes (recall the example of the Model T Ford). Productivity increases are in turn "distributed" in the form of higher wages and reduced working time through a whole complex system of wage negotiations and work regulations. Thus it is erroneous to treat advances in knowledge and technology as independent from the labor component in the production function. Such an approach significantly underestimates the impact of technological change, no matter how large the "residual" remains in the analysis. The causal relationship also goes from the direction of labor to technology.

After all, without a qualified labor force, no new technology can be developed and applied.

Capital and technology can be considered in a similar manner. In as much as technological advances are embodied in a physical plant and equipment, it is impossible to separate increases in the capital stock from increases in technology. Unless one maintains that technical progress occurs only in consumption, increases in the stock of capital should not be treated as though they were independent of increases in productivity and efficiency, i.e., technology. The relationship between capital and technology goes in two directions: an accelerating growth of the capital stock means its mean age is reduced, and new technologies become incorporated into the capital stock faster.

Additional criticism on the production function framework focuses on "the failure to account for quality change in the consumption basket, the disamenities of modern growth and the valuation to be placed on enhanced leisure time" (Metcalfe, 1987:619). Another important weakness of production function models is that they focus on economic output rather than on economic welfare. Nordhaus and Tobin (1972) have addressed this weakness in an ingenious effort to assess economic welfare growth in the USA from 1929 to 1965. For our purposes their analysis provides some zero order estimates of, first, economic welfare associated with leisure activities and nonmarket activities such as child and household care, and, second, welfare losses associated with the "disamenities" of urban life and pollution. Nordhaus and Tobin highlight the impact of technological change on leisure and nonmarket activities. Their results are summarized in *Table 3.3*. Where ranges are shown, they reflect the range of alternative assumptions considered when assessing the impact of technological change.

The summary results of Nordhaus and Tobin are presented below.

- The gross national product (GNP), the value of all goods and services produced and exchanged in the formal economy, constitutes only half of the total economic welfare.
- After subtracting capital consumption, intermediary goods and services, and "regrettables" such as military expenditures, the final output reflected in national income and product accounts equals about 75% of GNP, and only 38% of total economic welfare.
- By far the largest component of economic welfare (about half) is leisure.
- Valuation of other nonmarket activities (e.g., family and household work) puts them at about one-quarter of total economic welfare.
- Environmental disamenities subtract about 3% from total economic welfare, and nearly twice that percentage from GNP.

Table 3.3: National income versus total economic welfare by component estimated for the USA, 1965.

	10^9US(1958)^a	%
Gross national product	617.8	49.8
Capital consumption	−148.8	
Final output from national income and product accounts (NIPA)	469.0	37.8
Estimates of items not included in NIPA		
Leisure	626.9–712.8	50.5–53.00
Nonmarket activities	259.8–295.4	21.6–23.8
Other consumption and services	−115.6	
(Environmental) Disamenities	−34.6	2.8
Measure of economic welfare	1241.1	100.0
	(1205.5–1327.0)	

^{a}Values are calculated for 1958 US$, i.e., at prices of 1958.
Source: Nordhaus and Tobin (1972:10–11).

- Because income and product accounts capture only half of economic welfare, GNP growth rates tend to exaggerate the increase in material welfare through economic growth. For instance, the GNP of the USA increased between 1929 and 1965 by 3.1% per year, and the population by 1.3% per year. Thus per capita GNP grew by 1.7% per year (Nordhaus and Tobin, 1972:56). However, per capita economic welfare increased by only 1% per year, with a range from 0.5% to 2.3% depending on how the impacts of technological change on leisure and nonmarket activities are factored into the analysis.

If we accept these zero order results, we have a model where economic growth has positive welfare implications as we would expect, while, at the same time, leisure and nonmarket activities dominate both economic welfare and welfare gains. Given technology's role in expanding leisure time, these results reinforce the importance of technological change. Not only is technology the major "residual" in explaining economic growth, it is also a major source of the productivity gains that reduce working time and thus expand overall economic welfare. We may also add that without technology, environmental disamenities (pollution) from economic growth would subtract yet a larger part of the welfare gains.

One response to the critique of classical production function models comes from "endogenous growth" theory in which the classical production function is enlarged by additional endogenous factors representing advances in knowledge. Romer (1986), for example, introduces into the production function the notion of human capital as a complement to physical capital. Human capital essentially grows exponentially, reflecting the cumulative nature of knowledge, and yields positive returns that offset diminishing marginal returns from physical capital. (In the classical Solow model these were compensated for by exogenous technological progress.) A similar endogenous growth model has been suggested by Grossman and Helpman (1991). In their model, knowledge capital serves as an input to R&D activities that generate innovations that are in turn a main engine of economic growth.

Both knowledge capital and R&D exhibit a number of important properties. First, like physical capital, R&D shows diminishing returns (i.e., twice as much R&D money does not yield twice as much research or innovation output). Second, because knowledge capital is a public good and nonrival (i.e., one's "consumption" of knowledge does not reduce the amount available to others) there is an additional positive external effect of knowledge generation beyond the originating individual or firm. Knowledge can "leak out" and generate additional positive effects both nationally and internationally. Knowledge, both public and private, is also an important internationally traded good and an additional mechanism for the international diffusion of technical change. (For further discussion, see Dosi *et al.*, 1990; for a simplified model, see Silverberg and Verspragen, 1994; or Sentance, 1996.) Most importantly, the production of knowledge capital exhibits increasing returns – the more knowledge the better. This is an essential reason why firms, and society at large, invest in knowledge capital and perform R&D.

One important weakness in the approaches described here for endogenizing technological change is that they somewhat simplistically equate increases in knowledge with technological "progress" (i.e., productivity increases). This ignores both the investments required for technological change, without which most increases in knowledge capital and R&D would remain "blueprints", and the high risk and uncertainty involved in translating new knowledge into applied technological advances. By lumping all knowledge and technologies together, the models also ignore the fact that there is a dichotomy between increasing and diminishing returns in investments into new knowledge (e.g., R&D) and its application in the form of improved technology (e.g., learning effects). Upon closer examination, everything depends critically upon the life cycle stage of any particular technology, or sectoral activity, whether improvement potentials are large (possibility of increasing returns) or largely exhausted (diminishing returns).

Sectoral Production Function Models

Production function models have been applied to specific economic sectors as well as overall national economies. This section discusses models in two sectors: energy and agriculture. Production function-based energy models have recently had an important influence in the policy debate surrounding climate change. Agricultural models have yielded significant insights into the important influence of resource endowments (e.g., land) on technological change and on the persistence of technological change over long historical time periods.

Sectoral production function models basically enlarge the macroeconomic production function by adding additional factor inputs that provide greater sectoral detail. In most cases, however, the added sectoral detail comes at the expense of limiting the problem to a partial equilibrium analysis explaining the dynamics of only one sector of the economy. The modeling of the rest of the economy effectively assumes "business as usual". The sources of productivity gains are exogenous, modeled either as a direct external input, or by a trend ("residual") parameter. Within the sector of interest, however, more technological detail can be added.

Energy. In the energy sector, "top-down" production function models have gained particular prominence in recent years in the climate policy debate. Particularly notable are the seemingly contradictory findings about the possibilities and costs of emission reductions from these pessimistic "top-down" models compared to (optimistic) technology-rich "bottom-up" modeling approaches. A representative, state-of-the-art top-down model is the Global 2100 model developed by Manne and Richels (1992). In the Global 2100 model a macroeconomic production function with exogenous labor productivity increases is nested with additional energy factor inputs. The additional energy factor inputs are separated into two different categories – electric and nonelectric. The overall productivity (or efficiency) of energy use in this model depends first on prices. If relative prices change, energy use changes (compared to other factor inputs such as labor and capital), based on the assumed elasticity of substitution used in the model. Second, energy productivity in the model depends on a parameter labeled AEEI, an acronym for autonomous rate of energy efficiency improvement. AEEI captures all the remaining factors that might alter energy productivity such as economic structural change, changes in lifestyles, and changes in technology. Of course none of these stylized model parameters is directly observable. Therefore, all rely on either intricate econometric techniques or subjective judgements for their parametrization, and there is continuing debate on how they change across different economic and social settings. The USA has been the most intensively studied economy, but the applicability of these results to formerly centrally planned economies or China remain questionable. In its basic structure, Global 2100 and the AEEI provide a prototype of a "residual" representation of technological change.

However, Global 2100 also contains an optimization submodule to analyze specific energy supply technologies such as alternative electric power plants, refineries, synthetic fuel production facilities, and so forth. This includes both existing technologies and technologies that are not yet commercially proven but may be employed in the future. Fossil energy resources are also modeled. Therefore, depending on various factors such as future demand growth, the availability of energy resources, and other exogenously imposed constraints such as limitations on greenhouse gas emissions, the model calculates the least-cost energy supply structure. Should this lead to increased energy prices, the model endogenously determines macroeconomic energy demand adjustments and any reductions in GDP growth that might result. This constitutes its main advantage over "bottom-up" engineering models where demand and economic growth are usually treated exogenously, or dealt with through successive iterations between separate supply and demand models.

Thus the Manne–Richels model combines features of both macroeconomic "top-down" and sectoral "bottom-up" models. But in the end, technology improvements remain exogenous, despite all the technological detail and dependence on prices, resource availability, and other constraints. Two basic concepts are used. First is a residual time-trend parameter (the AEEI). Second is the concept of a "backstop" technology. A backstop technology (a term introduced by Nordhaus, 1973a) is a kind of generic technology that is available today but too costly to be economically competitive. An example would be a coal mine with an associated coal liquefaction facility as an alternative to cheap Middle Eastern oil. If prices rise high enough, the backstop technology can be taken "off the shelf" and start to diffuse into the energy market. Development costs are assumed to be negligible and do not enter the model calculations. In Global 2100, one backstop technology for electric energy and one for nonelectric energy are defined parametrically by their future dates of availability and costs. The sensitivity of model results to variations in these and other important model parameters is shown in *Figure 3.1* for an illustrative scenario of stabilizing global CO_2 emissions. The range of parameter values analyzed was derived from a Delphi-type expert poll (Manne and Richels, 1994). The five parameters that are varied in *Figure 3.1* are the following:

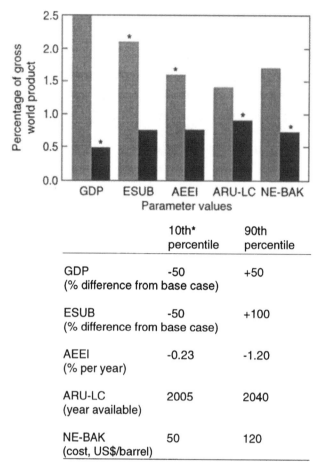

Figure 3.1: Sensitivity of loss in gross world product for stabilization of global CO_2 emissions as a function of variations in important input parameters for the Global 2100 model. (See text for an explanation of the abbreviations.) 10th percentile points and resulting model values are denoted by an asterisk. Source: adapted from Manne and Richels (1994:52).

- GDP growth, i.e., the product of increases in the labor force and in labor productivity;
- ESUB, the elasticity of substitution between energy and other factor inputs (labor and capital) in the macroeconomic production function;
- AEEI, the autonomous (i.e., nonprice induced) rate of energy efficiency improvement;
- ARU-LC, the availability date for the electric backstop technology; and
- NE-BAK, the cost of the nonelectric backstop technology.

In their comprehensive sensitivity analysis based on expert poll techniques, Manne and Richels (1994) have demonstrated how the model outcome, and the economic costs of stabilizing global CO_2 emissions, critically depend on productivity growth and technological change assumption. Here, the results serve to remind us that these critical variables and parameters are exogenous in the modeling framework.

The use of both a time-trend parameter and of backstop technologies leave the model with serious weaknesses in its representation of technological change. First, the model follows the classical "residual" approach for modeling nonprice-induced productivity increases. Second, both the assumed exogenous productivity improvements and the backstop technologies come at no effort and cost and are also insensitive to variations in exogenous and endogenous variables (such as environmental constraints or even straightforward productivity growth). The R&D and investments necessary to accumulate experience and realize technological learning are also entirely ignored. Third, the technological dynamics of the assumed backstop technologies are unrealistic. Once price increases trigger the deployment of backstop technologies, the economy and the energy system are assumed to settle into a new equilibrium where they remain forever. Technological change comes to a standstill. This is particularly unrealistic for a model that is usually run with extremely long time frames, such as to 2100 and, in some applications, to 2200.

Agriculture. In the agricultural sector there is a particularly large variety of products, soils, and climatic conditions, and techniques and technologies, from traditional to high-tech. As a result production function models developed for agricultural studies are among the most elaborate and complex. As a typical example the Basic Linked System (BLS) of national agricultural models (Fischer *et al.*, 1988, 1994) models 11 different agricultural commodities through individual production functions, covers a variety of agro-climatic zones, and incorporates additional physical input parameters such as soil quality and precipitation. In addition to the traditional factor inputs of capital, labor, and land, technology inputs such as seeds, fertilizers, and mechanization are explicitly modeled and included in the disaggregated production functions. Nonetheless, basic weaknesses in the representation of technological change remain. Productivity increases at the macroeconomic level remain exogenous, a "residual" parameter. Improvements in specific technologies (e.g., increased crop yields from new strains) are represented as time-trend functions requiring no prior R&D or investment. And there is no explicit recognition of the diversity of decision agents. Farmers in particular have proven to be an exceptionally diverse group, as has been confirmed, for instance, in anthropological, sociological, and technology diffusion studies.

However, agricultural studies using production function models have provided key insights through detailed consistent international and longitudinal studies of the evolution of agricultural productivity. Such studies (e.g., Hayami and Ruttan, 1971, 1985) have convincingly demonstrated several factors. First, they have demonstrated the large impacts of technological change on agriculture. Second, they have shown the importance of differences in factor endowments (e.g., land or labor availability) in shaping the evolution of technological change. Third, they have demonstrated the persistence and stability of technological change "trajectories", or their "path dependency", giving rise to the theory of induced technological and institutional innovation (see e.g., Binswanger and Ruttan, 1978).[4]

Figure 3.2 illustrates the diverse patterns of productivity increases in agriculture. Agricultural land productivity (Y/A) is plotted against agricultural labor productivity (Y/L) over time. The arrows show, in most cases, how both measures improved between 1960 and 1980. The arrows for the USA, Denmark, France, and Japan include data back to 1880. The numbers in parentheses give the percentages of the work force employed outside agriculture and thus mirror the structural impacts of improved agricultural labor productivity. Hayami and Ruttan (1985:124) group these productivity increase trajectories into three categories: Asian, European, and New Continent. They are related to the initial relative endowments of the factor inputs land and labor with starting values around 1,000, 10,000 and 100,000 ha per agricultural worker, respectively.

The figure shows trajectories ranging from those characterized by high land productivity, albeit labor intensive (e.g., Japan), to those characterized by high labor productivity, but with land-intensive agricultural systems (e.g., Australia). The variation arises from differences in initial factor endowments and resulting differences in relative prices, patterns of technological change, government policies, and so forth. What is important is the *consistency* and *stability* of the productivity increase trajectories. This implies that different agricultural systems and their technologies evolve along stable and mutually exclusive development paths. This is an example of path-dependent processes of technological change.

---

[4]Note that the term "induced innovation" is used differently from earlier microeconomic theories of induced technical change (Kennedy, 1964; Ahmad, 1966) formulated under the "demand pull" hypothesis of technological change (Habakkuk, 1962). For a critique, see Nordhaus (1973b); for a concise review of the debate, see Ruttan (1996).

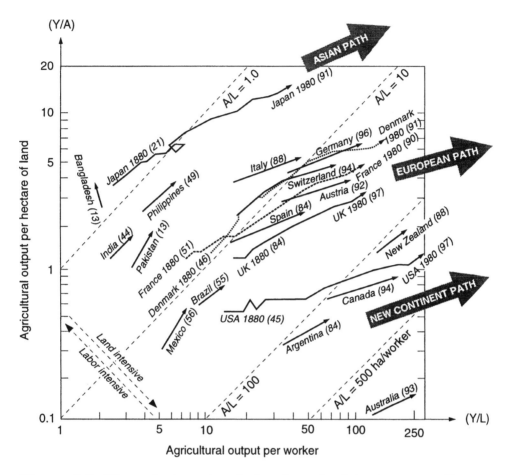

Figure 3.2: Three clusters of agricultural productivity increases. For an explanation of the figure and its units refer to the text. Source: adapted from Hayami and Ruttan (1985:121, 131).

"Path dependency" is used here in a dynamic rather than a static context.[5] Rather than simply being influenced by a discrete historical event, a given technological configuration is constantly reproduced and extended

---

[5] Our use of the term "path dependency" is different from its use in the technological selection literature (Arthur, 1988). There path dependency refers to historical events that constrain subsequent technological choices. Examples are the adoption of a particular standard (e.g., the DOS operating system in computers) or a particular technological artifact (e.g., the internal combustion engine). The initial random, and sometimes even suboptimal, selection of a particular standard or artifact gets "locked-in" and predetermines much of the future. We label this phenomenon "lock-in" rather than "path dependency".

in a particular direction, building on past achievements, R&D strategies, incentives, and distinct institutional and organizational settings. Instead of standards and artifacts per se, it is the pattern of technological change and the different directions it can take that is path dependent. See *Figure 3.2* and the examples of persistent differences in energy intensities given in Part II.

This stability has disadvantages. If a technology trajectory is "locked in" to evolve in a particular direction and relative prices or environmental standards change all of a sudden, it can be very difficult, costly, and time consuming to change course. This is very different from notions of equilibria employed in much of economics, where a system that is disturbed settles down more or less autonomously and rapidly into a new equilibrium condition. Consider, for example, energy efficiency in different economies. The USA has traditionally developed along a high energy-intensity trajectory. Primary energy use per unit of GNP in the USA has historically been about twice the level of Japan or the industrialized countries in Europe. Arguably, this reflects the historical abundance of domestic energy resources and low taxation resulting in relatively cheap energy. During the oil shocks of the 1970s and early 1980s energy was no longer cheap. But the US economy, its industrial structure, its settlement patterns, and its high use of energy for transportation were (and still are) locked into a path of "high intensity" development. Conversely, countries that were on a "high efficiency" path were much better prepared for rising energy prices. As the case of Japan illustrates, countries already on high-efficiency paths also managed the greatest efficiency gains.

To understand how such macrolevel path-dependent processes emerge requires a more detailed look inside the "black box" of technology (Rosenberg, 1982). Rather than imagining that technology evolves autonomously or is available (in the case of backstop technologies) "off the shelf" as required, we need to ask the following questions:

- By what processes is technological variety generated?
- How do different varieties acquire importance for an economy?
- How does the process of acquiring economic importance in turn shape the development of technological variety?

Tentative answers to such questions are given by microeconomic models of technological change. These focus on those who either generate or implement new technologies, and on their interdependent strategies for dealing with the large uncertainties and high risks inherent in any technological change.

3.1.2. Microeconomic perspectives

A Conceptual Model

From a neoclassical economic perspective, the agent of technological change – a firm – is characterized by a single decision agent possessing perfect foresight and acting rationally under a clear optimization criterion: maximizing profits. The decision agent reacts to outside market signals (price changes) but cannot affect the market itself.

Conversely, managerial or behavioral theories of the firm, particularly "evolutionary" theories (e.g., Nelson and Winter, 1982),[6] draw a much more complex picture of decision making and how uncertainty is handled. *Figure 3.3* gives a conceptual "wiring diagram" of the driving forces of technological change within a firm. It suggests that what matters most are management, organization, information, products and clients, and finance, which all interact with each other.

Management matters because in a typical modern large corporation, decisions are no longer made by the owner (the classical Schumpeterian entrepreneur) but by managers with diverse interests and motivations. Management also matters because it typically involves strategic decisions defining the long-term future of the firm. Instead of simply reacting to price signals from the environment, managers actively try to shape the environment. Different managerial strategies, motivations, and behaviors are at the core of models designed to explain why some firms engage in the risky business of technological innovation while others do not.

Organization matters because it affects attitudes toward innovative opportunities and the risks associated with pursuing them. Organizational theories emphasize that firms exhibit behavior that is independent of the individuals in the organization. The whole – sometimes referred to as "corporate culture" – is more than the sum of its parts. The collective ability to gather, interpret, and absorb information and to apply decision rules is constrained and shaped by different organizational traditions. As a result, different firms display very different attitudes toward innovation opportunities. A firm's organization also affects how it acquires information and knowledge, both from internal and external sources. For technology, a firm might adopt a strategy of internal learning. This may be costly, but it can quickly reduce uncertainty and keep new information under proprietary control. Alternatively, a firm might opt for external learning, waiting for competitors

---

[6]These theories are called evolutionary in the sense that economic behavior does not follow simple *ex ante* rules but is essentially *adaptive* in response to uncertainty, learning, and interdependence between decisions and their outcomes among economic agents.

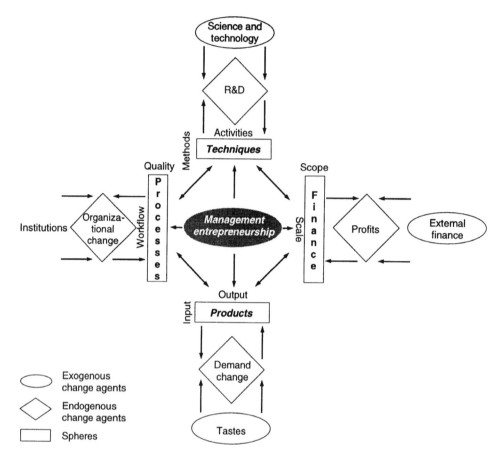

Figure 3.3: A conceptual diagram of factors influencing technological change at the firm level. Source: adapted from von Tunzelmann (1995:391).

to take the initiative and then imitating them through various legal and il-legal mechanisms like "redesign" or straightforward espionage. Although it might cost less, external learning can also entail greater uncertainty, time lags, and other intangible "transaction costs" (see also Williamson and Masten, 1995). Moreover, learning is never a one-way street. Good professors admit that they learn considerably from their students. External learning is only possible if you allow others to also learn from you. A firm that is hypersecretive cannot hope for open access to information from other firms. The importance and success of professional meetings and symposia derive from this reality.

Information matters because information is limited, incomplete, and often false. Even *if* decision makers wanted to behave according to the decision rules of neoclassical economics, it is not clear that they could. In practice rationality is "bounded" (Simon, 1957, 1992) by two important constraints. First, no manager could ever assemble all the information required for an optimal decision. Second, even if one could, decisions are usually so complex that no simple algorithm exists for evaluating all possible courses of action. Third, the success of whatever strategy adopted is not independent of the strategies and decisions of competitors. The outcome depends critically on what they do. Information also depends on the actions of others as well as being inherently uncertain. For example, a decision to invest in new technology depends on information used to project future performance and costs. These can be influenced by strategies (R&D, learning curve effects) both internal and external to the firm.

Products matter because new products are an essential mechanism for influencing a firm's environment: the market. Clients matter because they are both important sources of ideas and partners (or guinea pigs) in the development of new innovations. Studies ranging from high technology to the hamburger business have repeatedly concluded that most ideas for new products come from users (e.g., Peters, 1986; von Hippel, 1988). Firms with a "good ear" for customer suggestions gain a competitive edge and move into markets that others never knew existed. A market for "railways" did not exist when Stephenson built his first locomotive plant. The mass market for automobiles was "developed" by Henry Ford and his cheap Model T. Ironically, even firms that develop a new technology are sometimes ignorant about its possible markets (cf. Rosenberg, 1996). IBM initially assessed the market for mainframe computers at about a dozen. The CEO of Digital Equipment Corporation in the 1970s considered the idea that every household could own a computer as complete baloney. When Rank Xerox invented the photocopier, their first marketing study identified the total US market potential at about 10,000 machines.

Finally, finance matters. Technological change, from R&D to new plant and equipment investments, is a costly and risky business. Internal financial capabilities and external financial organizations are consequently key factors affecting technological change. Although this has long been recognized in microeconomics and by business people, financing has received relatively little attention in macroscopic technology studies, or in long-term scenarios of environmental change. A notable exception is the work of Nakićenović and Rogner (1996).

It would be unreasonable to expect any one model to incorporate all these microeconomic features in a single analytical framework. However, in the next section we do present briefly a representative microeconomic simulation model of technological change, that – while not incorporating all the features mentioned above – nevertheless contains sufficient microlevel rationale and detail to enable a better understanding of the drivers and their highly complex interplay that underlie empirically observed macropatterns of technological change.

An Illustrative Model

In a series of papers Dosi *et al.* (1986), Silverberg *et al.* (1988), and Silverberg (1991) have developed a "self-organizational" microeconomic simulation model that dismisses two central assumptions used in neoclassical models – the assumption of equilibrium and the assumption of "economic rationality". Economic rationality assumes that decisions are always optimal and based on perfect information. Rather than being based on optimization techniques, the model therefore uses simulation techniques.

The model includes various firms that have different capabilities and follow different strategies. Suddenly, the competitive equilibrium of the industry is disturbed by a new technology whose exact properties are both unknown and a function of the R&D strategies followed by different firms. The subsequent decisions taken by the different firms draw on specific knowledge bases that contain both freely available external information and localized internal information in the form of expertise and skills. Decisions are affected by the inherent uncertainty surrounding the new innovation, the strategic interactions among firms, and their interdependencies through the market. For any particular firm, the outcome of a decision depends, among other things, on the unknown strategic responses of other firms. Both existing and new technologies are dynamic (i.e., may realize productivity increases). Learning curve effects are explicitly modeled. Technology dynamics can result from either internal learning processes, external learning processes, or both.

Investment decisions must therefore weigh subjectively the potential for improving existing technologies (i.e., the remaining improvement potential of their respective learning curves) against the promise of the new technology. The exact rates and ultimate potentials for improvements are unknown. Different strategies are pursued by different firms. Some invest in in-house development. Others seek to avoid initially high development and learning costs by waiting for competitors to move first. This leaves open the possibility that an innovation might never be introduced if nobody takes the initiative to develop it. On the other hand, any one firm might acquire a new technology very cheaply after it is developed, improved and demonstrated by a competitor.

Runs with the model show an interesting pattern; that is, if the subjective assessment criteria used by a firm that adopts an innovation and becomes the net "winner" in one simulation are then applied uniformly to all firms in a subsequent simulation, the innovation is *not adopted* at all because no firm is ready to incur the costs of developing the technology and bringing it to commercial maturity. This supports the argument that diversity in expectations about the future, diversity in assessments of risk, and diversity in entrepreneurial strategies are the real drivers of technological evolution. Without diversity there is no evolution. If all firms have the same expectations, or share perfect foresight into the future, technological change does not occur. No firm is prepared to incur the risks of innovation. Firms innovate and invest because they have different expectations and hope to outperform their competitors (or prepare for not being outperformed themselves).

Figure 3.4 illustrates the microeconomic drama going on in a competitive environment "disturbed" by innovation. In this simulation, very early adopters of the new technology do not perform as well, in terms of market share, as firms adopting the technology a little later. However, note in particular what happens to the innovation "laggard" (firm 10 in the top panel of *Figure 3.4*): it is completely driven out of the market even if, with a lag, it manages to eventually catch up with the increasing productivity levels of its innovating competitors (bottom panel of *Figure 3.4*). The results confirm that the ultimate outcome of innovation strategies depends on factors partly beyond the control of an individual firm.

The bottom panel of *Figure 3.4* shows the productivity of each firm's capital stock as well as the industry average (dotted line). It shows that successful adopters reach productivity levels above average, which helps increase their market shares, while late adopters stay below the average and only gradually catch up, with the risk of being completely driven out from the market. This reflects how internal efficiency (skill levels) evolve within firms. While the innovation pioneers build up their skill levels, later adopters can benefit from this experience via external learning and, in this simulation, eventually may even overtake the earliest adopters. Conversely, waiting too long may mean that learning cannot catch up fast enough, productivity lags behind, and the "laggard" is driven out of the market altogether.

Thus independent of the ultimate potential technological merits of an innovation, its *appropriability* is crucial, as its ultimate potential can only be realized through "hands on" experience in the form of investments and learning. There is no such thing as "autonomous" technological learning either in the model or in real life. Indeed, the key to understanding technological dynamics is understanding the sources of both internal and external technological learning.

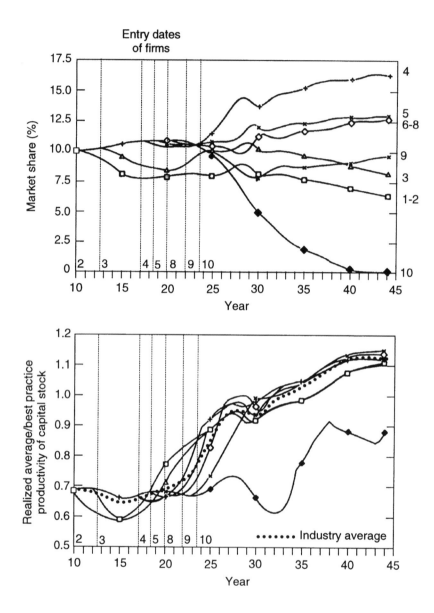

Figure 3.4: Market share (top panel) and productivity of capital stock (bottom panel) for firms applying different adoption strategies to a new innovation. The dotted line in the bottom panel is the industry average. Note that the figure only shows the modeled entry dates (of those applying a new technology) of seven out of ten firms. Source: adapted from Silverberg *et al.* (1988:1044–1045).

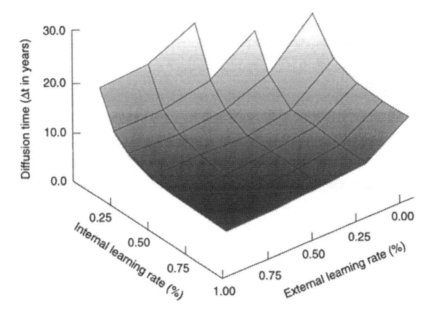

Figure 3.5: Time (in years) for a new technology to diffuse from 10% to 90% of industry capacity as a function of sources of information and learning: internal (private) versus external (public). Source: Silverberg (1991:220).

The model shows that internal and external sources of technological learning are only partially substitutable in their effects on innovation diffusion (*Figure 3.5*). Indeed the two are partially complementary. Without internal experimentation and development focused on specific applications, even the most ambitious mass media efforts to diffuse knowledge about an innovation are unlikely to cause rapid diffusion. Conversely, without at least a minimum of public learning and freely available information, even the most ambitious internal learning efforts remain isolated. Earlier we regarded the distinction between basic and applied research as a false dichotomy. Here, the evolutionary model indicates that there is no "exclusion principle" between the effects of public and private sources of advances in knowledge on technological diffusion.

3.1.3.   A preliminary conclusion

No model, macro or micro, as sophisticated as it may be, can predict when a new innovation materializes.[7] No model whatsoever can predict what particular new technological "combination" will give shape to an innovation.

---

[7]The exceptions are phenomenological models based on cyclical, or long wave theories of economic development. See e.g., Mensch (1979) or Marchetti (1980:274).

But all models, and all the empirical evidence, agree on the importance of technological change as a driving force in productivity increases, as an engine of economic growth, and as a constant cause of perturbations in the economic, material, and social structures of humankind.

Although technology is frequently modeled as something "external" to the economic and social system, it is in fact an endogenous product of the economic and social system, and technological change is constantly reshaping this system. On the one hand technological change provides the means to alleviate adverse environmental impacts and overcome resource constraints and scarcities. On the other it also creates new impacts and scarcities. We hope to have convinced the reader that models treating technology as exogenous, and technological change as "manna from heaven", are seriously deficient. We hope also to have provided sufficient examples to illustrate that "patterns" do exist that can be addressed by a whole range of theories from economics and other fields. The main lesson is that dynamic interactions between the micro- and macrolevels lead to the emergence of spatial and temporal patterns that are *driven* rather than dissipated by microlevel diversity. Moreover, these patterns can undergo a sequence of evolutionary instabilities at the microlevel of economic agents while still maintaining structure and evolutionary stability at the macrolevels of industrial sectors and the overall economy.

What emerges is the importance of *uncertainty, diversity,* and *learning* phenomena in technological change. Without taking into account diversity and learning, technological change cannot be satisfactorily modeled. This is especially important when interpreting empirical long-term regularities in diffusion and technological substitution dynamics. Regularity in evolutionary paths at the macrolevel does not contradict the existence and importance of uncertainty and diversity in the behavior of economic agents, in technological "expectations" and designs, and in appropriability conditions. Rather, it is just the opposite. Such regularity is the direct *consequence* of uncertainty and diversity.

Another key lesson is the importance of focusing not on invention and innovations as discrete events, but rather on the diffusion of individual technologies and of entire technological systems. It is only through diffusion that proposed new solutions become incorporated into the capital stock, into the economy, and into everyday practice.

As discussed in Chapter 2, the sources and impacts of technological change are multiple and diverse. In trying to capture this diversity, both micro- and macrolevel analyses have a common drawback: they are unable to capture the essence of structural change. Neither an aggregate production function approach, nor the most detailed evolutionary microlevel model, can adequately describe the transition from a rural, agrarian society, to an urbanized industrial one. An intermediate level of analysis is required. We believe it is provided by the concept of technology "families" and "clusters", even if such a categorization is largely qualitative. We will touch briefly on the empirical aspects of their identification, but are unable at this stage to suggest an entirely new class of model describing how many individual technologies interact, enhance one another, and give rise to entirely new sectors, forms of production, and consumption. The first vigorous stab at the problem was Marvin Frankel's seminal paper published in 1955, which focused on how technological interrelatedness in a mature economy (the UK, in Frankel's case) can slow down the diffusion of new technologies and technology systems. Unfortunately, not much new seems to have been written along these lines or incorporated into formal models since. However, it is precisely this "bundling" of technological change that makes technology such a powerful force in transforming economic structures, society, and the natural environment.

Chapter 4

Technology: History

Synopsis

Technological changes since the onset of the Industrial Revolution are summarized. The concept of technology clusters, i.e., a set of interrelated technological, infrastructural, and organizational innovations driving output and productivity growth during particular periods of time is used to explain these changes. Four historical technology clusters are identified, with a prospective fifth, emerging one. The most salient characteristics of each cluster are discussed with illustrative examples. The chapter concludes with a discussion of quantitative and statistical approaches that corroborate the concept of technology clusters.

4.1. A Long View of Technology Development: The Last 200 Years

This section is a synoptic *tour d'horizon* of 200 years of technological change. It provides a historical overview and identifies distinct periods of technological change in order to set the stage for more detailed discussions in Part II of individual technological changes and their global environmental change implications. Our principal organizing concept is that of technology "families" or "clusters". A technology cluster is a set of interrelated technological and organizational innovations whose pervasive adoption drives a particular period of economic growth, productivity increases, industrialization, trade, and associated structural changes.

Technology clusters do not follow one after the other in a rigid temporal sequence. Various clusters coexist in any given period, although the relative importance of each keeps shifting. Older technological and infrastructural vintages coexist with the dominant technology cluster. In some cases older clusters are perpetuated by government policy even after more modern technologies are well established in other parts of the international economy. Post-World War II industrial policy in the former USSR is a good example. Elements of an emerging cluster initially develop within specialized applications or in specific market niches. Eventually they emerge as a new dominant technological mode after an extensive period of experimentation and cumulative improvements.

At any given time economic growth is driven primarily by the dominant technology cluster, which is frequently associated with the most visible technological artifact or infrastructural system of the time. This provides the focus of studies by economic historians using the leading sector hypothesis, e.g., of the "railways era" (Fogel, 1970) or of the "age of steel and electricity" (Freeman, 1989). We emphasize the concept of technology clusters because any dominant individual technology or infrastructure studied under the leading sector hypothesis can explain only a fraction of economic growth.[1] It is impossible for a single leading sector, or a few individual industrial or infrastructural innovations, to account fully for growth, important as they might be. Only the combination of many innovations in many sectors and technological fields into entire technology families or clusters can adequately account for overall economic growth and the expansion of human activities that are the core driver of global change.

The four historical clusters we distinguish, plus a possible fifth cluster now emerging, all have important implications for economic growth and development. New products and markets emerge; transportation infrastructures widen existing markets; and new process technologies and forms of organization and management make it possible to raise industrial productivity. Macroeconomic and social policies help in distributing productivity gains, and rising incomes create a powerful demand-induced stimulus for industrial output growth. At the same time, energy, transportation and communication infrastructures facilitate changes and adjustments in agriculture, industry, and consumer markets.

As the dominant technology cluster expands, many technological elements of its successor are developed through scientific discoveries, innovation, and small-scale applications. However, considerable time is required before isolated developments converge and develop the interconnections that foster forward and backward multiplier effects characteristic of a distinct technology cluster. Eventually a new cluster emerges after a period of crisis that involves sometimes painful structural adjustments in both economic activities and in the social and institutional domains. Freeman and Perez (1988) emphasize the importance of a "mismatch" between established institutions and changing conditions created by technological change. Such concepts go back to the early 1920s, when they were first advanced by the eminent

---

[1]For case studies of coal, steel, and railways, see e.g., Fishlow (1965), Holtfrerich (1973), Fremdling (1975), von Tunzelmann (1982), O'Brien (1983), and Freeman (1989).

American sociologist William Fielding Ogburn (e.g., 1950). No technology cluster can emerge without new forms of organization and institutions. From that perspective there is indeed a dialectical relationship between technological and institutional/organizational change.

At this point, a word of caution is necessary. There exists no simple, linear cause–effect relationship between technological, institutional/organizational, and *global* change even if we have tried to synthesize historical development patterns through the concept of technology clusters. Both historical – and future – developments are characterized by a multitude of synergies and feedback mechanisms. The appropriate model, therefore, is one of coevolutionary processes, rather than one of linear cause–effect relationships. Technological changes across sectors interdepend; e.g., without advances in agriculture there is limited scope for industrialization; without appropriate institutional/organizational settings there is little incentive for experimentation (innovation) and diffusion of new technologies; without productivity increases output growth cannot be sustained; and finally, negative externalities like environmental impacts need to be addressed, even if we have to recognize the limitations of our knowledge and that successes of today may eventually turn out as a mixed blessing in a more distant future.

In terms of global (environmental) change it would perhaps suffice to concentrate on the last three clusters, because it is only since the turn of this century that human numbers and activities have grown to such an extent, and technologies diffused worldwide to result in both ubiquitous and truly planetary-scale impacts. (A few examples of these were summarized in *Table 1.1* in the introductory chapter). However, such a perspective would be too restricted. Having embraced the concepts of path dependency and mutually reinforcing changes (in the form of technology clusters), we have to acknowledge that many developments are deeply rooted in the past. They need to be understood first, before a speculative look into the future can be ventured. Long-term scenarios, reaching 100 years into the future (or even more) are very much *en vogue* today, in particular in connection with possible climate change. Perhaps the simple precept of control theory – that one should never extrapolate beyond half the length of the observational record – adds additional justification for our long historical perspective.

4.1.1. A qualitative account

The Rise of Industrialization

In the 18th century, a series of innovations (notably the spinning jenny, the flying shuttle, and the power loom) transformed the manufacture of cotton in

England and gave rise to what eventually became a new mode of production: the factory system. Innovations in the fields of energy (stationary steam engines) and metallurgy (replacement of charcoal by coal in the iron industry) were similarly revolutionary. All these reinforced one another and drove an Industrial Revolution that made England the world's leading industrial and economic power well into the late 19th century. As summarized by Mokyr (1990), technology embodied in machinery, leading to new forms of production, products, and markets, has been the *lever of riches*. Landes (1969:41) summarizes the many innovations of the Industrial Revolution under three principles. These three principles apply also to later stages of industrialization and equally to modernization and economic growth in developing countries today. The three principles include:

- The substitution of machines for human effort and skill;
- The substitution of fossil fuels (coal) for animate power, allowing for the first time an unprecedented consumption density and almost unlimited supply of energy; and
- The use of new and more abundant raw materials in manufacturing.

While important technological innovations can be identified in earlier historical periods, the impact and scale of the Industrial Revolution were due to a new qualitative characteristic of innovations. Innovations mutually enhanced each other and became embedded in profound transformations of the social and organizational fabric of society. The steam engine, the coal industry, railroads, and new steel production processes cannot be considered separately. They depended on one another, enhanced each other, and together contributed to economic growth via a multitude of "forward and backward linkages", to use economic terminology. Today, the same can be said about the internal combustion engine, the oil and petrochemical industries, synthetic fibers and plastics, etc.

Important social and organizational changes associated with the Industrial Revolution occurred in many areas, e.g., in the generation of new knowledge through science; the application of new knowledge to innovations; incentives for innovation generation and technology diffusion; and new modes of production, enterprises, and organization of market relations. Rosenberg and Birdzell (1986) emphasize the decisive role of new institutional arrangements such as the early separation of political and economic activities. It is "the interplay of people, economic institutions, growing markets and technology" (Rosenberg and Birdzell, 1990:25) that is the key to explaining the Western economic "miracle". Cameron (1989:163–182) also emphasizes the importance of social and intellectual, commercial, financial, agricultural, and even political developments. However, he cautions against the terminology of an "Industrial Revolution", which implies a pronounced discontinuity, and stresses that industrial technology and innovation changes were all rooted in earlier developments. In the "seamless web" (Hughes, 1988) of historical change it is difficult to assign relative weights to different factors. But the intellectual, institutional, and organizational changes were arguably the most fundamental. They provided the critical favorable environment for systematic experimentation (invention) and commercial application (diffusion) of innovations. This new qualitative characteristic is the principal reason for starting the technology cluster "clock" by the mid-18th century, the usual dating of the onset of the Industrial Revolution.

Technology Clusters

Table 4.1 groups industrial and economic developments since 1750 into four technology clusters. Also included is a speculative fifth cluster that might now be emerging. The top half of the table lists major technologies and products of the dominant cluster. Those of the emerging cluster that will dominate in the subsequent period are listed in the bottom half of the table. Key technologies are exemplified in the areas of energy, transportation, materials, industry, and consumer products. Each cluster lasts about five decades and can overlap the preceding or succeeding cluster by 20 years.

Table 4.1 provides only a rough sketch of the five technology clusters. In Part II a more detailed analysis of technology developments is given, following a simple three sector model of economic structural change. In that section these technology clusters are examined in terms of their importance for the agricultural, industrial, and service sectors. *Table 4.2* characterizes briefly each technology cluster in these three sectors and identifies the dominant "organizational style" of each cluster, largely following the taxonomy developed by Freeman and Perez (1988). The table also gives a stylized "innovation geography" showing the different locational patterns by which innovations originated and diffused internationally in each of the successive technology clusters. The "core" countries are at the center of technological developments and diffusion. "Rim" countries are rapidly catching up and likely to join the core in the subsequent cluster. All countries not listed separately for a particular technology cluster are effectively on the technological "periphery". They have little or no endogenous technological developments, and only isolated adoption of key technologies and infrastructures. They are generally kept in "economic backwardness" (Gerschenkron, 1962) with economies dominated by agriculture and raw materials production, both for domestic needs and exports to "core" countries.

Table 4.1: Five important technology clusters, 1750–2000.

	1750–1820	1800–1870	1850–1940	1920–2000	1980–
Dominant cluster					
E	Water, wind, feed, wood	Wood, feed, coal	Coal	Oil, electricity	Gas, electricity
T	Turnpikes	Canals	Railways, steamships, telegraph	Roads, telephone, radio & TV	Roads, air transport, multimedia comm.
M	Iron	Iron, puddling steel	Steel	Petrochemicals, plastics, steel, aluminum	Alloys, speciality materials
I	Castings	Stationary steam, mechanization	Heavy machinery, chemicals structural materials	Process plants, NC machinery, consumer goods, drugs	Environmental technologies, disassembly & recycling, consumer services
C	Textiles (wool, cotton), pottery	Textiles, chinaware	Product diversification (imports)	Durables, food industry, tourism	Leisure & vacation, custom-made products
Emerging cluster					
E	Coal, coke	City gas	Oil, electricity	Gas, nuclear	Hydrogen(?)
T	Canals	Mobile steam, telegraph	Roads & cars, telephone, radio	Air transport, telecommunication, computers	Hypersonic aircraft(?), high-speed trains
M	Puddling steel	Mass produced steel	Synthetics, aluminum	"Custom-made" materials, composites	Recyclables & degradables
I	Stationary steam, mechan. equipm.	Coal chemicals, dyes, structural materials	Fine chemicals, drugs, durables	Electronics, information technology	Services (software), biotechnology
C	Chinaware	Illuminants	Consumer durables, refrigeration	Leisure & recreation products, arts	Integrated "packages" (products & services)

Abbreviations: E, Energy; T, Transport and communication; M, Materials; I, Industry; C, Consumer products; NC, numerically controlled.

Table 4.2: A summary of technology clusters, 1750–2000.

	1750–1820	1800–1870	1850–1940	1920–2000	1980–
Technology cluster					
Agriculture	Agricultural innovations		Mercantilistic agriculture	Industrialization of agriculture	–
Industry	Textile	Steam	Heavy engineering	Mass production	Total quality
Services	–	–	–	Mass consumption	–
Organizational "style"					
Plant/company level	Individual entrepreneurs, local capital, small-scale manufacture	Small firms, joint stock companies	"Giants", cartels, trusts, pervasive standardization	Fordism/Taylorism, multinationals, vertical integration	"Just-in-time", TQC, horizontal integration
Economy and society	Breakdown of feudal & medieval economic structures	"Laissez-faire", Manchester liberalism	Imperialism, colonies, monopoly & oligopoly regulation, unionization	Social welfare state, Keynesianism "open" society	Economic deregulation, environmental regulation, networks of actors
Innovation geography					
"Core"	England	England, Belgium	England, Benelux, France, Germany, USA	USA, Canada, JANZ, EC-6, England	OECD
"Rim"	Belgium, France	France, Germany, USA	Central Europe, Italy, Scandinavia, Canada, JANZ, Russia	USSR, Central & Eastern Europe, Southern Europe	Asian Tigers, Russia, Eastern Europe, ??

Abbreviations: JANZ, Japan, Australia, New Zealand; TQC, total quality control (in manufacturing); EC, European Community.

4.1.2. A quantitative account

The usual approach to describe the quantitative rise of technology clusters is to calculate the growth of representative products, technologies, or systems. Such analysis is inevitably partial. Unless appropriate meta-systems that are important to more than one economic sector are used (e.g., energy or transportation), it can generate misleading overall inferences from too limited a set of examples. Part II will review in detail individual technology examples. Abundant research also exists on the growth of individual technologies, products, and infrastructures. Woytinsky and Woytinsky (1953), Hoffmann (1958), Landes (1969), Rostow (1978), Mitchell (1980, 1982, 1983), and Mokyr (1990) contain valuable historical data and easily available output statistics of the principal industrial commodities produced.

Here we will discuss two indicators that are aggregate representatives of many processes of technological and economic change. The first represents the diffusion histories of many innovations in one country, the USA. The second describes the growth of the "mass production/consumption" cluster after World War II based on a principal component analysis of a large number of individual indicators.

Figure 4.1 shows the results of a diffusion analysis of 117 processes of technological change in the areas of energy, transport, manufacturing, agriculture, consumer durables, communication and military technologies in the USA since the 19th century (for details see Grübler, 1990a). The figure presents the weighted average diffusion rate over time. That is the sum of the first derivatives of empirically estimated diffusion functions divided by the number of diffusion processes at any given time. (In most cases the estimated diffusion function is logistic.) The result is the diffusion equivalent of the annual GNP growth rate.

A rising average diffusion rate indicates the emergence of a whole technology cluster. The curve then tapers off as more and more diffusion processes that sustain a particular cluster tend toward saturation. A trend reversal indicates the progressive emergence of a new cluster, whose initial diffusion rates are low, perpetuating a period of slow growth and painful adjustments and structural change. It is no coincidence that the troughs in *Figure 4.1* coincide with periods of economic depression and recession (1870s, 1930s, and the period since the early 1970s). The figure illustrates the rise and fall of three successive technology clusters. It does not show the first technology cluster from *Table 4.1* because before the 1830s the USA was basically an agrarian society and classified as being on the periphery of technological innovation in *Table 4.2*.

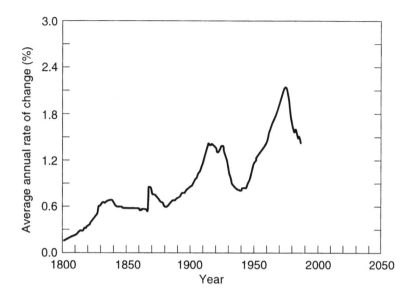

Figure 4.1: Average annual rate of technological, economic, and social change based on a sample of 117 diffusion processes in the USA (%/year).

A second representative aggregate indicator presented in this section follows a more conventional statistical approach in the international comparison of economic development and structural change: principal component analysis. Glaziev (1991) analyzes 50 indicators over the period 1950 to 1986 in the areas of agriculture, construction, the chemical industry, energy, electricity, transportation, and private consumption. He calculates, first, aggregated principal components of the overall evolution of growth and intensity of the seven areas. Second, he calculates the principal component of the seven principal components of the first level of analysis. The results, shown in *Figure 4.2*, are presented as a growth/intensity indicator for each country – the USA, Japan, the Federal Republic of Germany (FRG), the UK, and the former USSR.

The figure shows the evolution and intensity of development of the "mass production/consumption" cluster of the post-World War II period. The USA has most intensively developed this particular cluster and its associated industrial base and consumption patterns. Japan and Western Europe have followed suit, albeit at a lower intensity level. In all the OECD countries the cluster starts to decline in the 1970s and 1980s, indicating a slowdown in growth and possible transition to a new cluster.

It is important to note the decisive differences in the intensity of the development path of the USA, compared to Japan and Western Europe.

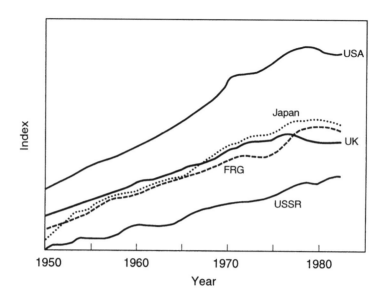

Figure 4.2: Index of the evolution and intensity of the "mass produc-
tion/consumption" technology cluster since World War II. Source: Glaziev
(1991:304).

While the OECD countries all develop along similar lines, with Japan catching up since the late 1950s, the intensity of development is quite different.
This result is consistent with spatial theories of innovation diffusion and
the spread of industrialization as discussed previously. Early starters such
as the USA have the longest growth phase and develop the mass produc-
tion/consumption technology cluster most intensively. Late starters catch
up, but realize lower intensity levels. The former USSR is below even the
Western European trajectory. It developed most of the mass production
technologies, but few of the mass consumption equivalents. Glaziev (1991)
describes the former USSR development path as one of "multi-modeness".
It simultaneously reproduced an outdated technological mode ("heavy engineering") alongside a more modern one. The failure of this industrial
development strategy is now well established, and its environmental legacy
is becoming increasingly clear.

Some Suggestions for Further Reading on Part I

Two books, available in paperback, and giving a good overview of empirical and theoretical aspects of technology are:

Mokyr, J. 1990. *The Lever of Riches: Technological Creativity and Economic Progress.* Oxford University Press (349 pp.). Paperback: ISBN 0195074777, price: US$19.20.

Rosegger, G. 1996. *The Economics of Production and Innovation: An Industrial Perspective.* 3rd edition. Butterworth-Heinemann (313 pp.). Paperback: ISBN 0750624337, price: US$48.

The former won the Schumpeter prize and provides a holistic account of technological evolution since antiquity. It is concise, fun, and insightful with an excellent writing style and illustrations. Mokyr avoids the usual Western technology bias and also gives a good account of developments outside Europe, particularly in China.

Unlike most economics textbooks, Rosegger's book is eminently readable. It provides a competent treatment of technological change theory within neoclassical economics, but also has sympathy with alternative (e.g., evolutionary) approaches and takes a wider disciplinary view, e.g., discussing the management science perspective on technology. Particularly useful are the numerous examples incorporated in the discussion of the customary economists isoquant curves. The book is already in its 3rd edition.

An excellent book integrating both historical and economic perspectives of technology, rich in its contextual breath, and remaining a "classic" in its field is:

Rosenberg, N., 1982. *Inside the Black Box: Technology and Economics.* Cambridge University Press (304 pp.). Paperback: ISBN 0521273676, price: US$28.70.

As a reference book for studying particular technologies and developments (obviously the sheer size and detail of a reference book series precludes any customary "reading"), we draw attention to the monumental Oxford History of Technology:

Singer, C., Holmyard, E.J., and Hall, A.R. 1954 to 1979. *A History of Technology.* Volumes I to VII. Oxford University Press.

Price quotations are from *British Books in Print*, February 1997. Conversion rate used: £1 = US$1.60.

Part II

Technology and the Environment: Natural and Human

Chapter 5

Agriculture

Synopsis

An overview of agricultural output and productivity growth is outlined. Three broad historical periods are distinguished. In the first, agriculture improves primarily through biological innovations in the form of new crops and new agricultural practices. In the second, new transport technologies enable agricultural production and trade to expand to a continental and then a global scale. In the third, mechanization, synthetic factor inputs, and new crops, all developed through systematic R&D, push agricultural output and productivity to unprecedented scales. Throughout all three periods labor productivity rises, requiring ever fewer farmers to feed growing populations both at home and abroad. The reduced demand for farmers precedes a related migration from rural to urban areas, labeled urbanization. Progress in agricultural technologies and techniques also progressively decouples the expansion of arable land from population growth and food consumption growth. Initially, this decoupling simply slows down the expansion of agricultural land. Subsequently, international trade effectively transfers the expansion of agricultural land to other countries, limiting further expansion in the industrialized countries. Finally, agricultural productivity increases to such an extent that agricultural land in the industrialized countries can be reconverted to other uses. Thus technological change, combined with saturating demands for food, translates into *absolute* reductions in agricultural land requirements. Technology begins to spare nature. In contrast with its decreasing land requirements, the overall expansion of agricultural production has more problematic impacts on global water use and global nutrient and geochemical cycles. The chapter concludes with a discussion of urbanization and urban environmental impacts. These can be seen as an important indirect impact of the productivity increases in agriculture driven by technology.

5.1.   Introduction

The history of agriculture over the last 200 years is one of tremendous increases in production to sustain an ever growing population and, for a privileged one-third of the global population, an ever more affluent diet. The greatest environmental impact has been the expansion of arable land by large-scale conversion of natural into managed ecosystems. The speed and manner in which this has happened at different times and places depends on changing patterns of population growth, food consumption, agricultural productivity in different countries, and the international division of agricultural production.

Throughout most of history population increases could only be sustained by enlarging the land area devoted to agriculture and increasing agricultural productivity per unit land area. Today, the expansion of arable land is most visible in parts of the developing world (see *Figure 5.1*). Similar patterns occurred in the northern hemisphere in the past. However, the forces driving today's changes are far greater than in the past. Particularly important is the tremendous absolute numbers with which our global population grows.

Reference to technology's impact on land use usually conjures up images of land covered by city skylines, sprawling suburbs, factories, roads, dams, pipelines, and other human artifacts. In reality, although detailed statistics are lacking, the area covered by such technological artifacts is most likely less than 1% of the earth's total land area.[1] In contrast, the percentage of the total global land area that is used for agriculture and pasture is close to 40% (FAO, 1991:47).

Technological changes in agriculture therefore directly affect much larger areas than other technological changes. They also affect a large share of the global labor force. Just as the productivity of land determines the land requirements for a given population, the productivity of labor determines the percentage of the population that is required to cultivate the land. As we will show later, the impact of technological change on raising agricultural labor productivity has been yet more dramatic than its impact on raising agricultural land productivity. This has enabled an ever increasing fraction of the population to engage in economic activities outside agriculture and to migrate from rural to urban areas.

---

[1]As explained in Section 5.8 we estimate the amount of land devoted to buildings and infrastructures at 250 m^2 per capita. This value is representative of densely populated countries like Japan and the Netherlands. Because most countries are less densely populated, the true worldwide percentage of land covered by human artifacts should be significantly less than the 1% cited here.

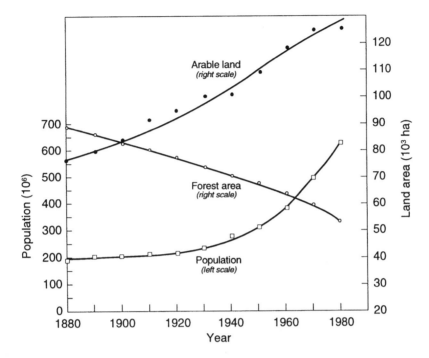

Figure 5.1: Population (millions) and land-use changes (thousands of ha of arable land and forests, respectively) in Pakistan, Bangladesh, Burma, Malaysia, Brunei, and Northern India. Source: Marland (1989:205).

5.2. Technology, Agricultural Land, and People

How does the succession of various agricultural technology clusters since the early 19th century (discussed in Section 5.3) relate to global change? First, there can be no doubt that without technology-driven improvements in agricultural productivity the global population could not have increased to its present size of almost six billion. Technological change led to far-reaching increases in land and labor productivity, thereby decoupling population growth from the expansion of agricultural land, and freeing people for other economic activities and enabling urbanization. Agriculture has become truly gigantic. About 1,500 million ha of arable land are cultivated globally. Grain production is around 2,000 million tons,[2] and the number

---

[2]Source: *FAO Yearbook: Production* (various volumes). For 1990, the tonnage of grains included 600 million tons of wheat, 520 million tons of rice, 480 million tons of maize, and 180 million tons of barley.

of domesticated animals surpasses 15 billion,[3] i.e., about three animals for each person.

Second, the successive transportation revolutions from steam locomotives and ships to today's systems of road and air transport have increased the spatial division of labor. This has enabled the expansion of large-scale export-oriented production and trade, including agricultural products, and the increasing concentration of people in urban areas. Perhaps the most pervasive changes brought about by the Industrial Revolution are the increases in spatial density and productivity arising from transportation systems that cover ever larger distances at ever lower costs (see Chapter 7).

Until recently, patterns of technological change have been geographically diverse. It has only been during the last 50 years that technologies have become truly global, and also only in this period that agricultural land productivity increases have outpaced population growth. To feed the world in 1980 with 1950 agricultural productivity levels would have required an additional 500 million ha, an increase of 33% over the 1,500 million ha actually required in 1980. This is an indication of how technology – through productivity increases – can sometimes "spare" nature. It is necessarily hypothetical as it is unlikely that in the absence of technological change the global population and economy would have expanded as much as they have.

5.3.   Three Clusters of Change in Agricultural Technologies

5.3.1.   An advance summary

In this section we examine three agricultural technology clusters covering the period from the early 19th century up to the middle of the 19th century, the period from then to the 1930s, and the period from the 1930s to the present.[4] We will consider technological and mechanical innovations (e.g., tractors, manufactured fertilizers), biological innovations (e.g., new crops from other continents, new high-yield varieties), and social and organizational innovations (e.g., land reforms).

The categorization into three agricultural technology clusters starting in the early 19th century seems justified when considering how global

---

[3]The number of domesticated animals includes 1.3 billion cattle; 0.3 billion horses, mules, buffalos, and camels; 2.7 billion pigs, sheep, and goats; and 11.6 billion chickens, ducks, and turkeys.

[4]For the sake of brevity, this discussion omits the first technology cluster shown in *Table 4.1*, which is primarily of interest in connection with the origins of industrialization (see Chapter 6).

agriculture was transformed through successive waves of systematic deployment of numerous interrelated biological, mechanical, and institutional/organizational innovations. However, many of these innovations had much earlier origins.

It would be impossible to understand developments in agriculture since the 19th century without mentioning the important diffusion of numerous new crops and animals that accelerated after the 16th century. Most biological innovations that started to transform European and subsequently global agriculture had their roots in the exchange of crops and animals across continents, for example, the potato and maize introduced to Europe from the Americas, and wheat and horses introduced to the Americas from Europe. However, as long as these exchanges were not integrated into an entirely new system of agricultural production, their impact on diets and agricultural productivity remained localized and comparatively limited. Things started to change dramatically in the 19th century, summarized here under the heading of the first agricultural technology cluster of "agricultural innovations".

In the period of "agricultural innovations", agriculture is transformed through the widespread diffusion of new species imported from other regions, new agricultural techniques such as complex crop rotation patterns, and new institutional innovations affecting various operational practices. The resulting increases in agricultural output and productivity sustain rising populations and rapid urbanization in the core regions of industrial take-off, particularly in England. Agriculture thus sustains early industrialization, which is characterized by an emerging factory system (especially textiles) and the widespread application of stationary steam power in industry.

The second technology cluster is centered around heavy engineering industries, particularly steel, and new transport and communication infrastructures (e.g., railways, steamships, and the telegraph) based on the widespread diffusion of mobile steam engines. Agricultural changes come mainly in the form of further diffusion of innovations from the previous period, plus biological innovations (e.g., new higher yield crops) outside the industrialized countries. Agricultural changes also involve improvements in mechanization and the introduction of mined phosphate fertilizers. The most important characteristic of the period, however, is the global spread of transport infrastructures and the resulting expansion of world trade. In agriculture this marks the beginning of large-scale export-oriented crop production in the tropical countries (e.g., grains, cotton, rubber, sugarcane, coffee, and tea). The political term for this increasing spatial division of agricultural production is colonialism. We refer to this period as "mercantilistic agriculture".

The third technology cluster is characterized by Fordist mass production (and consumption) of consumer goods, by petroleum as a primary energy carrier and feedstock for industry and transportation, and by new communication systems – the telephone, radio, and television. Agriculture is revolutionized by the widespread application of industrial innovations: mechanization, synthetic factor inputs in the form of fertilizers and pesticides, and new high-yield plant varieties developed through agricultural R&D. These innovations result in unprecedented increases in output and productivity. From their beginnings around the turn of the century in Europe and North America, they spread after World War II to become truly global. Globally, land-use conversions fall significantly behind the rate of population growth, and the most developed regions begin to reconvert agricultural land to forests. We refer to this period as the "industrialization of agriculture".

5.3.2.   Important developments prior to the 19th century

During the first of the three technology clusters discussed here, covering the period starting from the beginning of the 19th century, European agriculture was radically transformed by biological innovations in the form of new crops and by new farming practices. None of these innovations was entirely new. All had been used on a smaller scale in parts of Europe, particularly the Netherlands, or were imports from the Americas, such as corn and potatoes. Particularly important were increasingly complex crop rotation patterns in conjunction with new fodder crops such as clover and, later, lucerne. These eliminated fallow periods and previous constraints on the feed supply for animal stock, particularly during the winter. Typically a new crop rotation pattern involved wheat, turnips, barley, and clover, but in some parts of Europe more complex patterns (e.g., the Flemish seven-course rotation) were introduced. Fertilizer became more available due to the increased stock of animal husbandry, and guano imports from Peru starting in the 1820s, and nitrate imports from Chile starting after the 1840s. Grigg (1987:100) characterizes the new agricultural system as a greater integration of livestock and arable husbandry. The new system did not significantly improve overall land productivity, but it enabled better utilization of fallow and grassland areas. Increases in cropland could draw on these fallow and grassland areas. Europe is the only region where such conversions took place before the mid-19th century at a noticeable scale. They are estimated at 25 million ha from 1800 to 1850 (see Section 5.5) and noticeably slowed down the rate of deforestation.

The introduction of new staple food crops from the Americas was also important. These were corn (maize, the basis of polenta) and the potato.

Table 5.1: Originating areas for the worldwide diffusion of agricultural crops and animals.

Southeast Asia	Europe	Americas	Africa
Aubergine	Barley	Avocado	Coffee
Banana	Bee	Cocoa	Hard wheat
Lemon	Cattle	Maize	Sorghum
Lime	Horse	Manioc	
Orange	Oats	Peppers	
Rice	Pig	Pineapple	
Spinach	Rabbit	Potato	
Sugar cane	Sheep	Pumpkin	
	Wheat	Rubber	
		Sisal	
		Squash	
		Tobacco	
		Tomato	

Source: Ponting (1991:110).

Table 5.1 gives an overview of the origins of the worldwide diffusion of agricultural crops and animals. While Europe probably gained more new crops and animals from other continents than vice versa, other continents did gain from Europe too. Perhaps the best example is the horse, introduced in North America by the Spanish. Horses subsequently became an integral part of the "indigenous" Indian culture on the North American continent.

Agricultural diffusion also had its own precursors. In antiquity the Romans introduced the cherry tree from Asia to Europe, and wine growing diffused from southern to northern Europe. But these remained isolated developments. The first more systematic diffusion of agricultural crops corresponds with the rise of Islam and runs from about the seventh to the tenth century. India was the principal origin of new crops that diffused to the Middle East, North Africa, and southern Europe. The new crops included hard wheat (which became an important staple in North Africa in the form of couscous, and in Italy in the famous form of pasta), rice (reaching northern Italy by the 15th century), lemons, limes, and vegetables such as spinach and aubergines. Sugar cane was introduced from India to Mesopotamia and the eastern Mediterranean. From there it diffused to the West Indies as part of the second wave of the global agricultural diffusion triggered by Christopher Columbus' journey to the Americas in 1492. This second wave of diffusion led to a much wider transformation of global agriculture as staple foods like corn and potatoes were to substantially increase the food supply in areas of rapid population growth such as Europe. It also prepared for the first large-

scale specialization of agricultural production in particular parts of the world
and the first large-scale production of export crops. The first of these export
crops was sugar cane.

5.3.3. The period of agricultural innovations

In the period of "agricultural innovations", the developments described
above were finally broadly diffused, especially in Europe. New staple foods,
combined with intensified animal husbandry, resulting in increased fertilizer
availability (manure), enhanced by increasing fertilizer imports, and generally intensified land use, increased European agricultural output significantly. Perhaps the best single indicator for the radical change in European
agriculture is the fact that the last famine affecting all of Europe was in 1816–
1817. It was caused partly by dislocations from the Napoleonic wars and
partly by dismal weather conditions, due probably to volcanic dust ejected
into the atmosphere by the Tomboro volcanic eruption (Ponting, 1991:106).
To be sure, it took quite some time before the agricultural innovations diffused to every part of Europe. Vulnerability stemming from the overreliance
on any one individual of the new biological innovations also continued to
persist (as illustrated in the case of the Irish potato famine discussed in
Box 5.1). Nevertheless, continent-wide famines ceased to occur as a result
of a more productive agricultural system.

Next to the new major staple foods of maize and potato, additional economically important new crops were introduced: the sugar beet and tobacco
(although the latter had no nutritional value whatsoever). The sugar beet
diffused widely in Europe after the discovery of sugar refining and the opening of the first factory in Silesia in 1801. The sugar beet experienced a particularly strong boost due to the continental blockade during the Napoleonic
wars and the loss of sugar (cane) imports from the Caribbean.

The pervasive diffusion of new crops characteristic for the period of
"agricultural innovations" implies a more diversified and enlarged dietary
base. However, the diffusion of new crops is accompanied by the diffusion
of new pests and species that become a nuisance in new ecosystems where
their growth is unchecked by natural predators. Pests introduced from the
Americas such as potato blight and later the Colorado beetle threatened the
success of the potato "revolution" (see *Box 5.1*). The Irish potato famine
particularly illustrates how isolated agricultural innovations can overcome
constraints only temporarily, and the vulnerability of expanding food production based on a single staple crop. Another example is the vine disease
phylloxera, which was imported from North America in the 1860s and nearly

---

Box 5.1: The Irish Potato Famine*

A major famine struck Ireland in 1845–1846, and is generally referred to as the "Irish potato famine". Population pressure (the population of Ireland increased from about 0.8 to 8.5 million between 1500 and 1846) and extremely uneven distribution of agricultural land (the average plot size amounted to less than an acre with 650,000 landless laborers living in a state of permanent destitution) led to the adoption of the potato as almost the exclusive staple food. At the beginning of the 19th century potatoes were planted on about half of the entire cropland area, and about half of the population relied on the potato as its sole food. In fact, without the successful introduction of the potato from the Americas, it would have been impossible to sustain a population of more than eight million even at a minimum level of nutrition given the tiny farmland plots and widespread poverty precluding purchase of food.

Poor harvests had caused recurrent widespread starvation ever since the middle of the 18th century. But the arrival of the potato blight disease from America in 1845 triggered straightforward catastrophe. The 1845 harvest was lost partially and that of 1846 almost entirely, as was the case in most of Europe.

The human consequences of this crop failure were exacerbated largely by policies adopted by the British government. These involved not interfering with the workings of the free market in food and halting all public relief works such as road construction (and thus the last income possibilities of the rural poor) in order to stop people becoming dependent on government welfare. Large quantities of grain were imported, but even larger quantities of the Irish grain harvest were exported to England (where the crop harvest had been bad), often under armed guard. The food market worked freely, but the rural poor had no income (not to mention no government aid) to purchase corn or maize.

It is estimated that about one million people died either directly from lack of food, or from the outbreak of diseases that affected the undernourished population. Another million people emigrated during the famine or immediately afterwards. By the end of the 19th century another three million had left the island.

The Irish famine illustrates the dangers of being overdependent on a single successful innovation, in this case the potato. However, the real failure was not technological but rather political, institutional, and social: starvation amid plenty, even under large-scale food exports. There was no absolute food shortage in Ireland. Those who died could simply not afford to buy food and the authorities were neither prepared to distribute food for free nor to provide income possibilities for the poor to purchase food. This distributional issue also forms the central part of the interpretation of more contemporary famines in the developing world (cf. Newman, 1990).

*Based on Ponting (1991:106–108).

---

drove European wines to extinction.[5] Diffusion also creates opportunities for pests to "migrate" across crops. The Colorado beetle, for instance, did not originally thrive on potatoes but only "discovered" its taste for potatoes

---

[5]Phylloxera is an insect that attacks vine roots in one phase of its reproductive cycle (Jackson, 1994:133–134). It illustrates both the importance of quarantine regulations and the effectiveness of biological methods of disease control. The vines in current European vineyards are all grafted onto phylloxera-resistant American root-stocks. Only in some very isolated spots (including the vineyard of the author) have pre-phylloxera European vines survived.

once potato farming was introduced to Colorado. From there the Colorado beetle spread globally.

The diffusion of pests affects unmanaged ecosystems as well as agriculture. An example of an indigenous species wiped out entirely by imported predators is the dodo. Large numbers of dodos once inhabited the island of Mauritius. The dodo is a large bird that cannot fly, but never needed to as it had no predators on Mauritius. Once humans began hunting the dodo its population dropped quickly, and it was finally driven to extinction by rats inadvertently imported by European sailing ships. The current connotation of "dodo" is definitely unfair. The bird was far from being stupid. It was simply not adapted to evade predators that it never had, and was thus defenseless when predators were introduced from different ecosystems.

All things considered, however, the global diffusion of agricultural crops has had far more benefits than drawbacks. Despite some losses from imported pests, the new agricultural crops and animals enabled substantial expansion of agricultural output, enhanced food security, and diversified diets during the period of "agricultural innovations".

Organizational and institutional changes also played a decisive role in increasing agricultural output. The most important such changes were the abandonment of peasant serfdom in Europe during the 18th century, plus a number of land reforms[6] allowing the concentration of farmlands and resulting economies of scale. New fodder crops and the abandonment of fallow lands used previously for communal pasture implied important institutional changes concerning land rights and usage. To keep grazing animals off cropland, farmland became increasingly enclosed. Between 1760 and 1840 over 6 million acres in England were redistributed in separate holdings by private Enclosure Acts (Fussel, 1958:17). Agricultural practices also changed, although the change was not always smooth. The introduction of horse-powered threshing machines, for example, faced opposition in the form of the violent Captain Swing movement in England in 1830 (cf. *Figure 2.20*),[7] and the machines diffused slowly. Through the middle of the 19th century, advances in farming techniques followed similar patterns in many European countries, with some time lags between countries (France was a particular laggard). Yields in England increased slowly from about 16 bushels per acre in the late 16th century to 20–22 bushels 200 years later (Fussel, 1958:31).

A final important institutional development was the establishment in all European countries by the mid-19th century of centers of agricultural

---

[6] For an account of Sweden see e.g., Anderberg (1991:403–426).

[7] For an excellent account of the causes, events, and consequences of this first manifestation of agricultural Luddism see Hobsbawn and Rudé (1968).

research and education. In the USA public sector R&D in agriculture became institutionalized with the founding of the US Department of Agriculture in 1862. These institutions and systematic R&D efforts paved the way for both biological and mechanical innovations leading to even more spectacular improvements in agricultural output in the subsequent periods.

5.3.4. The period of mercantilistic agriculture

From approximately the mid-19th century to the 1930s, agricultural practices introduced earlier in England and some European countries spread to vast peripheral regions of Europe such as Russia. This spread was characterized not so much by the transmission of agricultural techniques used in the more densely populated areas of Europe, as by new developments in transport, manufacturing, and science that accompanied the process of industrialization (Boserup, 1981:116–117). These could spread only after the iron and chemical industries were developed and their products had become sufficiently cheap to be used in agriculture. Commercial fertilizer and large-scale imports of food and fodder could not be introduced until a railway network was in place and the steamship was widely diffused. Imports of animal products required refrigeration, and transportation over longer distances required new methods of food preservation. The interconnections and mutual reinforcement among advances in transport technologies, new methods for refrigeration and food preservation, and growing world trade in manufactured goods (to pay for food imports) were central to the emergence of "mercantilistic agriculture".

The transport revolution combined tremendous improvements in accessibility with rapidly falling transport costs.[8] This enabled unprecedented regional specialization and the opening of vast new agricultural areas in the Canadian provinces, the American mid-west, the Argentine pampas, the Russian steppes, and the interior of Australia. Thus as the food hinterlands of industrialized core regions shrank, these regions relied increasingly on external food sources to provide for diversified diets including products produced only in distant climatic zones.

The introduction of industrial innovations (i.e., mechanization) was important particularly for raising labor productivity. In North America agricultural labor was scarce relative to land, and mechanical innovations for stationary applications intensified. These included the mechanical reaper (1831; see e.g., David, 1975), the transportable threshing machine (1850),

---

[8]Typically rail freight costs declined by up to a factor of five between the 1870s and the 1920s (Grübler, 1990b:117–119).

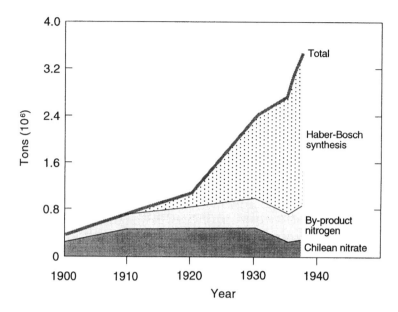

Figure 5.2: World nitrogen production by process: 1900–1938 (in million tons, cumulative totals). Data source: Zimmermann (1951:789).

and the milking machine (1850). However, in the absence of a light high-output movable power source (the 20th century tractor), the impact of these innovations remained limited, particularly outside North America.

For raising land productivity the discoveries of synthetic fertilizers, superphosphates (invented in 1841 and the only chemical fertilizer of the 19th century), nitric fertilizers (1906), and ammonia synthesis for nitrogen fertilizers in 1912 (the Haber-Bosch process) were all very important. Stimulated by military requirements during World War I (the production of TNT explosives required fixed nitrogen), the Haber-Bosch synthesis quickly became the dominant source[9] of synthetic nitrogen, supplying 75% of fixed nitrogen fertilizers by the eve of World War II (*Figure 5.2*).

Nitrogen fertilizers were applied widely in European and world agriculture after the 1920s (cf. *Figures 5.5* and *5.6*). Combined, fertilizers, pesticides, fungicides, and the breeding of new plant varieties enabled significant increases in yields. These improvements were most pronounced in Europe where land conversions were reduced to half the value of the previous five decades (cf. Section 5.5).

Agricultural land productivity increased mostly in Europe, although some new plant varieties were introduced outside Europe [for example, new

---

[9]The other two sources were mined nitrates from Chile, and nitrogen produced as a by-product of coke production.

high-yield rice varieties introduced in Japan doubled yields between 1880 and 1930 (Hayami and Ruttan, 1985:468)]. The increase in European production together with large-scale food imports provided for an ever increasing population while reducing land conversions from forests and grasslands to cropland. Outside Europe land productivity did not increase noticeably, with the exception of Japan mentioned above. In North America advances in labor productivity were not accompanied by comparable advances in land productivity. As a result cropland expansion continued vigorously. Wire-fencing facilitated the conversion of pasture to cropland. An estimated 100 million ha of grassland were converted to cropland in North America between 1850 and 1920 (cf. Section 5.5).

Finally, innovations in food preservation proved very important. These included tin cans, concentrated milk, the invention of absorption refrigeration in 1850, and the invention of ammonia compression refrigeration in 1876. Refrigeration technology remained cumbersome, suffering from frequent leaks of the highly reactive ammonia. This problem would be solved only in the 1930s through the introduction of chemically inert chlorofluorocarbons, or CFCs. (Today we know that CFCs have environmental problems of their own, being the major contributor to the depletion of the earth's stratospheric ozone layer.) Refrigerated steamships allowed meat to be imported from as far away as Australia, New Zealand, and Argentina. These improvements in food preservation, coupled with the dramatically decreasing transport costs of the railway and steamship era, enabled an unprecedented expansion of agricultural trade. By the 1870s, England – at that time the world's leading economic power – had net imports of agricultural products exceeding the net export value of all manufactured goods (Woytinsky, 1927:212). By the end of the 19th century Russia was sending large-scale grain exports to Central Europe and England. World trade in agricultural products doubled between the 1870s and 1913.

5.3.5. The period of agricultural industrialization

Since the 1930s global agriculture has been transformed from a resource-based to a technology-based industry. The transformation has occurred partly through the development of new technologies, but more importantly through a series of institutional innovations that accelerated both technological innovations and their diffusion. Examples include new public and private sector suppliers of innovative plant varieties and agricultural technology, institutions and services for transferring technical knowledge to farmers, public and private sector R&D, input supply and marketing organizations, and the development of more efficient labor, credit, and commodity markets.

The industrialization of agriculture is characterized by three developments: biological innovations, new cheap factor inputs, and mechanization. Advances in all three have reinforced each other and led to spectacular increases in both labor productivity and, for the first time since the Industrial Revolution, land productivity in developed as well as developing countries (*Figure 5.3*). In the industrialized countries, productivity increases have been so large as to entirely decouple the expansion of agricultural land from population growth. Indeed, "industrialized" agriculture has allowed significant reconversion of cropland back to grassland and forest cover (*Figure 5.4*).

The first set of developments, biological innovations, included the introduction of new crops and broad diffusion of new high-yield plant varieties developed through systematic agricultural R&D. Diffusion in previous periods involved the spread of varieties already in use in some other parts of the world. In the period of agricultural industrialization, there has been an additional sort of diffusion, that of new varieties systematically "engineered" and actively diffused in order to raise agricultural productivity. These include new hybrid corn and rice varieties, perhaps the most important contributions of applied biology in the 20th century. It is no coincidence that the first detailed economic studies of agricultural diffusion focused on hybrid corn (Griliches, 1957).

While the new plant species increased yields, diffusion of crops between continents both opened new export markets (examples are soybeans in the USA over the last 30 years and, more recently, in Brazil) and improved and diversified local diets. Maize and manioc, for example, have become important food supplements in Africa, while sweet potatoes, maize, and peanuts have supplemented Asian rice and wheat diets.

The second set of developments concerned cheap new factor inputs in the forms of fossil energy and synthetic fertilizers and pesticides. Fertilizer availability was no longer limited by animal production and naturally occurring deposits. Already prior to World War II, ammonia synthesis became the dominant source of nitrogen fertilizers, and since then global fixed nitrogen use has risen from some 3 million tons to 80 million tons (*Figure 5.5*), with increasing shares for Eastern Europe, the former USSR, and particularly the developing countries. The growth of phosphates has also been tremendous. Over 150 million tons of phosphate rocks are now mined globally (Smil, 1990:431). Today nitrogen and phosphate fertilizers affect nearly every major biospheric flow of nitrogen and phosphorus nutrients on the planet (cf. Section 5.6). *Figure 5.6* shows the increases in total fertilizer application around the world. Regional disparities today are comparatively small with the exception of Europe and Africa. The former is significantly above the world average. The latter is significantly below. [Note however, the recent

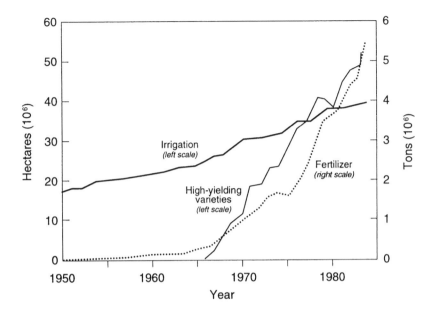

Figure 5.3: Increasing agricultural land productivity in India since 1950 through planting more high-yielding varieties, irrigation (both in million ha), and the application of fertilizers (in million tons). Data source: Sarma and Gandhi (1990:36).

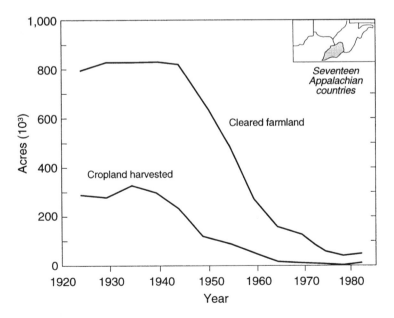

Figure 5.4: Decrease in cleared farmland and harvested cropland since the 1930s on the Appalachian Plateau, USA. Source: Hart (1991:66).[1]

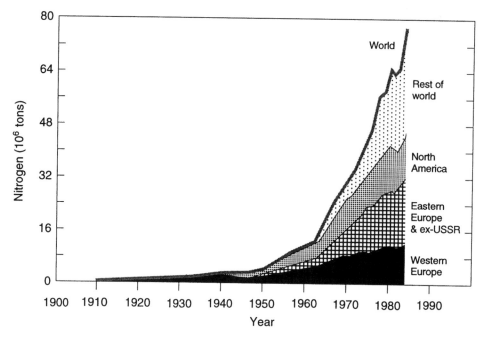

Figure 5.5: World nitrogen fertilizer use by region (cumulative totals, in million tons nitrogen). Data source: Kolmhofer (1987:Table 3).

drastic decline in fertilizer use in Europe (data include Western and Eastern Europe) and the former USSR shown in *Figure 5.6*, as a result of economic recession and restructuring of the formally centrally planned economies.]

Pesticide use has also grown enormously in the period of agricultural industrialization. Currently, over 3 million tons of formulated pesticides are produced annually (Brown *et al.*, 1994:92–93). The adverse environmental impacts of long-lived pesticides have been well documented, starting with Rachel Carson's *Silent Spring* in 1962, and the use, particularly of DDT, has been limited since. There have been significant technological improvements in terms of degradable pesticides, and new methods of biological pest control and integrated pest management. However, we remain in a race against the increasing resistance of insects and plant diseases to pesticides. Currently over 500 insects and mites, more than 150 plant diseases, and over 100 weeds are estimated to have developed resistance to one or more pesticides (Brown *et al.*, 1994:92).

The third set of developments was characterized by mechanization, symbolized by the farm tractor. The substitution of inanimate power (and fossil energy) for animal and human power (and renewable energy in the form of feed) alleviated another constraint on agricultural output: labor. *Figure 5.7*

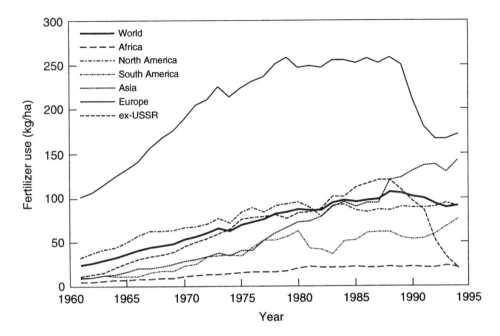

Figure 5.6: World fertilizer application (in kg/ha cropland) by major regions. Numbers include nitrogen and phosphate fertilizers as well as ground rock phosphate. Source: based on *FAO Yearbook: Production* (various volumes). Graphic courtesy of Gerhard Heilig, IIASA.

shows the evolution of direct human labor and energy inputs in US agriculture since 1920. Labor inputs decreased by approximately a factor of five while fossil energy inputs increased by a factor of five. The increase in total agricultural energy requirements (by about 50%), however, is much lower than the fossil energy increase. Initially, much of the energy consumed in agriculture was nonfossil energy, such as traditional animal energy and fuelwood. Much of the increase in fossil energy has replaced these other energy forms, not just human labor.[10] Agricultural consumption of nonfossil energy

---

[10] Our calculations for energy requirements in the form of feed energy, wind and water power, and fuelwood are based on estimates of Fisher (1974:158–159). Because Fisher's fuelwood estimates are not disaggregated among uses, our figure may be an overestimate. The error should be small because, by the 1920s, fuelwood use in the USA was confined largely to rural, agricultural areas (Schurr and Netschert, 1960). Note, however, that the figure includes only direct energy inputs, not the *embodied* energy required for the manufacture of tractors, fertilizer, food processing, etc. There are no reliable historical estimates for embodied energy needs in agriculture. Contemporary estimates for intensive wheat farming indicate energy needs of 2.7 MJ per ton, 20% of which are directly consumed on the farm in the form of fuel for tractors and harvesters, and 80% are off-farm consumption, mostly for fertilizer manufacturing (Bonny, 1993:59).

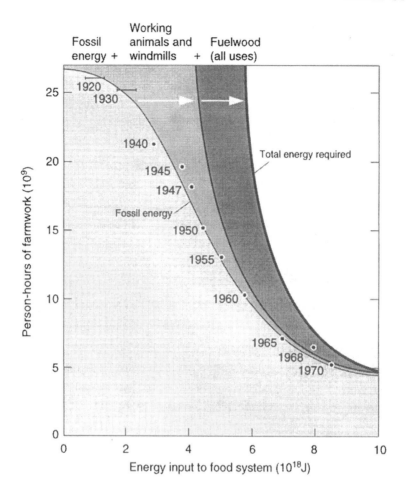

Figure 5.7: Direct human labor (billions of hours) and energy inputs (10^{18} joules) to US agriculture, 1920 to 1970. Data Source: Fisher (1974:158–159) and Steinhart and Steinhart (1974:51).

in the USA is estimated to have peaked at around 5×10^{18} joules (J) in the 1920s. As a result, total energy consumption in US agriculture increased from about 6×10^{18} J in the 1920s to only a bit less than 10×10^{18} J in the 1970s, while total output more than doubled over the same time period (Hayami and Ruttan, 1985:482). Thus energy requirements per unit of output dropped. The principal reason is the higher end-use efficiencies of modern technologies fueled by fossil energy. While a horse typically converts 3% of the energy embodied in feed to useful work, the energy efficiency of a farm tractor is 30%.

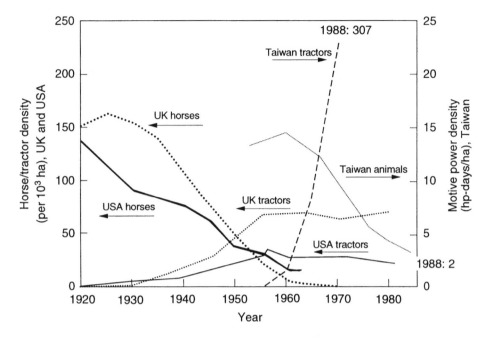

Figure 5.8: Displacement of animal labor in farming: UK, USA (number of horses and tractors per hectare cropland), and Taiwan (horse-power-days per hectare cropland). Source: adapted from Grigg (1982:133) and Jones (1991:626).

In addition, because of mechanization, large areas that had been required to feed working animals became available for crop use. In the USA, the area required to feed farm horses and mules was nearly 40 million ha in the 1920s (US DOC, 1975:510), twice as large as the area devoted to export products and about half the area used for domestic production. Mechanization made much of this available for crops.

Figures 5.8 and *Figure 5.9* show the spread of mechanization around the world. Over the last 20 years, the developing countries' share of farm tractors has been rising rapidly.

Industrialization also changed the demand for agricultural products. It created new demands for agricultural raw materials, but also partially replaced many agricultural raw materials with synthetic products. The first synthetic fibers were based on a reconstituted natural polymer: cellulose. Cellulosic fibers, made using a whole range of processes but all called rayon, carved out a modest market niche (less than 10% of global fiber production) even before World War II. Next came the true noncellulosic synthetic fibers like polyamides (nylon) and polyesters (dacron). While synthetic fibers have

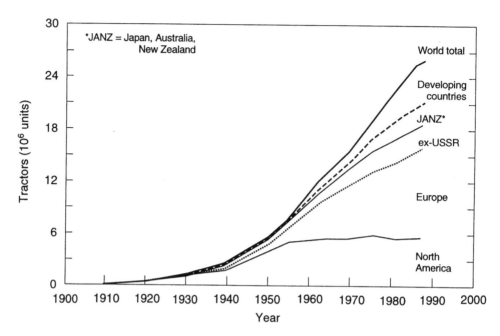

Figure 5.9: Diffusion of agricultural tractors worldwide and by region (cumulative totals, in millions). Data source: Woytinsky and Woytinsky (1953:515) and *FAO Yearbook: Production* (various volumes). For the data of this graphic see the Appendix.

not displaced either cotton or wool in absolute terms, they have captured virtually all fiber output growth in the textile industries since World War II. Since the 1960s they have also dominated fiber production in absolute terms (*Figure 5.10*).

With the development of the electrical industry and motor vehicles, rubber was transformed from a minor curiosity to a major raw material. Production of natural rubber rose rapidly to about 1 million tons in the 1930s, with over 90% of this production concentrated in Southeast Asia. Rubber plantations covered some 5.6 million hectares in Asia in the 1930s, split almost equally between large plantations and small holdings (Woytinsky and Woytinsky 1953:620). By the late 1980s world rubber production exceeded 14 million tons. Fortunately this 14-fold increase in rubber production since the 1930s did not translate directly into a 14-fold increase in land requirements. The increase in land requirements has been much smaller due to synthetic rubber – yet another product of the petrochemical industry. Currently synthetic rubber accounts for two-thirds of the world's rubber output, which also now includes a small contribution from recycled rubber (*Figure 5.11*).

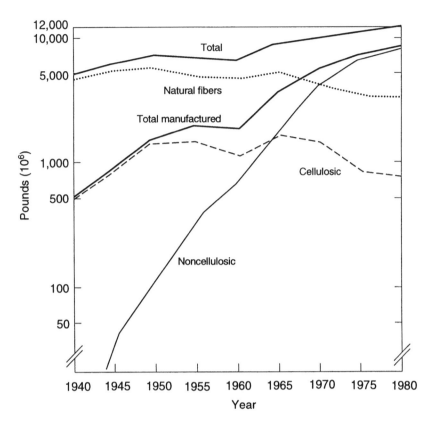

Figure 5.10: Global consumption of natural and manufactured fibers (in millions of pounds, logarithmic scale). Source: Ayres (1989a:48).

The industrialization of agriculture has increased the output of both food and raw materials dramatically. Output growth has more than matched population growth. In some areas, particularly Europe and North America, this has allowed the large-scale reconversion of marginal agricultural lands to forestry. And it has led to large and costly agricultural surpluses. Agricultural policies in the OECD countries subsidized agricultural production to the tune of US$300 billion in 1990 (OECD, 1991:5).[11] This is split almost equally between direct producer subsidies and transfers from consumers, who must pay above world market prices. This total subsidy is comparable to the total value of world trade in crude oil. At the extreme end (e.g., Switzerland, Norway, and Japan), subsidies equal about three-quarters of the value of agricultural output (OECD, 1991). Such subsidies have the effect of vastly raising agricultural output and favoring diets to become increasingly

---

[11]US$ in this book refers to constant 1990 money and prices, unless otherwise stated.

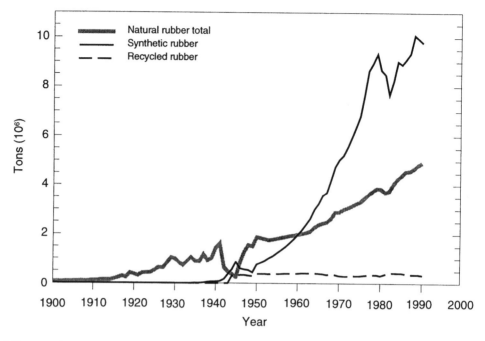

Figure 5.11: World rubber production: synthetic, recycled, and natural (in millions of tons). Data source: Woytinsky and Woytinsky (1953:621–623) and UN *Statistical Yearbook* (various volumes). For the data of this graphic see the Appendix.

dominated by animal products. In most western countries livestock products account for as much as two-thirds of the value of output (Grigg, 1987:102).

Table 5.2 summarizes the state of agriculture in selected world regions based on 1990 FAO data. The table shows that industrial innovations have diffused into agriculture on a global scale, although large regional variations remain. Differences also remain in land productivity in terms of food calories per arable hectare due to differences in the output mix and the intensity of cultivation, fertilization, and mechanization. The picture that emerges is one of opportunities to increase food production per hectare of arable land in many regions without engaging in forest clearing, thus providing additional food for continuing population increases.

The question remains whether such increases in land productivity in developing countries will be sufficient to keep pace with population increases. The history of Europe and of North America illustrates the potential of technology to fulfill such an objective. However, the technologies that will be applied in the future will depend mainly on economic and social policies addressing key potential constraints. The first is the shortage of capital.

Table 5.2: Agriculture: Land, people, and technology, 1990.

	Europe	Former USSR	North America	JANZ[a]	China	Asia	Africa	Latin America	World
Population (10^6)	497.7	288.0	275.7	143.6	1,135.5	1,855.9	647.5	448.3	5,292.2
Arable land (10^6 ha)	140.1	232.4	235.9	52.2	96.6	351.7	186.7	179.8	1,475.4
Irrigated area (10^6 ha)	17.3	20.8	18.9	5.0	44.9	94.9	11.2	15.6	228.7
Farm tractors (10^6)	10.3	2.7	5.4	2.4	0.9	2.2	0.6	1.4	25.9
Fertilizer use (10^6 t)	31.9	26.5	19.9	3.6	18.9	21.4	3.5	8.4	134.1
Food supply (10^9 cal)	1,723.9	976.6	994.6	414.3	2,946.0	4,222.2	1,421.0	1,204.4	13,903.0
Arable land-use intensity									
People/km^2	355	124	117	275	1,175	528	347	249	359
Fraction irrigated	0.12	0.09	0.08	0.10	0.47	0.27	0.06	0.09	0.16
Tractors/km^2	7.4	1.2	2.3	4.6	0.9	0.6	0.3	0.8	1.8
Tons fertilizer/km^2	22.8	11.4	8.4	6.9	19.6	6.1	1.9	4.7	9.1
Food output 10^6 cal/km^2	1.23	0.42	0.42	0.79	3.05	1.20	0.76	0.67	0.94
10^3 cal per person	3.5	3.4	3.6	2.9	2.6	2.3	2.2	2.7	2.6

[a]Japan, Australia, New Zealand.
Data source: *FAO Yearbook: Production* (various volumes).

The second is the environmental impact of increased land productivity and agricultural output.

Before speculating on the future, however, let us summarize quantitatively the impacts of the three agricultural technology clusters. First are the impacts on agricultural productivity, in particular land and labor (Section 5.4). Second are global change impacts in the form of land-use changes (Section 5.5). Third are other global environmental changes such as impacts on water resources, nutrient cycles, and greenhouse gas emissions. This category also includes local and regional impacts that are common to many regions (Section 5.6).

5.4. Impacts I: Productivity

Agricultural productivity can be measured in different ways. Here we focus on changes in the productivity of the factor inputs land and labor. Generally, agricultural systems that are highly land productive (i.e., require little land input per unit of agricultural production) are labor intensive. This is the case in Asia. Conversely, agricultural systems that are highly labor productive are *ceteris paribus* land intensive. These are the "new continent" agricultural systems in North America and Australia. Up to now the differences between the two extremes of "Asian" and "New Continent" agricultural systems have been persistent. The apparent stability in their respective land and labor productivities was shown in *Figure 3.2* in Part I and was explained on the basis of theories of "path dependence" and induced technological change. For a deeper understanding of why different agricultural systems have moved persistently along different frontiers of agricultural land and labor productivity we need to first understand their different initial starting conditions.

5.4.1. The importance of initial conditions

By 1100, China already had a population estimated at approximately 100 million and a population density of about 25 people per km^2. Europe only reached this population density 600 years later. In the 11th century Europe's population was about 30 million (McEvedy and Jones, 1978:19) and its population density less than seven people per km^2. Agricultural practice included long fallow periods, and agricultural productivity was correspondingly low. Typically, fields did not yield more than three to five times the seed sown. Only in exceptional harvests did the ratio rise to six or seven (Slicher van Bath, 1963:15).

By the end of the 17th century Europe's population had increased to about 100 million with a density of approximately 20 people per km². This expansion was hardly smooth. Plagues and wars caused substantial fluctuations in population and in agricultural output and land use (Abel, 1980). The overall population expansion, however, together with the emergence of the medieval city and an urban bourgeoisie, depended on a large number of innovations in agriculture, transportation, and energy. These innovations reduced physical toil and improved labor productivity. They had less effect on land productivity. Given the low population density, large virgin forests were a readily available resource for increases in agricultural output. Consequently, population growth caused large-scale conversion from forests to agriculture between the 11th and 15th century. This resulted from both inward colonization (e.g., France and England) and outward colonization (e.g., Germany). Agricultural settlements on cleared forest areas can be recognized even today in many parts of Europe by virtue of their distinctive land-use patterns (e.g., the "Waldhufenflur" in Germany and Austria, see Engel, 1979; and Grübler, 1992a).

Interruptions in this trend were only temporary. Areas depopulated by the Black Death or wars in the Middle Ages were later recolonized (Abel, 1956:52). Overall, throughout the Middle Ages and the Renaissance, Europeans used their forests in an "eminently parasitic and extremely wasteful way" (Cipolla, 1981:112). Many areas such as the maquis of southern France, the barren areas of central Spain, and the eroded coastlines of the Adriatic denuded by the Venetian ship-building industry are today testimony to the profound land-use changes and deforestation of Europe after the 10th century. *Figure 5.12* shows European albedo changes due to the large-scale conversion – mostly between the 11th and 14th centuries – of forests to agricultural land. Only since the 1950s have some reversals of the historical pattern started to become evident. More recent developments in other continents echo the earlier experience of Europe (cf. Woodwell, 1990).

The widespread disappearance of forests by the 17th century led to timber shortages, particularly in England, and rapidly rising energy prices. Charcoal prices tripled in the period from 1630 to 1690, and many attempts were made to introduce substitutes such as coal. Land became the limiting factor to population growth as exemplified by Malthus' pessimistic vision of increases in agricultural productivity falling far short of population increases (cf. Glass, 1953:140).

Labor productivity was just enough that 10–20% of the 17th century European population could engage in activities outside agriculture. The

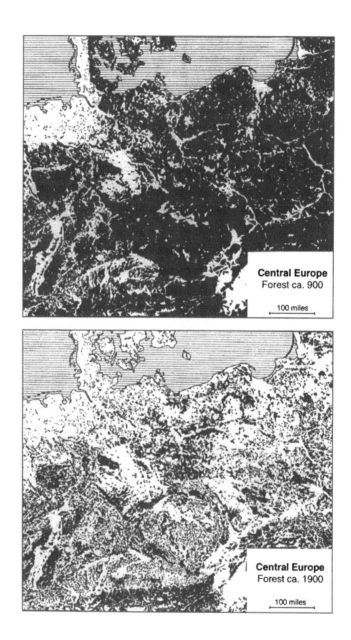

Figure 5.12: Forest cover in Europe 900 (top) and 1900 (bottom). Source: Darby (1956:202–203).[2]

overall productivity of agricultural land supported slightly more than one person (and two draft animals)[12] per hectare of arable land.

In contrast, China, whose population in 1600 was similar to Europe's (about 150 million people), was able to feed 15 people per hectare of cultivated land. This far exceeds even present European land productivity levels! Agricultural land availability had long been *the* principal constraint in China to increases in agricultural output. As a result, China developed an agricultural system characterized by labor-intensive, high-intensity rice cultivation with correspondingly high yields. For such a system both technological innovations and social and organizational innovations were extremely important. Wetfield rice cultivation required sophisticated civil engineering (terraced fields) and hydraulic engineering (dams, locks, water storage, etc.) for proper drainage and irrigation. Gates, pumps, and water-raising devices (norias) controlled the flow of water. The scale of these water control projects required elaborate organizational skills. Wittfogel (1957) referred to China as a "hydraulic civilization". Perkins (1969:61) reports that more than 50,000 water management projects can be identified in various government gazettes. Of the 5,000 water control projects whose construction can be dated, 94% were constructed between the 10th and 19th century. The agricultural system was supplemented by an elaborate transport system. In the case of food shortages, this and efficient social organization enabled effective relief. The related administrative techniques are reported in Yates (1990:164–165).

China was also ahead of Europe in agricultural technology. China replaced the scratch plow with the iron plow centuries before Europe. The adoption of wetfield rice cultivation has already been mentioned. Seed drills and other tools were introduced in China as early as the 11th century. There was early widespread use of diverse fertilizers (urban refuse, lime, ash) and methods of pest control (e.g., copper sulfates as insecticides). Finally, the Chinese published a large number of texts and handbooks on agricultural technology, thus furthering the diffusion of advanced agricultural knowledge and techniques (Mokyr, 1990:209).

This historical retrospective on agriculture in Europe and China explains their different starting points at the onset of the Industrial Revolution. These differences determined to a large extent the differences in their development paths over the next 300 years and their resulting environmental impacts. The distinction between Asian, European, and New Continent development paths is an important and powerful one (Hayami and Ruttan, 1985). The three

---

[12]Inventories from the 16th century in England indicate an average farm size of about 30 sown acres and 27 draft animals per farm (Langdon, 1986:208).

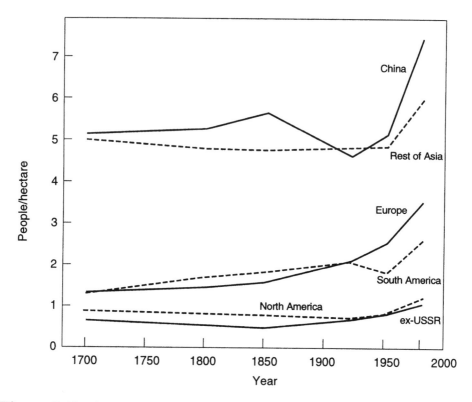

Figure 5.13: Agricultural land productivity (population per hectare crop-land) for different world regions. Source: derived from data in Durand (1967:137), Demeny (1990:49), and Richards (1990:164).

have different patterns of land and labor productivity increases, agricultural output and diets (e.g., dominance of rice versus grain and meat production), agricultural practices, and technology.

5.4.2. Land and labor productivity

Figure 5.13 presents first-order approximations of agricultural land productivity in terms of a region's population divided by its cropland. Differences in land productivity reflect different agricultural systems and different stages of agricultural development, and the figure still masks persistent subregional differences. For example, more than 200 million people still apply the simplest mode of agricultural production (shifting cultivation) resulting in 15–20 ha being required to feed one person. At the other extreme, there are areas where three crops per year are grown, and less than one-twentieth of a hectare produces enough food for one person (Buringh and Dudal, 1987:12).

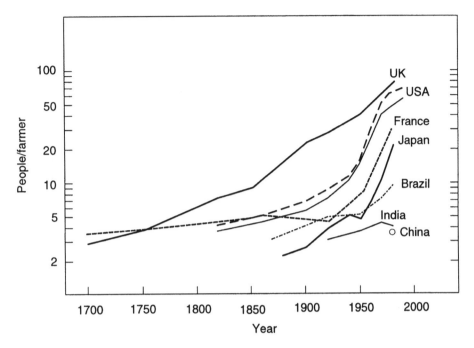

Figure 5.14: Agricultural labor productivity in terms of total population supported per agricultural worker. The range shown for the USA corresponds to domestic and total supported population (including exports). Source: Grübler (1994a:295).

Land productivity in Japan has been significantly higher than the Asian average.[13] Land productivity in France has been consistently below the European average.[14] *Figure 5.13* also does not reflect recent significant interregional trade in agricultural products which would increase land productivity estimates for regions that are net food exporters. This effect is illustrated by the range shown for the USA in *Figure 5.14*.

Despite these caveats, *Figure 5.13* still illustrates clear differences among regions in agricultural land productivity. Much of the variation is explained by differences in initial conditions, in the mix of agricultural products, and in diets, such as the difference between rice-oriented systems and those oriented toward grain and meat. It is particularly noteworthy that with the

---

[13]Land productivity in Japan already exceeded 8 people per hectare arable land in the 18th and 19th centuries and currently exceeds 20 people per hectare (Grigg, 1980:265).

[14]Values for France did not exceed 1.5 people per hectare cultivated land (excluding pastures) throughout the 18th century and well into the 1920s (Grigg, 1980:198–203) compared to values between 3 and 4 for England and Wales over the same time period (Grigg, 1980:165–177).

exception of modest productivity increases in Europe and perhaps South America (where data are much less certain), agricultural land productivity did not increase in the 18th and 19th centuries. Over this period, therefore, there was a direct one-to-one correlation between population increases and conversion of land to agriculture.

Increases in land productivity became noticeable in Europe by the second half of the 19th century. These were due to the host of technology changes characterizing the period of "agricultural innovations". European productivity increases accelerated in the period of "mercantilistic agriculture", surpassing two people per hectare cropland – one-third higher than at the beginning of the Industrial Revolution. By the second half of the 20th century, increases in land productivity became noticeable for the first time in all regions outside Europe. These improvements were due primarily to synthetic fertilizers and the diffusion of high-yield crops that are characteristic of the period of "agricultural industrialization".

In contrast, agricultural labor productivity measured in terms of the total population per agricultural laborer increased continuously from the onset of the Industrial Revolution (*Figure 5.14*). Note the semi-logarithmic scale of *Figure 5.14* to emphasize the improvement of nearly two orders of magnitude. These developments first took place in England. Other industrialized countries followed in the 19th century, although with some delay in the case of France. Values would be even higher than those shown in *Figure 5.14* if agricultural exports were included. The alternative data series in *Figure 5.14* illustrates the trade effect for the USA.

During the 20th century labor productivity has increased fast, and the overall trend is one of convergence toward an employment structure where only a few percent of the labor force is employed in agriculture.[15] Although in many developing countries like China and India about 70% of the work force is still employed in agriculture, similar structural shifts are very likely to occur in the future. The experience from the developed countries and their accelerating rates of change over time can serve as a guide for scenarios about future transitions in developing countries.

The long-term patterns described above are corroborated by shorter-term analyses of productivity increases by Hayami and Ruttan (1985). These were shown in *Figure 3.2* in Part I. Hayami and Ruttan identify three clusters of productivity increase trajectories – Asian, European, and New Continent. These are related to the relative endowments of land and labor.

---

[15]There is of course a definitional issue. Many activities previously performed in the agricultural sector now employ people in the industrial and service sectors. Hence, the percentage of the work force in all food-related activities (farming, food processing, distribution, etc.) is significantly above the few percent that remain on farms.

The next step is to understand the land-use changes that result from each of the three agricultural technology clusters.

5.5. Impacts II: Land-Use Changes

5.5.1. A quantitative account

Table 5.3 summarizes global land-use and population changes since 1700. These figures were calculated to give a general indication of the direction of change and are not presented as highly accurate assessments of land use in particular years. Of the green areas of our planet some 5 billion ha (38%) were covered by forests in 1980, close to 7 billion ha (51%) by grassland, and 1.5 billion ha (11%) by cropland. Compared to 1700, this represents a decrease in global forests by about 1.2 billion ha and an equal increase in cropland. This land-use change has altered ecosystems, destroyed wildlife habitats, changed regional climates, and released some 150 billion tons of (elemental) carbon into the atmosphere. It has left environmental "footprints" over a time-scale of centuries (see Section 5.6).

Most population growth is in the developing countries and consequently they dominate land-use transformations in both absolute and relative terms. The current developing countries account for three-quarters of global population growth since 1700, and for almost the same percentage of deforestation. They account for a somewhat smaller percentage (60%) of increases in cropland. Asia accounts for over half of the world's population growth between 1700 and 1980. The shares of other regions range from 7% to 10%. Deforested areas are largest in Africa and Latin America (300 million ha), followed by Asia (250 million ha), and the former USSR and Oceania (218 million ha). The expansion of cropland is more evenly distributed among regions. Changes have been greatest in Asia (+313 million ha between 1700 and 1980), followed again by Africa (+265 million ha) and the former USSR and Oceania (+253 million ha). Cropland increased by 200 million ha in North America and by 135 million ha in Latin America. Changes in Europe were comparatively smaller (+75 million ha).

Table 5.3 clearly illustrates substantial differences among regions in the impacts of population growth on agricultural land use. Whereas Asia accounts for 57% of global population growth between 1700 and 1980, it accounts for only 25% of increases in cropland. At the other extreme, the former USSR and Oceania account for only 7% of global population growth, but 20% of cropland increases.

To illustrate the different land intensiveness of the Asian, European, and New Continent development paths we use the data of *Table 5.3* to calculate

Table 5.3: Global and regional land-use and population change (in million ha and millions, respectively).

	1700–1800	1800–1850	1850–1920	1920–1950	1950–1980	Total 1700–1980	% of global change	1980 land use and population	% of world
Europe									
Forests	−15	−10	−5	−1	+13	−18	2	212	4
Grassland	−15	−25	−11	−3	+2	−52	—	138	2
Cropland	−30	+35	+15	+5	−15	+75	6	137	9
Population	+53	+63	+105	+79	+92	+392	10	484	11
North America									
Forests	−6	−39	−27	−5	+3	−74	6	942	19
Grassland	—	−1	−103	−22	+1	−125	—	790	12
Cropland	+6	+41	+129	+27	−3	+200	16	203	14
Population	+3	+20	+89	+52	+82	+246	7	248	6
Former USSR & Oceania									
Forests	−29	−42	−86	−38	−23	−218	19	1,187	23
Grassland	+2	+7	−12	−9	−22	−34	—	1,673	25
Cropland	+27	+35	+97	+47	+47	+253	20	291	19
Population	+19	+30	+62	+50	+95	+256	7	288	7
Africa and Middle East									
Forests	−11	−15	−68	−96	−118	−308	27	1,088	22
Grassland	—	+5	+23	+24	−9	+43	—	2,218	33
Cropland	+11	+9	+47	+71	+127	+265	21	329	22
Population	0/+1	+4	+39	+70	+250	+364	10	470	11
Latin America									
Forests	−6	−19	−51	−96	−122	−295	25	1,151	23
Grassland	+2	+11	+25	+54	+67	+159	—	767	11
Cropland	+4	+7	+27	+42	+55	+135	11	142	9
Population	+9	+15	+67	+63	+200	+354	9	364	8
Asia									
Forests	−38	−20	−50	−53	−89	−250	22	473	9
Grassland	−1	−8	−11	−12	−31	−63	—	1,202	18
Cropland	+38	+29	+61	+65	+120	+313	25	399	27
Population	+195	+171	+216	+372	+1,190	+2,144	57	2,579	58
World									
Forests	−105	−145	−287	−289	−336	−1,162	100	5,053	38
Grassland	−12	−11	−89	+32	+8	−72	—	6,788	51
Cropland	+116	+156	+376	+257	+331	+1,236	100	1,501	11
Population	+278	+603	+578	+686	+1,909	+3,755	100	4,433	100

Positive or negative sign indicates direction of change. Net land conversion may not add due to rounding errors.

Source: Land-use figures are derived from Richards (1990:164). Population data are from McEvedy and Jones (1978) and Demeny (1990:42).

Table 5.4: Land-use change per capita population growth ($\Delta L/\Delta POP$), hectare per head additional population.[a]

	1700–1800	1800–1850	1850–1920	1920–1950	1950–1980	1700–1980
Europe						
Forests	−0.28	−0.16	−0.05	−0.01	+0.14	−0.05
Grassland and cropland	+0.28	+0.16	+0.04	+0.03	−0.14	+0.05
Cropland	+0.57	+0.56	+0.14	+0.06	−0.16	+0.18
North America						
Forests	−2.00	−1.95	−0.30	−0.10	+0.04	−0.30
Grassland and cropland	+2.00	+2.00	+0.29	+0.10	−0.02	+0.30
Cropland	+2.00	+2.05	+1.45	+0.52	−0.04	+0.81
Former USSR & Oceania						
Forests	−1.53	−1.40	−1.39	−0.76	−0.24	−0.85
Grassland and cropland	+1.53	+1.40	+1.37	+0.76	+0.26	+0.85
Cropland	+1.42	+1.17	+1.56	+0.94	+0.49	+0.99
Africa[b] and Middle East						
Forests	(−11.00)	−3.75	−1.70	−1.37	−0.47	−0.85
Grassland and cropland	(+11.00)	+3.50	+1.80	+1.36	+0.47	+0.85
Cropland	(+11.00)	+2.25	+1.20	+1.01	+0.51	+0.73
Latin America						
Forests	−0.67	−1.27	−0.76	−1.52	−0.61	−0.83
Grassland and cropland	+0.67	+1.27	+0.78	+1.52	+0.61	+0.83
Cropland	+0.44	+0.53	+0.40	+0.66	+0.27	+0.38
Asia						
Forests	−0.19	−0.12	−0.23	−0.14	−0.07	−0.12
Grassland and cropland	+0.19	+0.12	+0.23	+0.14	+0.07	+0.12
Cropland	+0.19	+0.17	+0.28	+0.17	+0.10	+0.15
World						
Forests	−0.38	−0.24	−0.50	−0.42	−0.18	−0.31
Grassland and cropland	+0.37	+0.24	+0.50	+0.42	+0.18	+0.31
Cropland	+0.42	+0.26	+0.65	+0.37	+0.17	+0.33

[a] Source: Table 5.3.
[b] Figures for Africa prior to 1800 are highly uncertain.

marginal land-use changes, i.e., the increase in land use divided by the increase in population (*Table 5.4*). In addition to illustrating differences among development paths, the marginal land-use changes calculated in *Table 5.4* will also serve as reference points for later quantifying the land-use impacts of new technology and large-scale export-oriented agricultural production driven by new trading opportunities.

Table 5.4 shows an average of about 3,000 m^2 (0.3 ha) of forests converted to agricultural land (almost exclusively cropland) for each additional person since 1700. However, there are large variations over time and among regions. A representative value for the Asian development path is around 2,000 m^2 (0.2 ha) per capita. It is on the order of 5,000 m^2 (0.5 ha) per capita for a European-type path, and between 10,000 and 20,000 m^2 (1–2 ha) per capita for a New Continent path. These values serve as reference points for estimating likely changes in arable land in different regions in the absence of technological change and external trade. In regions with marginal land-use changes above these reference values agricultural land has expanded faster than the population. For instance, between 1850 and 1920 cropland grew faster than the population in the former USSR and Oceania and in Asia. This indicates large-scale land conversion for export crop production. The same is true for Latin America between 1920 and 1950.

Conversely, where marginal land-use changes are below the reference values defined above, and especially where marginal land-use changes are declining, land productivity is increasing due to technological change. Examples are Europe since 1850, North America since 1920, and all regions since 1950. As technology progressively decouples land-use changes from population growth, recent years have even seen instances of reconversion of croplands to grasslands and especially forests.

Figure 5.15 illustrates graphically the different land-use change intensities per head additional population since 1700. "Asian" versus "New Continent" (former USSR, Oceania, North America, and Africa) agricultural systems are clearly discernible from the marginal land-use change profiles. The intermediate "European" (and Latin American) trajectory is also clearly visible. Most notable, however, is the consistent declining trend across all regions. Over time, less and less land-use conversions are required to meet the growing food needs of an expanding population. Extrapolating these historical trends leads to a very different picture of future land-use changes from those suggested by "conventional wisdom" global scenarios, e.g., those developed within the framework of the Intergovernmental Panel on Climate Change (IPCC). Invariably in these scenarios, future population growth, combined with modest expected increases in agricultural land productivity, lead to enormous expansion of cropland areas and resulting depletion

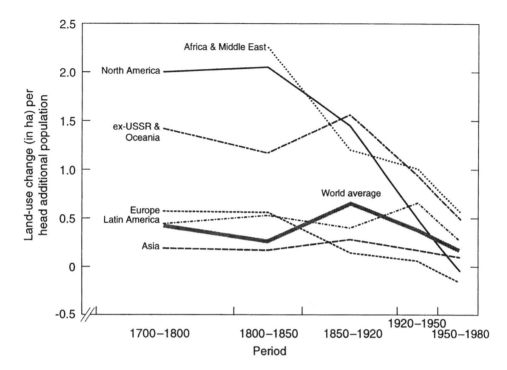

Figure 5.15: Land-use changes (in ha) per head additional population. Average for six world regions and the world for five successive time periods since 1700. Data source: *Table 5.4*.

of tropical forest cover (for a review see Alcamo *et al.*, 1995). Conversely, *Figure 5.15* suggests that if historical agricultural land productivity trends should continue, progress in agricultural technologies and techniques might, within the next three decades, enable a complete decoupling of land-use changes from population growth.

Based on *Table 5.4* and *Figure 5.15* the remainder of this section discusses the impacts of the three agricultural technology clusters on land productivity and thus on land-use patterns. In the mid-19th century global cropland expanded by about 270 million ha, almost entirely at the expense of forests, which lost some 250 million ha. With the exception of Europe, and to a lesser degree Asia, both grassland and cropland replaced forest cover. The intensification of agricultural production in Europe translated primarily into converting agricultural lands used "extensively" (i.e., grasslands) to "intensive" uses (i.e., croplands). This helped reduce the amount of deforestation. The estimates given in *Table 5.3* indicate that 25 million ha were converted from grasslands to croplands in Europe from 1800 to 1850, and an additional 11 million ha were converted between 1850 and 1920.

The development of the textile industry in Europe together with rising incomes and a growing population led to a large demand for cotton and wool that was satisfied by imports from abroad. However, expansion of cotton production in the USA and later on more widely in the subtropics appears to have had a large-scale impact on land use only after the 1850s. This conclusion is based on the period's relatively modest trade figures for cotton and wool, compared to the end of the 19th century. This means that the high land conversion figures in *Table 5.4* for regions outside Europe prior to 1850 cannot be explained by massive land conversions for export crop production. Nor can they be explained fully by population growth. The resulting residuals cast doubt on the land conversion estimates and population growth estimates in *Tables 5.3* and *5.4*, particularly for Africa and, to a lesser extent, North America.

The impact of technological change on land-use patterns first becomes noticeable in the period of "mercantilistic agriculture". The new transport technologies of the steam era caused transport costs to drop rapidly and opened the possibility of large-scale international trade in bulk agricultural products, raw materials for the textile industry, and luxury items such as sugar and spices. Although statistical records are scarce, we have tried to assemble in *Table 5.5* some zero-order estimates of land areas used for export crop production from 1850 to the mid-1920s.

Because increases in land productivity due to technological change are only discernible in Europe, we can infer that all additional land conversions outside Europe, beyond their respective reference values for marginal land-use changes, were for export crops. This leads to an estimate of 20–50% of all cropland expansion outside Europe being devoted to export crops. As agricultural trade prior to the mid-19th century was generally modest,[16] we can infer that nearly all these areas represent net land-use changes over the period 1850 to the 1920s. These rough estimates suggest that 20% of land-use changes in North America and the former USSR and Oceania between 1850 and the 1920s were for export crop production. The percentage for Asia, excluding China, is estimated at 30%, and for Latin America 50%.

In absolute terms North America dominates with some estimated 25 million ha converted to cropland for export (cotton and grains), followed by Asia (mostly India) with some 20 million ha, and the former USSR, Oceania, and Latin America with around 15 million ha each. The available trade statistics (Woytinsky, 1926:109–220; Mitchell, 1982:472–477) indicate that agricultural exports from Africa were comparatively modest. This reinforces

---

[16]Exceptions are cotton exports from the USA and Egypt as well as trade in sugar. Areas producing export crops by the 1850s are subtracted from the land-use change figures of the 1850–1920s period given in *Table 5.5*.

Table 5.5: Expansion of cropland for export crop production (zero-order estimates) in the period of mercantilistic agriculture (1850 to ca. 1920), in million ha.

Region	Products Luxury[a]	Grain[b]	Industrial raw materials[c]	Total	As % of increase in cropland area 1850–1920 (*Table 5.3*)
North America	0.2	17.0	7.7	24.9	19
Former USSR & Oceania	–	16.3	–	16.3	17
Africa	0.2	–	>0.8	>>1.0	?
Latin America	3.5	10.7	>0.1	>14.3	53
Asia (excl. China)	2.5	5.0	>11.8	>19.3	32
Total (5 regions)	6.4	49.0	20.4	75.8	21

[a]Sugar(cane), tea, coffee, tobacco.
[b]Barley, corn, oats, rice, rye, wheat.
[c]Cotton, flax, hemp, jute, rubber.

our doubts about the estimates in *Table 5.3* of African land conversions from 1850 to the 1920s.

Taking these estimates for export-oriented land-use changes into account, the marginal land-use changes of *Table 5.4* drop to about 12,000 m^2 (1.2 ha) per capita for North America and the former USSR and Oceania, and to about 2,000 m^2 (0.2 ha) per capita for Asia. These results are in agreement with the marginal land-use change values adopted above under *ceteris paribus* conditions, i.e., in the absence of technological change and export crop production.

Thus, in the period of "mercantilistic agriculture" some 380 million ha of additional cropland were brought into production worldwide. An estimated 290 million ha were converted from forests and 90 million ha from grasslands. The single largest change was the "conquering of the West" in North America, where an estimated 100 million ha were converted from prairies to croplands in which new drought-resistant wheat crops were planted. The ecological consequences of such a large-scale conversion of what we now recognize as a vulnerable ecosystem with loose and fragile soil were only fully felt in the 1930s (see the discussion in Section 5.5.2 on land quality).

Globally, some 80 million ha of cropland expansion can be attributed directly to export crops and thus serve as an indicator of the impacts of the steam age's transport revolution. Technology's contribution to increased land productivity seems to have been most pronounced in Europe. There land conversion to cropland was reduced to half the value (+15 million ha) of the previous 50 years. Given the near doubling of the population, perhaps

some 40 million ha were "spared" in Europe through increases in land productivity and food imports between the 1850 and 1920 period. The two
factors contributed roughly equally.

Since 1920, in the period of "agricultural industrialization", global cropland increased by close to 600 million ha, about the same increase as during
the entire period from 1700 to 1920. However, population growth since 1920
added nearly twice as many people to the global population as were added
between 1700 and 1920. This is indicative of the increasing decoupling of
land-use changes from population growth. Beyond that broad conclusion,
it is extremely difficult to disentangle the individual effects of agricultural
technology, its international diffusion, trade intensification, altered diets,
and changing outputs in different agricultural systems. Nevertheless, some
general observations can be made based on the data in *Tables 5.3* and *5.4*.

From 1920 to 1950 land productivity increases in Europe reduced
marginal land conversions (the amount of agricultural land needed for each
person added to the population) to half the value between 1850 and 1920.
In North America marginal land-use changes were reduced to one-third their
earlier level. This considerably reduced the expansion of cropland from
140 million ha in the previous period to approximately 30 million. In addition the substitution of farm animals by tractors freed large areas devoted to
feed production. In the USA alone some 40 million ha thus became additionally available for crops. Land productivity increases outside the industrialized countries, however, remained small. The single largest regional land-use
changes in this period were the deforestation of approximately 100 million
ha in Africa and another 100 million ha in Latin America. In Latin America
more than half of this area was converted to grassland and meat production
for export.

Changes since 1950 have been even more dramatic and widespread. Land
productivity has increased in all regions. Productivity increases in Europe
and North America allowed, for the first time since the Industrial Revolution,
agricultural lands to be converted to forests (approximately 18 million ha)
and, at the same time, to maintain tremendous increases in agricultural
output (and surpluses). In non-OECD countries land productivity increased
from 1950 to 1980 at an annual rate of about 1%. Although population
growth was roughly twice as fast (2.1% per year), the land productivity
increases that were achieved remain a formidable accomplishment of the
Green Revolution. Without these increases, cropland outside Europe and
North America would have had to expand by close to 400 million ha above
the 350 million ha increase that actually did occur between 1950 and 1980.

5.5.2. Land quality

Up to now we have focused on quantitative land-use changes since the Industrial Revolution. Also important are qualitative changes, particularly land degradation. Land degradation has been a perennial problem since antiquity. Silting of soils due to bad irrigation practices destroyed the agricultural base of the large empires of Mesopotamia, and is a continuing threat to the sustainability of productivity increases through irrigation. That irrigation and soil degradation can have negative environmental effects far beyond the negative effects of agriculture is perhaps best illustrated by the history of the US "Dust Bowl" and that of the Aral Sea in the former USSR discussed briefly below.

Soils[17] are susceptible to human influences in a variety of ways. First, agricultural crops remove nutrients from soils, and these must be replaced. Second, management practices influence soil quality. For example, conventional irrigation can lead to siltage or salinization, i.e., the deposition of minerals that remain after irrigation water evaporates. Conversely, droplet irrigation techniques developed in Israel reduce salinization risks and at the same time conserve water. Finally, erosion removes soils. The extent of erosion depends significantly on farming practices. Avoiding extended periods of soil exposure without vegetation cover significantly reduces erosion.

One of the most dramatic incidences of large-scale wind erosion occurred in the USA in the 1930s. Ploughing the vast grasslands of the American Great Plains and planting drought-resistant wheat greatly increased agricultural production. However, without appropriate countermeasures soils were exposed to erosion when not covered by vegetation. This happened on a grand scale following periodic droughts in the 1930s; the exposed soils were carried away in large quantities creating huge dust storms, hence the term "dust bowl". The first major storm in May 1934 is estimated to have removed some 350 million tons of topsoil (Ponting 1991:260), depositing it over the eastern United States of America, even over the Atlantic. Frequent storms followed in 1935 and again in 1938. By that time some three and a half million farmers had abandoned farms in the area. Oklahoma lost about one-fifth of its population, and in some counties nearly half of the population left, many of their cars inscribed "Oregon or bust".[18]

---

[17] As suggested by Buol (1994:227), although identifiable processes are common to all soils, their magnitude varies substantially in different settings. Hence, it is more appropriate to use the plural form and speak of soils.

[18] An impressive photographic account of the dust bowl generation and their descendants is given in Ganzel (1984).

There have been significant global losses of humus, the organic component of soils, across a variety of different ecosystems and agricultural areas. Humus losses over the last 300 years have averaged approximately 300 million tons per year. The rate has nearly tripled over the last 50 years to some 760 million tons per year (Rozanov *et al.*, 1990:213). The resulting loss of organic carbon is estimated at 90 GtC (billion tons carbon) over the last 300 years and 40 GtC over the last 50 years.

Soils are, in most respects, a renewable resource; the real problem lies in a temporal mismatch. Soil deterioration can occur within a few years while soil restoration can take decades. In the tropics, for example, planting crops on deforested land can reduce soil humus, nutrients and organic carbon by 20–50% within a few years. The soils are then exhausted, and cultivation must shift to new areas. Recovery through secondary vegetation and eventual forest regrowth takes many decades. As summarized by Buol (1994:228), human cultivation results in "chemical deterioration primarily in tropical areas where soils with low native fertility are abundant and human institutions have not developed to replace the nutrients removed in the food. This reduced chemical ability to support adequate vegetative cover invariably leads to accelerated physical soil damage via erosion".

5.6. Impacts III: Other Global Changes

5.6.1. Water

Water is the lifeblood of the biosphere. The annual flow of the global hydrological cycle is approximately 580,000 km^3 (all data from Shiklomanov, 1993). Evaporation over the oceans (510,000 km^3) exceeds precipitation (460,000 km^3). Precipitation over land (120,000 km^3) exceeds evaporation (70,000 km^3). Runoff into the oceans is approximately 50,000 km^3.

The water stored in the global hydrosphere is three orders of magnitude greater than the annual precipitation (and evaporation) flow. Total water vapor plus fresh and saline water are estimated at close to 1,500 million km^3. Freshwater makes up only a small fraction of this total, 85 million km^3. Of this, approximately 60 million km^3 is groundwater, 24 million km^3 is in ice sheets, 300,000 km^3 is in lakes, reservoirs and rivers, less than 100,000 km^3 is in soil moisture, and 14,000 km^3 is in the atmosphere.

Theoretically, the upper limit of renewable water resources is set by the total annual flow of the hydrological cycle (580,000 km^3). In practice, the upper limit of renewable water flows available for humanity is much smaller. From the total amount of freshwater runoff over the continents only about one-third (15,000 km^3) is available as stable runoff. The rest flows rapidly into the oceans as flood runoff. Of the 15,000 km^3 of stable runoff only about 10,000 km^3 is available in inhabited areas (Rogers, 1994:237).

Current worldwide water withdrawals are about 3,000 km^3, i.e., humankind uses one-third of the stable water runoff available in inhabited areas. This, together with the possibility of elaborate reservoir schemes to capture runoff in sparsely populated areas and flood runoff, suggests there is no immediate looming scarcity. However, both quantitative and qualitative trends in water demand urge for caution. Quantitatively, global water use since 1900 has increased by approximately a factor of five (from 600 to 3,000 km^3). The increase is due both to population growth (by a factor of 3.3) and to increased per capita water use (by a factor of 1.5) (Raskin *et al.*, 1995:2). In addition there has been a deterioration in water quality. It is estimated that up to 6,000 km^3 of water are required to dilute 450 km^3 of polluted wastewater (UNEP, 1991). This is effectively a demand multiplier of 13. Thus both increases in the efficiency of water use and improvements in water quality will be required to slow current trends in increased global water demand.

Global aggregates also mask regional differences, as water availability is essentially defined at the local and regional level (except for very large, capital-intensive water transfer projects). Falkenmark and Widstrand (1992) have developed a typology of water availability in which they define two levels of scarcity. Countries with water availability of less than 500 m^3 per capita annually are classified as facing "absolute scarcity". Countries with water availability between 500 and 1,000 m^3 per capita annually are classified as facing "scarcity." Raskin *et al.* (1995:11) list 13 countries in the category of absolute scarcity and 7 countries in the category of scarcity. Most are either in the Middle East or are small city and island states. Annual freshwater availability in Quatar is estimated at less than 50 m^3 per capita. In Bahrain and Kuwait it is less than 10 m^3 per capita. Fortunately these countries have large fossil fuel (natural gas) reserves that can fuel large-scale desalinization facilities. But desalinization is not cheap. As an indicator of relative local resource scarcity, gasoline in these countries[19] costs less than water!

Throughout the world, the largest consumer of water is agriculture. Agriculture uses approximately 2,000 km^3 of water annually. Households, services, and industry use about 1,000 km^3. Per capita annual agricultural water use equals approximately 400 m^3 (400,000 liters) per person. There are large regional variations, however, ranging from about 200 m^3 per person in

---

[19]Similar statements can also be made for countries with much larger water availability. For instance, bottled spring water in the USA is also more expensive than gasoline.

Africa and Eastern Europe to 800 m^3 per person in the former USSR (Raskin *et al.*, 1995:7).

Practically all agricultural water is used for irrigation. The rest, a very small amount, is consumed by livestock. Irrigation consumes on average 10,000 m^3 of water per ha (Raskin *et al.*, 1995:45–46). Worldwide irrigated areas have increased from 8 million ha in 1800, to some 50 million in 1900, 100 million in 1950 (Rozanov *et al.*, 1990:210), and over 320 million ha in 1990 (FAO, 1994). Thus, irrigation water use has increased by about a factor of 30 since the onset of the Industrial Revolution with significant environmental impacts (see *Box 5.2*). Since the early 1960s the greatest increases in areas being irrigated have occurred in China and India, where over 15 million ha have been put under irrigation (Heilig, 1995).

Irrigation is a key technology for increasing agricultural productivity and yields. Only about 16% of global cultivated land is irrigated, but that 16% produces approximately one-third of all crops. Only 10% of US cropland is irrigated, but it contributes about one-third to the total value of cropland production[20] (Raskin *et al.*, 1995:43). US yields on irrigated farms are about four times those on rain-fed farms. Given the importance of irrigation, plus the potential constraints on water availability as discussed, increasing the economic efficiency of water use will become increasingly important for agriculture. Currently in many countries – the USA and developing countries alike – agricultural water use is highly subsidized. In both California and China farmers pay only 10% of water supply costs (Leach, 1995:82) and enjoy hereditary water rights. Urban and industrial users face much higher prices. The result is often lavish and wasteful water use by agriculture.

A central component of irrigation systems – reservoirs – should be highlighted as an immense modern version of a basic technological artifact discussed at the very beginning of this book – containers. Today there are roughly 30,000 water reservoirs in the world covering 800,000 km^2 (WRI, 1990) with a total filled capacity of 6,000 km^3. That equals two years' worth of total global water use. About 1,800 of them are large reservoirs, holding volumes of over 100 million m^3 (0.1 km^3). Their growth since the turn of the century is shown in *Figure 5.16*. Prior to 1900 water reservoirs globally held 14 km^3 water. By 1950 this figure had grown to 528 km^3. By 1985 these large reservoirs held about 5,000 km^3 (or 5,000 billion tons) of water, i.e., an increase by more than a factor of 350 in less than a century. This is

---

[20]This is partly due to the higher value of agricultural products typically produced on irrigated farmland such as vegetables.

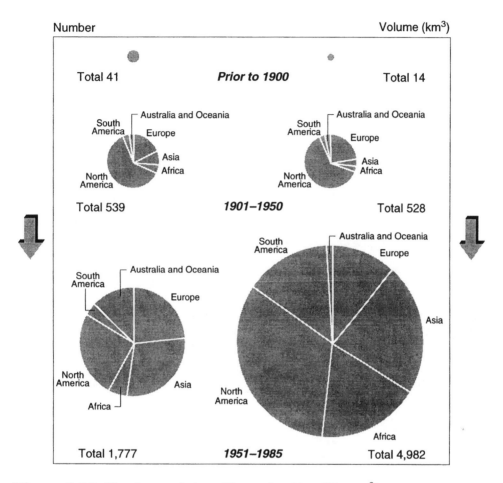

Figure 5.16: Number and size of large (>100 million m^3) water reservoirs by region, 1900–1985. Source: L'vovich and White (1990:239).[3]

humanity's largest material-handling enterprise. In comparison the tonnage of the seven heaviest industrial commodities is less than 2.5 billion tons per year (cf. Chapter 6). The size of water reservoirs is a powerful illustration of how "big" technology has grown and to what extent humanity has indeed become a "hydraulic civilization".

Water withdrawal for irrigation purposes can have a number of ecological impacts far beyond agricultural impacts. Perhaps the most dramatic illustration in this century is provided in the case of the disappearing Aral Sea (see *Box 5.2*).

Box 5.2: You Cannot Fill the Aral Sea With Tears

The Aral Sea, located between the former Soviet Republics of Kazakhstan and Uzbekistan, is disappearing. In 1960, the Aral Sea was the world's fourth largest inland body of water. Since then, its area has decreased by more than 50%, its volume fallen by 75%, and its salinity risen more than three-fold, approaching that of seawater (Miklin, 1996:4).

The desiccation of the Aral Sea is the most rapid and dramatic alteration of a major water body induced by human action. Plowing the Kazakhstan steppe and the establishment of vast areas of irrigated cotton fields requires enormous amounts of water that have been diverted from the Aral's main tributaries, the Amu Dar'Ya and the Syr Dar'Ya rivers. (The Aral Sea has no river outflow, its water balance was maintained by high evaporation rates.) More than 7 million hectares of irrigated agricultural land in the Aral basin divert almost all water of the rivers flowing into the Aral Sea. With continued evaporation this led to the progressive disappearance of the Aral Sea, illustrated quantitatively and graphically below.

Aral Sea statistics, 1960–2000.

	Water level (meters)	Surface area (km^2)	Volume (km^3)	Salinity (g/liter)
1960	53	66,900	1,090	10
1971	51	60,200	925	11
1976	48	55,700	763	14
1994	37	31,900	298	25–35[a]
2000 (estimate)	33	25,200	212	20–60[a]

[a] Range corresponds to small and large sublakes, respectively.

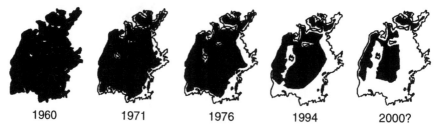

| 1960 | 1971 | 1976 | 1994 | 2000? |

Adapted from Miklin (1996:3–4).

The ecological consequences are severe: changes in regional climate patterns (rising temperatures and decreasing precipitation); salination (wiping out the sea's fish population and with it a thriving local fishing industry); deposition of salt in the 35,000 km^2 dried out seabed (that is being carried away by wind and contributes to further soil degradation of agricultural areas); serious health problems of the local population due to blowing dusts and salt and contaminated water, which are further compounded by poor health and medical infrastructures. Infant mortality in the region is the highest in the entire area of the former USSR, reaching 60 deaths per 1,000 live births (Miklin, 1996:7). The region is officially designated as an ecological disaster zone. Reviving the Aral Sea may be impossible; it certainly cannot be done without a complete change in water use and management in the entire region which might prove very difficult to implement. But as the Uzbek poet Mukhammed Salikh said: "You cannot fill the Aral with tears" (Gleick, 1993:5).

5.6.2. "Grand cycles" of nitrogen and phosphorus

Agriculture is the major anthropogenic influence on both the nitrogen (N) cycle and the phosphorus (P) cycle. The nitrogen and phosphorus cycles are summarized in *Tables 5.6* and *5.7* in terms of their major global reservoirs and annual fluxes.

Biotic nitrogen fixation by bacteria supplies the bulk of metabolizable nitrogen that limits photosynthesis in general and food production in particular. Farming therefore depends on assuring adequate nitrogen inputs. Early techniques included recycling organic wastes in the form of manure, and planting nitrogen-fixing legumes such as clover, soybeans, and alfalfa. The first inorganic nitrogen fertilizers were introduced in the 19th century in the form of Chilean nitrates and guano, but the real breakthrough came through ammonia synthesis using the Haber-Bosch process as discussed in Section 5.3.4.

Overall, human activity has approximately doubled the rate of global nitrogen fixation since preindustrial times (Ayres *et al.*, 1994:135–153). The main mechanisms have been synthetic nitrogen fertilizers, leguminous crops, fossil fuel combustion, and biomass burning associated largely with land-use conversion such as deforestation. Estimates for the end of the 1980s place these related fluxes at 90 Tg (million tons) from fertilizers, some 40 Tg through planting of leguminous crops (Ayres *et al.*, 1994:146), which is included in the terrestrial biotic fixation rate in *Table 5.6*, roughly 40 Tg through fossil fuel combustion, and perhaps an equal quantity through biomass burning (phytomass in *Table 5.6*). The balance of the phytomass burning line in *Table 5.6* (10–200 Tg) is from natural processes such as savanna fires.

The resulting large increase in nitrogen mobility creates both economic losses (e.g., fertilizer leaching) and environmental concerns. Nitrates can pollute groundwater resources, and NO_x emissions from combustion are a major cause of urban photochemical smog. Ammonia (NH_3) emissions from fertilizer application and from dense livestock populations add to nitrogen oxides as an additional source of acidification.

As a result all industrialized countries now have nitrogen emission controls in both stationary (power plants) and mobile sources (cars) (cf. the example of catalytic converters in *Figure 2.12* in Part I). Nitrogen oxides, together with sulfur oxides, are the principal precursors of acidic precipitation. These are compounded by additional nitrogen volatilization in the form of ammonia (NH_3) from fertilizers and intensive animal farming, particularly cattle. *Figure 5.17* shows European nitrogen emissions in 1990.

Table 5.6: Major global reservoirs[a] and fluxes[b] of nitrogen (N).

Reservoir		Estimated totals of N (in Pg)
Igneous rocks		$14–57 \times 10^6$
Atmosphere	N_2	3,800
	N_2O	1.8
	NO_x	0.0006
Hydrosphere	N_2, NO_3^-, organic N	20,000
Soils	NO_3^-, NH_4^+, organic N	100–760
Phytomass	Terrestrial	7.7–10
	Marine	300–500
Zoomass		200–370
Anthropomass		0.006

Flux		Estimated flows of N (in Tg/year)
Atmospheric fixation		1–30
Biotic fixation	Terrestrial	44–200
	Marine	1–130
Anthropogenic fixation		90
Fossil fuel combustion	NO_x	10–35
	NH_3	5–10
Phytomass burning		10–200
Biogenic NO_x releases		20–230
Denitrification	Terrestrial	40–390
	Marine	40–330
Volatilization		30–250
Atmospheric deposition	NH_3/NH_4^+	80–240
	NO_x	30–120
	Organic N	10–100
River runoff		10–40

[a]Measured in Pg (i.e., 10^{18} grams).
[b]Measured in Tg (i.e., 10^{12} grams).
Source: Smil (1990:424).

In the mid-1990s European nitrogen emissions totaled some 13 million tons elemental nitrogen. About half of this (6 million tons) came from agriculture, e.g., in the form of ammonia emissions from dense livestock populations. Four million tons were emitted from mobile sources (i.e., transport vehicles such as cars, buses, and aircraft), and an additional 3 million tons from stationary sources (mostly power plants burning fossil fuels). It is interesting to note that agriculture remains such a dominant source of anthropogenic nitrogen emissions even in a region with high fossil energy consumption and dense automobile and aircraft traffic.

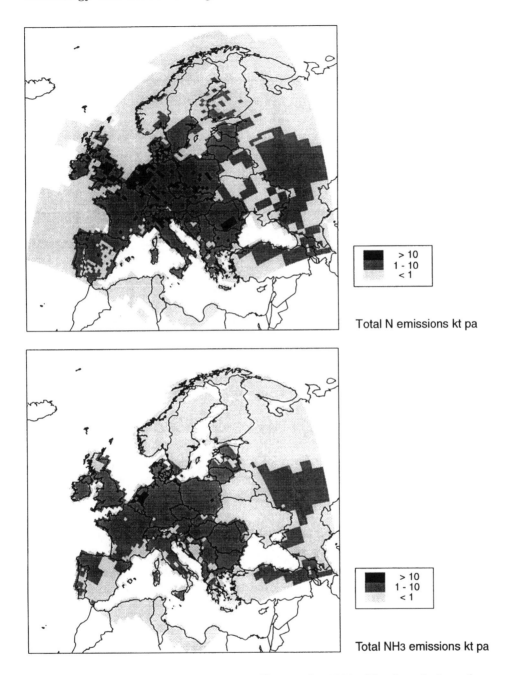

Figure 5.17: Nitrogen emissions in Europe in 1990. Total emissions from all sources (top) and from agricultural sources (NH$_3$, bottom; included in the total emission figures shown in the top panel). Three classes of emission levels are given, in 1,000 tons (kt) nitrogen per year. Source: Transboundary Air Pollution Project, IIASA. Graphic courtesy of Chris Heyes, IIASA.

Table 5.7: Major global reservoirs[a] and fluxes[b] of phosphorus (P).

Reservoir		Storage of P (in Pg)
Phosphorus rock	Resources	110–300
	Reserves	20–70
Soil		40–160
Ocean (dissolved P)		80–128
Phytomass		1–3
Zoomass		0.03
Anthropomass		0.003

Flux		Flows of P (in Tg/year)
River transport		14–18
Phosphate rock mining		20
Fertilizer applications		15
Uptake by biota	Terrestrial	200
	Marine	600–1,000
Detritus	Terrestrial	200
	Marine	2–10
Vertical ocean mixing		50–70

[a]Measured in Pg (i.e., 10^{18} grams).
[b]Measured in Tg (i.e., 10^{12} grams).
Source: Smil (1990:430).

Finally, nitrogen in the form of nitrous oxide (N_2O) contributes substantially to the global greenhouse effect. Since preindustrial times the atmospheric concentrations of N_2O have increased from 275 ppbv (parts per billion by volume) to over 310 ppbv. Current anthropogenic fluxes range from 3 to 8 TgN per year, mostly from agriculture and biomass burning. Nitrous oxide is particularly significant as a greenhouse gas because of its long atmospheric residence time – approximately 120 years – and its high radiative forcing – 180 to 320 times larger than CO_2 on a per molecule basis (IPCC, 1995).

Phosphorus (*Table 5.7*) has a less multifaceted set of environmental impacts. Phosphorus leaching can lead to "overfertilization" (eutrophication) in lakes and rivers, and this constitutes a major water quality problem in areas with intensive agriculture or untreated urban sewage discharges. Natural phosphorus inputs to agriculture are limited,[21] and most fertilizer comes

---

[21]Total global livestock wastes are around two billion tons per year. One-third is recycled on fields, corresponding to an input of 3 TgP out of an annual total of 15 TgP in phosphorus fertilizers (Smil, 1990:430–431).

Box 5.3: Ocean Island: The Ecological and Social Effects of Phosphate Mining Under Colonialism*

Ocean Island was a small Pacific island, coverered by tropical vegetation and inhabited by about 2,000 Banabans, following a typical Polynesian lifestyle. The island was unique (like its sister island Nauru) in that it consisted almost entirely of solid phosphate rock. Annexed by Britain in 1901, the mineral rights were sold for a payment of 50 pounds (!) a year to the British-owned Pacific Island Company, in a treaty of dubious legality. Mining and phosphate exports started on a large scale immediately.

By 1905 Ocean Island exported 100,000 tons of phosphate a year and Pacific Island Company made a profit of some 20 million pounds a year. The company was bought out by the British, Australian and New Zealand governments that established the jointly owned British Phosphate Commission. By the 1920s mining produced about 600,000 tons of phosphate a year and the native inhabitants saw their island rapidly ravaged by clearing of the topsoil, mining, and no subsequent land reclamation whatsoever. Despite protests, mining continued and part of the modest mining royalties (in the meantime increased to 6 pence per ton) were used by the Commission to purchase Rambi Island from Fiji. (However, 85% of the royalties were used to cover the costs of colonial administration of Gilbert and Ellice Islands to which Ocean Island had been incorporated. None of the royalties was disbursed to the Banabans.)

Occupied by the Japanese during World War II, the native population was deported to the Caroline Islands. After the war, the Banabans were not allowed to return, but resettled on Rambi island instead, some 2,000 km away from their original home and with a very different climate. The islanders were finally offered 500,000 pounds by the Phosphate Commission as compensation for all the mining damage the island had incurred. The Banabans refused, litigating the British Government in the 1970s in the longest civil case ever heard. They failed, mainly because the court ruled that the original (50 pounds a year) mining rights were a legally binding contract. However, the court did find that the Government had breached its obligation to care for the islanders' future, but refused to award any compensation. Ultimately the islanders and the Phosphate Commission settled on a compensation payment, that covered merely the costs of the legal case. By 1980, 20 million tons of phosphate had been extracted, the deposit exhausted, Ocean Island left uninhabitable, and the Banabans had received pitiful compensation for their loss.

The inhabitants of Nauru fared a little better. Allowed to return to their island after the war, Nauru gained independence in 1968, and the management of phosphate mining was transferred to them in 1970. The islanders now live with high incomes and Western lifestyles along a narrow coastal fringe, the only part of the island not devastated by mining, which took cumulatively some 60 million tons of phosphate. Despite a story of small islands, Ocean Island and Nauru illustrate the high price that had to be paid locally for increasing agricultural productivity and lowering food prices abroad.

*After Ponting (1991:218–221).

from mining over 150 million tons annually of phosphate rock.[22] This creates the usual environmental problems associated with large-scale surface mining. These are well illustrated by the destructive impact of phosphate mining on the Pacific islands of Nauru and Ocean Island (see *Box 5.3*). These two tiny islands have provided close to 100 million tons of phosphate since the turn of the century.

Clearly, Ocean Island illustrates a worst-case scenario. Mining impacts can be mitigated and in fact land reclamation technologies are highly developed. Perhaps the most impressive examples of land restoration and reclamation are provided in the giant German opencast lignite mines.

5.6.3. Carbon and methane

Agricultural and cropland expansion interfere substantially with global flows of both carbon dioxide and methane, the two most important greenhouse gases (next to water vapor). Agriculture dominates anthropogenic methane emissions. For carbon, the impact of agriculture and land use is secondary compared to other industrial activities like energy use, but still important.

Current biotic carbon emissions occur largely in the tropics where most biomass burning and land-use changes are concentrated. Annual biotic carbon emissions are estimated at 1.1 GtC (billion tons of elemental carbon) with an uncertainty range from 0.6 to 2.6 GtC. This compares to industrial emissions, mostly from fossil fuel combustion, of some 6 ± 0.5 GtC (IPCC, 1995). Throughout history, however, carbon emissions from land-use change have also been substantial. From 1800 to 1990 some 150 ± 40 GtC were released globally into the atmosphere as a result of land-use change (Grübler and Nakićenović, 1994; IPCC, 1995), while approximately 200 GtC were released from fossil fuel combustion during the same period (IPCC, 1990; Grübler and Nakićenović, 1994).

Methane (CH_4) is the second most important greenhouse gas after carbon dioxide (CO_2). Atmospheric concentrations of methane have risen from 700 ppbv in preindustrial times to over 1,700 ppbv today, a clear indication of anthropogenic interference with natural fluxes. The total annual methane flux from natural and anthropogenic sources is about 535 Tg per year. Less than 200 Tg comes from natural sources, mostly wetlands but also other sources such as termites. About 100 Tg comes from mining and fossil fuel combustion, and close to 300 Tg comes from biotic sources (IPCC, 1995). Of the biotic sources, the largest is agriculture. Cattle produce some 100 Tg from enteric fermentation and animal waste, and rice paddies contribute some 60 Tg. The remainder is generated by biomass burning (40 Tg),

---

[22]The best introductory text on phosphates remains Sheldon (1982).

landfills (40 Tg), and sewage (25 Tg) where methane is released during the anaerobic decomposition of organic matter. Although such estimates have large uncertainties, they leave no doubt that agricultural activities are the main source of anthropogenic methane emissions.

Technological measures (most of them with considerable secondary benefits) include, for instance, leak-plugging in natural gas pipeline and distribution systems, degassing of coal mines from the methane present in coal seams (dangerous, and explosion-prone), changing feeding practices of livestock, or modified wetfield rice cultivation methods. Because the atmospheric residence time of methane is only 12 years, emission reductions would translate relatively quickly into slower growth or stabilization of atmospheric concentrations.

5.7. Global Changes in Human Occupations and Residence

Technological change, through vastly increasing agricultural labor productivity, has permitted an increasing share of the growing rural population to transfer to urban employment. While this has numerous long-term advantages, it can also have disadvantages, evident today in many rapidly growing megacities of the developing world. Agricultural industrialization has also led to an increasing division of labor. Farming has evolved from a vertically integrated activity where farmers produced their own inputs such as seeds, fertilizer in the form of manure, and traction power supplied by livestock, and where they also took responsibility for storing, conserving, and marketing their products. The shift has been toward horizontal integration with increasing specialization. The shift from vertical to horizontal integration offsets to some extent the dramatic shifts away from agricultural employment patterns that are discussed below. Many activities previously performed within the agricultural sector are now performed in the industry and service sectors. Many jobs on the farm have moved either to industrial manufacturing plants that produce seeds, fertilizer, tractors, and other farm machinery, or to food processing industries and the service sector (e.g., food retail and restaurants).

This section therefore examines global changes in occupational structure and residence associated with the overall historical shift from a rural, agrarian society toward an urban, industrialized one.

5.7.1. Moving away from agriculture

In most advanced industrialized countries today, less than 3% of the work force works on farms. Prior to the Industrial Revolution, and still in many

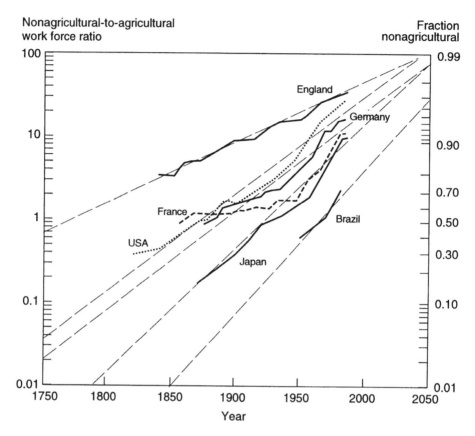

Figure 5.18: Ratio of nonagricultural to agricultural work force (logarithmic scale). Source: Grübler (1994a:316).

developing countries, that figure was 75% or more. *Figure 5.18* summarizes the shift away from agricultural employment. It shows the ratio of nonagricultural to agricultural workers for different countries over time. The figure indicates a long-term convergence in the employment structures of industrialized countries. The few long-term data available for developing countries indicate a similar secular trend. Nonetheless, the current share of agricultural employment in developing countries still spans a wide range, reflecting different levels of economic development and diverse agricultural policies. *Figure 5.18* suggests that the shift away from agricultural labor can be approximated by a simple logistic curve.

Thus, countries are converging toward only a minor percentage of their work forces being directly employed in agriculture. In the industrialized countries, this trend began more than 200 years ago. In developing countries that are now "catching up", this transition happens at an accelerated pace.

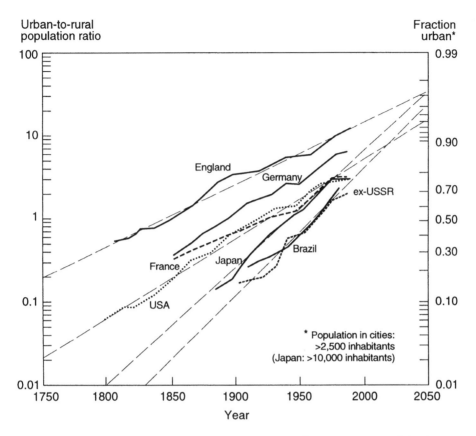

Figure 5.19: Ratio of urban to rural population (logarithmic scale). Source: Grübler (1994a:317). For the data of this graphic see the Appendix.

5.7.2. Moving into cities

Figure 5.19 provides a similar picture for the rural to urban population shift. Again the data show convergence among the countries sampled. The similar dynamics in the employment shift and the residence shift point to their close relationship. The two have happened sequentially in all industrialized countries – the shift out of agricultural employment has preceded urbanization. However, in countries now in the midst of these transitions such as the former USSR and Brazil, both processes are occurring simultaneously. This subsequently places enormous stress on infrastructures, financing, and policy capability.

The urbanization patterns in *Figure 5.19* are similar to those for other technological and economic structural change processes that we have discussed. There is a certain convergence between different countries, but there also remains great diversity. This is a function of how long ago the process

Table 5.8: Urban populations living in informal settlements, 1980.

	Total population (thousands)	Population in informal settlements	
		Thousands	Percent
Addis Ababa, Ethiopia	1,668	1,418	85
Luanda, Angola	959	671	70
Dar es Salaam, Tanzania	1,075	645	60
Bogota, Colombia	5,493	3,241	59
Ankara, Turkey	2,164	1,104	51
Lusaka, Zambia	791	396	50
Tunis, Tunisia	1,046	471	45
Manila, Philippines	5,664	2,666	40
Mexico City, Mexico	15,032	6,013	40
Karachi, Pakistan	5,005	1,852	37
Caracas, Venezuela	3,093	1,052	34
Nairobi, Kenya	1,275	421	33
Lima, Peru	4,682	1,545	33
São Paulo, Brazil	13,541	4,333	32

Source: WRI (1990:76) based on UN (1980:125–154) and HABITAT (1987:77).

started in different countries. For example, England (including Wales) has been consistently more highly urbanized than Germany and the USA where these transitions began later. In industrialized countries future urban growth will be comparatively modest given the overall low population growth rates in these countries and the fact that over 80% of their populations already live in urban areas.

Conversely, developing countries with their much higher overall population growth rates are also in the midst of the rural–urban transition process. Thus the exceptionally rapid growth of urban agglomerations in many developing countries results from a three-fold structural change process: the transition away from agricultural employment, high overall population growth, and increasing rural to urban migration. How to provide adequate housing, sanitation, health, and transportation services for cities in developing countries is perhaps the biggest challenge for technology in the 21st century. The scope of the challenge is shown in *Table 5.8* which estimates the populations of various shanty-towns or *favelas* (euphemistically referred to as informal settlements) around the world.

Technology's largest contribution to urbanization has been in the form of infrastructural developments. The key to urban growth has been the capacity to provide clean water, to dispose of sewage, to provide clean grid-dependent energy carriers (city gas, and later electricity and natural gas), and to transport people and goods. Other technological changes have been important – such as those in construction and housing, from air conditioning and elevators to domed stadiums and even indoor skiing in Tokyo. However, it is also lamentably clear from both *Table 5.8* and Friedrich Engels' writings on England's early industrial working class that a "technology push" in housing has never been a driver of urbanization.

Infrastructural advances, however, have proved vital even though their establishment lagged behind urban growth in industrialized countries and continues to do so in developing countries owing to capital shortages (for an excellent history see Tarr and Dupuy, 1988). Without clean water and sanitation, mortality in urban agglomerations would be impossibly high, and indeed this was the case in the frequent epidemics that plagued European and American cities in the early phases of industrialization. Without clean energy, air pollution would render large cities nearly uninhabitable, as has been shown by the recurrent coal smog episodes in London (see Brimblecombe, 1987) or in Pittsburgh until the middle of this century. Without modern transport infrastructures, large cities could not be supplied adequately with food, materials, and energy; nor could they handle their internal traffic volumes.

Berry (1990:103–119) has illustrated very convincingly the link between transport infrastructure development cycles and urbanization in the USA.[23] Grübler (1990a) has demonstrated similar linkages for other countries. For example, through 1860 the development of a canal network proved essential for supplying food and energy (coal) to a rapidly growing London. High European and North American urbanization growth rates from the mid-19th century until the 1930s correlated with the development of both intercity and intracity rail systems. Currently it is the internal combustion engine that underlies strong correlations between urbanization ratios and high transport intensities, both for individual modes of transport (e.g., cars) and public modes (e.g., buses and aircraft).

Due to the preponderance of short- to medium-distance trips (i.e., *urban* trips) in total travel demand and the high traffic flows between main urban centers, transport infrastructures are more a function of population density and urbanization rather than of absolute country size. This can be easily conceptualized based on the so-called gravity model of spatial interaction. In this, the propensity of communication (expressed, for instance, through the number of telephone calls or traffic volumes) between two cities is directly

---

[23]The causal relationship, characteristic for such coevolutionary processes, goes both ways. New transport infrastructures have enabled further urbanization. In turn urbanization provided powerful incentives for transport infrastructure investments both within cities as well as between cities.

proportional to their respective "masses" (e.g., population size, purchasing power) and inversely proportional to their intervening "distance" (measured in terms of travel time and costs, rather than simple geographical distance). Thus, urbanization and transport infrastructure developments both within and between cities are tightly linked.

5.8.   Environmental Problems of Urbanization

5.8.1.   Urban land use

Leaving aside small islands and city states (such as Hong Kong with 5,400 inhabitants per km^2), the countries with the highest population density in the world are, in decreasing order: Bangladesh, South Korea, the Netherlands, Japan, and Belgium. All have population densities above 300 inhabitants per km^2. Summary data for three of the five are shown in *Table 5.9*. Land-use patterns in these countries are particularly interesting because of either their high population growth rate (Bangladesh) or their high population density and high degree of industrialization (the Netherlands). Japanese statistics also make it possible to look at the three largest metropolitan areas (MA; Tokyo, Osaka, and Nagoya), which are shown in *Table 5.9*. The table also gives data for a single city, specifically Vienna, Austria, where population densities are much higher (approximately 4,000 people per km^2) than in the larger administrative regional or national divisions for which aggregate land-use statistics are usually available.

Perhaps the most surprising fact to emerge from *Table 5.9* is that even in the most densely populated countries the dominant use of land is for semi-natural forests and managed ecosystems such as water bodies and agricultural land. These account for over 90% of all land. Even in the Netherlands, with its high population density and long history of industrialization, built-up areas account for less than 10% of all land. Moreover, even in metropolitan and urban areas, 25–50% of land is still covered by forests. This feature of urban land is reinforced by the inclusion of a separate section on urban agriculture in the recent excellent report of the second UN Conference on Human Settlements (HABITAT, 1996).

At the regional scale, built-up areas range from 120 to 220 m^2 per capita. For the example city, Vienna, the built-up area – with around 50 m^2 per capita – has significantly smaller values than the regional aggregates. The reason for this is simple. Zoning plans delineate areas where buildings can be erected. Within these zones, only a fraction of the entire area can be physically "filled" with structures (buildings, roads, etc.). The larger the administrative unit, i.e., from a city district, to the entire city, to metropolitan

Table 5.9: Green versus built-up land in densely populated areas.

	Bangla-desh	Nether-lands	Japan	Japan: 3 major MA	Vienna
Population					
(1) Density (people/km^2)	766	388	322	1,411	3,925
Land-use (%)					
(2) Rivers and lakes	9.6	9.0	3.5	3.6	3.4
(3) Forests	13.6	9.7	67.0	52.7	24.6
(4) Grassland	4.2	31.2	1.1	0.0	26.5
(5) Cultivated land	64.4	22.6	14.3	16.0	
(6) Parks and recreational areas	n.a.	6.3	n.a.	n.a.	9.4
(7) Subtotal (4–6)	68.6	60.1	15.4	16.0	35.9
(8) Infrastructures[a]	n.a.	1.8[b]	0.5	2.8	12.3
(9) Residential buildings	n.a.	7.4	1.9	2.4	14.2[c]
(10) Industry and commerce	n.a.	1.3	0.3	0.4	4.3
(11) Office and public buildings	n.a.	n.a.	0.9	1.1	2.1
(12) Subtotal (9–11)	n.a.	8.7	3.1	3.9	20.6[c]
(13) Other uses	8.2	10.8	1.1	7.4	3.9
(14) Built-up land (8+12)	n.a.	10.5	3.6	6.7	32.9[c]
Per capita land use (m^2/capita)					
Forests (3)	170	244	2,049	373	63
Green areas (4–6)	854	1,513	471	114	91
Infrastructures (8)	n.a.	45[b]	86	37	31
Building area (12)	n.a.	218	121	211	53

Abbreviations: MA = Metropolitan area; n.a. = data not available.
[a]Roads, railways, airports.
[b]Only roads.
[c]Includes private gardens and parks.
Sources: Bangladesh: FAO (1991:52), Netherlands: van Lier (1991:386), Japan: Japanese Statistics Bureau (1987:7), Vienna: ÖIR (1972:I-XXIII).

areas, etc., building areas (zones) become aggregated in the spatial statistics. This results in larger per capita building-area statistics at higher levels of aggregation. Lower level aggregates are therefore closer to actual physical land utilization. Thus, land that is actually used (i.e., physically covered) by buildings in a city like Vienna does not exceed some 25 m^2 per capita, to which an equal area for infrastructures has to be added, bringing the total to 50 m^2 per capita.[24] This suggests that the urban area covered by physical structures (buildings, highways, etc.) may not exceed 20% of the total area classified as built-up area zones.

---

[24]Low per capita land requirements imply high settlement densities. This applies both to settlements in industrialized and developing countries alike. Settlement and population densities between high-rise residential areas in Europe and *favelas* in Latin America are surprisingly similar (even if their transport, energy, and sanitary infrastructures are very different). Urban density is therefore not conditioned on high incomes.

In the built-up areas, land requirements for infrastructures are considerable, ranging from 14% to 17% at the national level (Japan and the Netherlands) and from 37% to 42% in urban agglomerations (Vienna and the three largest metropolitan areas in Japan). Thus the land requirements for moving people and goods in urban areas are not far below those for housing, industry, and commerce.

Applying the value of 250 m^2 per capita for built-up areas and 50 m^2 for areas actually covered by built structures to the global population of 5.3 billion people, gives an estimate of 130 million ha of built-up land worldwide. This is only about 1% of the total global land area. Physical structures like buildings and infrastructures most likely do not cover more than 25 million ha globally, or less than 0.2% of the total global land area.

However, such global percentages mask potentially serious land-use conflicts over usable land, as settlements impinge on agricultural and forested land. Heilig (1996:6), for example, estimates that urban and rural settlements account for 4% of China's usable land, where the definition of usable land excludes deserts, mountains, snow cover, and so forth. In 1988–1989 alone, close to 100,000 ha of cultivated land in China was lost to settlements, transport, and water infrastructures (Heilig, 1996:11). Such changes are quasi-permanent as settlements and infrastructures have extremely long lifetimes.

Former fields may convert to forests within a few decades. Settlements and infrastructures last almost forever. Since antiquity very few cities have disappeared altogether (e.g., Troy, Carthage, and Ch'ang-an, the Chinese capital of the western Han and T'ang period). There is perpetual reconstruction and reuse of urban land for different urban uses (Weinberg, 1985; Marland and Weinberg, 1988). The land that urban structures occupy is effectively permanently excluded for alternative uses.

5.8.2. Urban environmental quality

Urban populations generally enjoy higher incomes than rural populations, and this reduces their vulnerability to environmental stresses. Urban infrastructures such as water and sewerage systems further improve environmental conditions. Nonetheless, there remains substantial urban poverty around the world and, with it, substantial urban environmental stress. Large urban population concentrations also create environmental stresses independent of their wealth. In a rural village, smoke from domestic fires quickly disperses. Multiplied by millions in a large metropolis that smoke translates into urban smog and serious health hazards. Emissions from a single truck supplying a rural community go largely unnoticed. In a city emissions from dense road traffic produce photooxidant smoke and lead poisoning.

Urban poverty remains widespread. Thirty-four percent of the urban population in Asia (excluding China) live below the absolute poverty line. This is less than the 47% of Asia's rural population living below the poverty line, but it is still substantial. In Latin America the percentage of the urban poor is 32% (45% in rural areas), and in Africa 29% (58% in rural areas) (HABITAT, 1996:113). Urban poverty is compounded by higher food prices in urban areas and the unavailability of subsistence activities drawing upon common property resources, such as collecting fuelwood from forests and grazing domestic animals on common land. For the rural poor these are important nonmonetary income supplements, especially in developing countries.

Urban poverty usually means inadequate access to infrastructures providing clean drinking water, sanitation, and garbage disposal. At a global level 1 billion people have no access whatsoever to a functioning safe water supply (HABITAT, 1996:265). Some two billion people lack sanitation (HABITAT, 1996:269). The consequences in terms of disease and human mortality are well documented. They constitute a prime example of environmental problems, particularly in urban areas, arising from "too little" technology rather than "too much". Comprehensive statistical data for urban areas do not exist, but survey data (HABITAT, 1996:266) suggest that for the cities with the worst water supply performance (e.g., Abidjan, Nairobi, and Delhi), 50–70% of all dwellings have no piped water. All the cities with more than 50% of dwellings without piped water are in the poorest of developing countries. This deficiency is due to both capital shortages and poverty – the poor could not pay for these services even if the infrastructures were available. Subsidies do not necessarily solve the problem. In Mexico City, where water prices are below costs, the resulting subsidy to those privileged citizens who receive municipal water equals US$1 billion in direct costs alone (HABITAT, 1996:407). The poor gain little. Municipal water supplies in developing countries also suffer from high losses due to pipe leakages and "other water unaccounted for", i.e., water withdrawn through illegal connections without payment. "Unaccounted losses" equal 20–50% of public water supplies in many cities of the developing world (WRI, 1996:66).

Urban environmental problems due directly to high population concentrations are most noticeable as air and water pollution. We return to these later. However, the large appetites of cities for water have other consequences beyond pollution. Where water cannot be piped in, regional water resources can be degraded significantly. Mexico City, for example, has close to 20 million inhabitants and is located in a high, naturally closed basin. The water supply comes almost exclusively from a local aquifer. Its depletion causes significant land subsidence. Over the last 100 years the central area of Mexico City has subsided by close to eight meters. In areas of heavy water withdrawal children mark their height on water well casings to see if they are growing faster than the ground is sinking (WRI, 1996:64). Venice provides another example. Heavy groundwater withdrawal for industry in the Mestre area, located along the lagoon of Venice, has led to subsidence and repeated flooding of the city's historical center. Gigantic, costly flood control projects have been proposed to protect the city center, but the overall environmental impacts remain problematic.

The principal environmental problems of high population concentrations, however, come from the large amounts of solid, liquid, and airborne waste that they generate, overstretching the assimilative capacity of the environment. Globally, total solid and liquid urban wastes amount to approximately 1 billion tons per year. Annual urban waste generation ranges from 200 to 400 kilograms per capita in developing countries to as high as 1,250 kg/capita in some OECD countries (WRI, 1996:70). Total urban waste generation exceeds 400 million tons in the OECD countries (OECD, 1993:137) or some 500 kg/capita on average. Average country values range from 300 kg/capita in many European countries to over 700 kg/capita in the USA. For developing countries statistical data are scarce. If we assume a lower range estimate of 100 kg/capita per year (HABITAT, 1996:271), we can estimate a total waste flow of a similar order of magnitude as in the OECD countries. Thus, despite the uncertainty in some of the data, it is clear that the amount of urban waste represents formidable challenges in terms of collection, disposal, and overall waste management. For solid waste the options are landfills, incineration, and recycling. All require considerable technological or organizational investments. And they increasingly require new consumer habits and ethics and new forms of social control. We return to this point in Chapter 7.

Turning to air pollution, *Figure 5.20* provides an overview of air quality in 20 megacities worldwide. The most disappointing feature of the figure is the large "white spots" indicating missing data. Using technology as a tool for diagnosing and systematically collecting data on environmental quality has not yet diffused even to many large metropolitan cities. However, even with sparse data a significant "North–South" divide in urban air quality is noticeable. Cities in industrialized countries generally fare well with respect to traditional air pollutants such as particulates and sulfur dioxide. Yet in common with many megacities of the developing world, they experience urban air pollution problems from dense motor traffic. Despite the

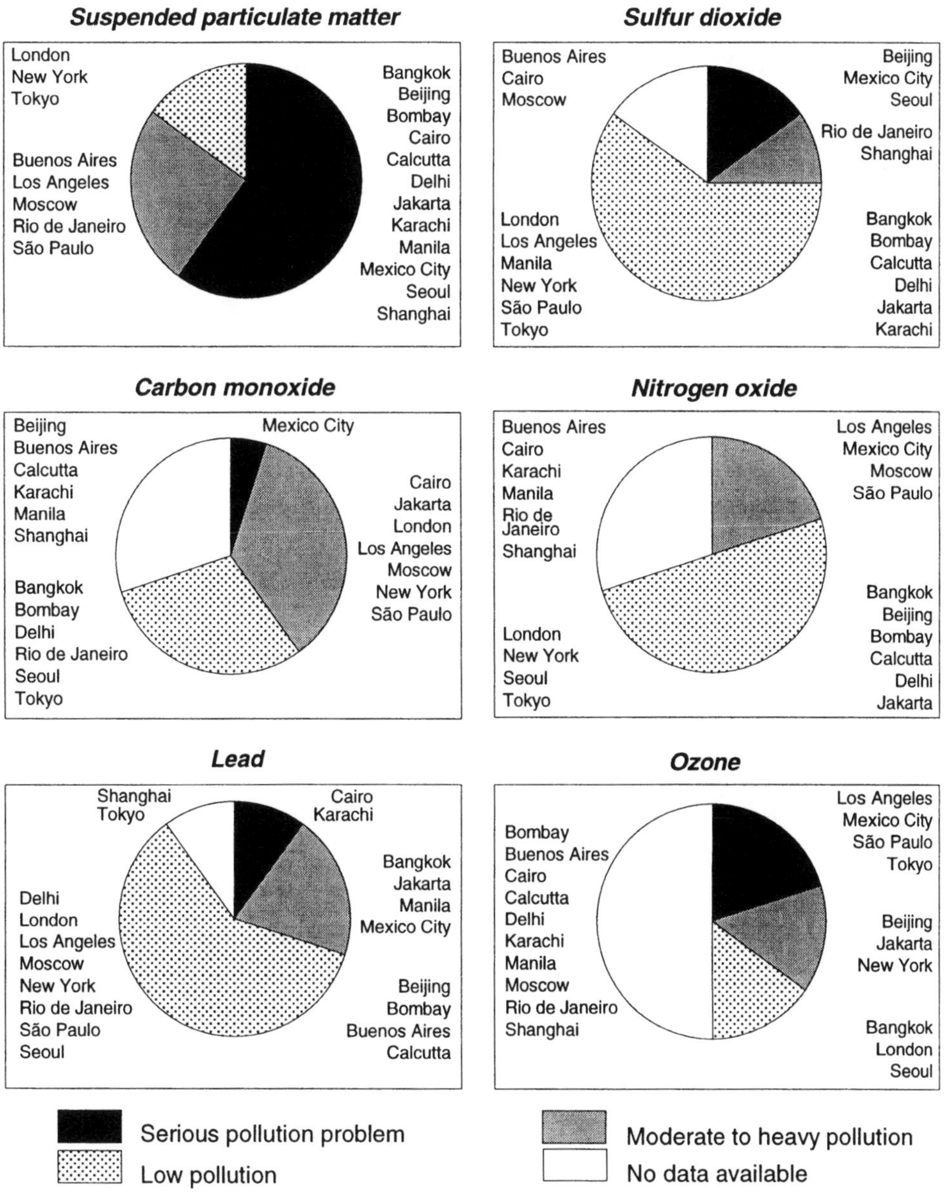

Suspended particulate matter

London
New York
Tokyo

Buenos Aires
Los Angeles
Moscow
Rio de Janeiro
São Paulo

Bangkok
Beijing
Bombay
Cairo
Calcutta
Delhi
Jakarta
Karachi
Manila
Mexico City
Seoul
Shanghai

Sulfur dioxide

Buenos Aires
Cairo
Moscow

London
Los Angeles
Manila
New York
São Paulo
Tokyo

Beijing
Mexico City
Seoul

Rio de Janeiro
Shanghai

Bangkok
Bombay
Calcutta
Delhi
Jakarta
Karachi

Carbon monoxide

Beijing
Buenos Aires
Calcutta
Karachi
Manila
Shanghai

Bangkok
Bombay
Delhi
Rio de Janeiro
Seoul
Tokyo

Mexico City

Cairo
Jakarta
London
Los Angeles
Moscow
New York
São Paulo

Nitrogen oxide

Buenos Aires
Cairo
Karachi
Manila
Rio de Janeiro
Shanghai

London
New York
Seoul
Tokyo

Los Angeles
Mexico City
Moscow
São Paulo

Bangkok
Beijing
Bombay
Calcutta
Delhi
Jakarta

Lead

Shanghai
Tokyo

Delhi
London
Los Angeles
Moscow
New York
Rio de Janeiro
São Paulo
Seoul

Cairo
Karachi

Bangkok
Jakarta
Manila
Mexico City

Beijing
Bombay
Buenos Aires
Calcutta

Ozone

Bombay
Buenos Aires
Cairo
Calcutta
Delhi
Karachi
Manila
Moscow
Rio de Janeiro
Shanghai

Los Angeles
Mexico City
São Paulo
Tokyo

Beijing
Jakarta
New York

Bangkok
London
Seoul

■ Serious pollution problem   ▨ Moderate to heavy pollution

▩ Low pollution   □ No data available

Figure 5.20: Overview of air quality in 20 megacities as indicated by concentrations of particulate matter, sulfur dioxide, carbon monoxide, nitrogen oxide, lead, and ozone. Black denotes a serious problem (concentrations exceed WHO guidelines by at least a factor of two). Grey areas denote moderate to heavy pollution (concentrations exceed WHO guidelines by up to a factor of two). Dotted areas denote low pollution (WHO guidelines are met), and white indicates that no data are available. Source: adapted from WHO and UNEP (1993:40–42).

Table 5.10: Particulate concentrations and exposures in eight major human microenvironments.

Group of nations	Concentrations (μg/m^3)		Exposures (GEE)^a		
	Indoor	Outdoor	Indoor	Outdoor	Total
Developed					
Urban	100	70	5	<1	6
Rural	60	40	1	<1	1
Developing					
Urban	255	278	19	7	26
Rural	551	93	62	5	67
Total			87	13	100

^{a}GEE = Global Exposure Equivalent.
Source: Adapted from Smith (1993:545).

fact that urban energy consumption is generally much higher in industrialized countries than in developing countries, differences in the structure of energy consumption (see Chapter 6) and active abatement measures have more than compensated in the case of particulate matter and sulfur dioxide. The industrialized countries are now relatively free of their earlier chronic problems of urban smog caused by using wood and coal as principal fuels.

Conversely, megacities in developing countries face a double challenge. They still have substantial air pollution from "traditional" sources (i.e., burning large quantities of biomass fuels and coal for basic energy needs), while increasingly sharing with cities in developed countries "modern" sources of pollution, i.e., motorization. We will return to sulfur dioxide in Chapter 6 and to motorized traffic pollution in Chapter 7, and will focus here only on particulates. In focusing on particulates, it is important to first correct the dominant image that the largest threat to human health is urban smog. We tend to think of air pollution as an "outdoor" urban phenomenon, but in fact some of the highest concentrations of pollutants occur *indoors* and in *rural* areas. *Table 5.10* estimates particulate concentrations and exposures in eight different major human microenvironments covering the eight possible combinations of indoor and outdoor settings, urban and rural settings, and developed and developing countries.

Table 5.10 illustrates the importance of considering exposure to pollution rather than just the amount of pollution. Exposure is the product of a pollutant's concentration times and the hours people are exposed to that concentration. To aggregate different exposures to a common unit Smith (1993) introduced the concept of the Global Exposure Equivalent (GEE) [see also Smith (1988) on the methodology underlying such calculations].

We have renormalized Smith's GEE index with the global exposure equivalence index being 100 at the world level. The respective relative exposure indexes thus give a more straightforward insight into the relative rankings of indoor versus outdoor, rural versus urban, and developing versus developed countries' population exposures to pollution.

The differences in particulate concentrations between developed and developing countries are striking. On average, developing country urban indoor particulate concentrations exceed those in industrialized countries by a factor of close to three. The difference for urban outdoor concentrations is a factor of four. Given the larger populations in developing countries this results in an overproportional exposure to pollution.

In developing countries average indoor and outdoor urban particulate concentrations are quite similar (250–280 μg/m^3), but because people spend more time indoors, exposure is much higher indoors than outdoors. By far the largest exposure to particulate pollution occurs in rural indoor environments (62% of GEE), followed by urban indoor pollution exposure (19% of GEE). The reasons for this are the same in both rural and urban areas of developing countries: traditional biomass fuel use in inefficient cooking and heating stoves. The results are considerable health risks, particularly for women and children, who spend more time in indoor environments. Conversely, the particulate emissions we usually associate with deteriorating urban air quality, i.e., in outdoor environments, correspond to only 13% of total GEE, with most of it in developing countries.

This analysis supports those who argue that poverty is the biggest "polluter". Economic resources determine access to services delivered by different technological means. Those without resources must use "cheap and dirty" technologies. Those more fortunate use "expensive and clean" technologies. This turns out to be more important for the human environment than where people live (rural versus urban) or the impacts of large population concentrations in cities. This argues for social and economic development as the preferred strategy in reducing pollution (cf. World Bank, 1992). This logic is correct and powerful for traditional environmental pollution caused by inadequate water supplies, inadequate sanitation, and particulates and sulfur dioxide. But it is less true for "modern" forms of pollution that increase with affluence, such as municipal waste, motor vehicle emissions, and energy-related emissions of carbon dioxide. In the next chapter we look at these environmental challenges and their spectacular rise since the onset of the Industrial Revolution.

Copyright acknowledgments

Chapter 6

Industry

Synopsis

The chapter starts with a brief quantitative overview of global industrial expansion and the disparities that remain between centers of industrialization and those regions that are catching up. Overall expansion has been enormous. It has been possible only through successive replacements of manufacturing technologies, materials, and energy sources, and through continuing improvements in the organization of industrial production. These changes have yielded enormous productivity gains in labor, materials, and energy use per unit of production. Such productivity gains have sustained increasing levels of industrial output, increased work force incomes, and reduced working time. Productivity gains have also eased the demands on natural resources and reduced traditional environmental impacts such as indoor and urban air pollution. At the same time, however, new environmental concerns have emerged at the global level. Synthetic substances are depleting the ozone layer, and increased concentrations of greenhouse gases, mostly from fossil energy combustion, are causing global warming. Historically, environmental productivity gains have been outpaced by output growth. Only in the last two decades have gradually saturating demands in bulk materials combined with continued productivity increases resulted in near stabilization of materials and energy use in the most advanced industrial countries. The history of energy and carbon use illustrates the predominant pattern. Energy use per unit of economic output has declined by 1% per year, and carbon emissions per unit of energy use has declined by 0.3% per year. This is a combined carbon productivity increase of 1.3% per year. However, economic growth has averaged 3% per year. Thus carbon emissions increased in absolute terms. What is more promising is that until now environmental productivity gains have been the only unplanned side effects of overall technological productivity gains. That these gains follow classical technology learning curve patterns suggests there is a large potential for future environmental productivity gains once they become an explicit objective.

6.1.   Introduction

Industrialization is a process of structural change. *Figures 1.1* and *6.1* illustrate the industrialization process, i.e., the shift from agriculture toward industry, particularly manufacturing, both in employment and in productivity and output growth. Industrial productivity and output growth are important drivers of overall economic growth and increase both individual and national incomes. This steadily enlarges the market for industrial products. Underlying historical productivity and output growth are a host of interrelated organizational and technological developments. Economies of scale, division of labor, and the expansion of markets through trade are the main economic drivers of industrial growth. These reflect essential changes in the *organization* of the production and distribution of goods, and how new technologies are developed and deployed. It is these organizational developments that are central to industrialization. Indeed, the term *industrial society* has come to mean a particular way of organizing economic and social relationships, from science and industrial management to the fine arts. An industrial society is based pervasively on the economics of specialization and standardization in order to produce final products in ever greater numbers and, paradoxically, ever greater variety.

Our emphasis on organizational change is not meant to belittle technological "hardware" changes in the form of new production processes and industrial products. It is simply a reminder that without "software" changes the impact of technological innovations is necessarily limited. Industrialization is *more* than just adding new technology components to an otherwise unchanged preindustrial system of production, distribution, and the societal division of labor. This may explain the limited success of ambitious large-scale, but isolated industrialization projects, in many developing countries. Industrialization is not just more and newer technological hardware. It entails a deep transformation of the social, economic, and spatial organization of production, and thus of the social fabric at large.

6.2.   Industrialization: Output and Productivity Growth

Since the onset of the Industrial Revolution[1] in the middle of the 18th century, global industrial output and productivity have risen spectacularly. *Figure 6.2* compares three sets of data. The three do not account for

---

[1]This term (coined by Toynbee, 1896) may in fact be a misnomer in that its implicit concept of discontinuity ignores important developments in protoindustrial societies that paved the way for accelerated rates of change after the mid-18th century. For a concise discussion see Cameron (1989).

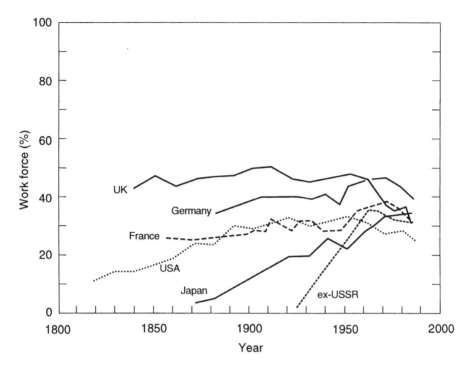

Figure 6.1: Percentage of work force employed in industry for selected countries. Source: Grübler (1995a:40).

improvements in quality and variety of industrial output and differ also in their estimates of growth during early industrialization. There are also inherent methodological and data uncertainties; however, all three estimates agree on the basic dynamic pattern of global industrialization since the middle of the 19th century: exponential growth. The Bairoch (1982) data include a special attempt to estimate industrialization levels outside Europe and North America. They therefore provide a more realistic picture of the dynamics of early industrialization.

Based on the Bairoch (1982) data, global industrial output has risen by approximately a factor of 100 since the 1750s. Over the last 100 years, where all the three data sets agree reasonably well, output has grown by a factor of 40 – an average growth rate of 3.5% per year. Per capita industrial production has increased by a factor of 11 – equivalent to a growth rate of 2.3% per year. Increased per capita output has therefore been a larger contributor to output growth than simple population increases.

It is important to note that *Figure 6.2* measures output in terms of monetary value. The picture for industrial material output growth will be different as the material intensity of industrial value added has varied over

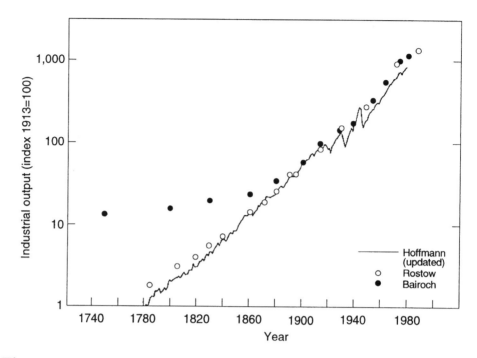

Figure 6.2: Growth of global industrial output (index 1913=100), a comparison of three estimates. Data sources: updated (UN, 1970 to 1995) from Rostow (1978:662), Bairoch (1982:292), and Haustein and Neuwirth's (1982:27–30) update of Hoffmann's (1955) estimates.

time. In the early phases of industrialization, rapidly falling material costs (and prices) caused industrial material output to grow even faster than the monetary output growth shown in *Figure 6.2*. Over the past few decades, however, material intensity has been declining in the OECD countries (cf. Williams *et al.*, 1987), a point to which we return in Section 6.6.

Figure 6.2 also masks significant variations in growth across regions and countries. In addressing various patterns of growth in different countries, perhaps the best known theory of industrial and economic growth is that developed by Walt Rostow (1960, 1978). Rostow distinguishes an early "take-off" phase, followed by a "drive to technological maturity" that ultimately leads to "high mass consumption". Rostow established the timing of the various phases in different countries by an extensive empirical analysis of growth rates of leading industrial sectors (textiles, railroads, steel, electricity, etc.), embedded in a broader analysis of macroeconomic and overall industry growth.

Rostow's theory has been criticized because it postulates essentially a linear development model in which latecomers to the industrialization process follow the development patterns of early industrializing countries. Because there may, in fact, be development options available today, for example, to China, that were not available to early industrializing countries, past development is not necessarily the best guide for the future. Nonetheless, Rostow's theory remains an important contribution for understanding the historically uneven industrialization paths in the highly industrialized countries of the OECD.

A second approach to the analysis of historical variety of industrialization patterns stems from geography, especially the spatial distinction between "core" and "periphery." Regions are classified along a continuum between core and periphery that reflects either straightforward geographic proximity or functional classifications based on, for example, industrial activity, trade, or communication flows. Here we adopt a functional classification as follows.

Industrialization starts in the "core". Core countries subsequently display the highest degree of industrialization and take the lead in introducing new technology systems and entire technology clusters. These subsequently spill out to the industrial "rim" and eventually to the industrial "periphery". Process and product innovations constantly enlarge the industrial base of the core and steer structural changes in its industry. The core also leads in the transition to a postindustrial, service-dominated economy. In other words, the industrial "core" is defined by those countries having developed and adopted a particular technology cluster most pervasively. The "rim" generally adopts a particular technology cluster later, and/or only partially.

Levels of industrialization are therefore lower in the "rim". Its technology base is quite diverse. In some areas it is as sophisticated as the core, but other areas are characterized by more outdated technological vintages. Overall, the exchange of information and goods is less intensive than that within the core, but the rim still participates significantly in the international division of industrial production, particularly in manufacturing. The industrial sector is quite important and growing steadily, as reflected in structural indicators of value added, employment, and the material intensiveness of industry.

Finally, regions in the "periphery" have the weakest industrial and technological base and are remote from international flows of information and goods. Exports are dominated by primary commodities, and industry produces mostly for local markets although islands of high-tech and heavy smokestack industries do exist. Structurally the economy is dominated by agriculture and service activities operating largely outside the formal economy. The degree of industrialization and urbanization remains comparatively low.

Table 6.1: The global geography of industrialization. Index of level of industrialization (industrial output) of England in 1900 = 100.

	1750	1830s	1870s	1920s	1980	1980/1750
Index level						
Core	2	20	180	950	7,400	3,080
Rim	5	20	40	190	2,300	430
Periphery	120	145	100	220	1,300	11
World	127	185	320	1,360	11,000	87
Growth rates (%/year)						
Core		2.6	4.6	3.6	4.0	3.6
Rim		1.7	1.3	3.3	5.0	2.7
Periphery		0.5	−0.7	1.7	3.5	1.1
World		0.5	1.1	3.1	4.1	2.0
Regional shares (%)						
Core	2	10	56	70	67	
Rim	4	11	12	14	21	
Periphery	94	79	31	16	12	

All figures are rounded. Regional shares and factor increases calculated from the original data may differ from the rounded figures.
Source: Bairoch (1982).

Table 6.1 summarizes the distribution of industrial output growth using the geographical taxonomy just described. The industrial output of England in 1900, based on Bairoch's estimate, is used as a reference value. *Table 6.1* shows that England's industrial output in 1900 approximated that of the entire globe 150 years earlier. Conversely, global industrial output in 1980 was more than 100 times greater than England's output 80 years earlier, and an equal order of magnitude larger than global industrial output at the onset of the Industrial Revolution. Industrial output in the core grew at persistently higher rates than in the rim and periphery. Only since 1920 has the rim begun to catch up to the core (cf. its higher growth rates shown in *Table 6.1*).

Taking the world as a whole, *Table 6.1* shows that the weight of the industrialized core countries in the early phases of industrialization was comparatively low. By the mid-19th century, however, the industrial core countries had achieved global dominance and accounted for over half of the global industrial output. Ever since, global industry has been dominated by a comparatively small number of countries – the core persistently accounts for about two-thirds of global industrial output. The relative decline of the periphery, and its absolute decline between 1830 and 1870, is the inverse of the industrial core's rise. Despite growth rates of 3.5% annually over the last five decades, the periphery has fallen further behind the core, which has grown at 4% per year. Thus the gap between the two has widened in both absolute and relative terms.

This analysis shows that while industrialization has become a global phenomenon, there are comparatively few examples of countries successfully catching up to the industrialized core. (Japan is the best example.) Instead, catching up appears to occur rather within a region or between regions with similar degrees of industrial development. While Austria and Finland have indeed caught up to the European core, disparities in income and industrialization between North and South America, or between Europe and India, have not narrowed. In some cases, notably Africa, they have widened.

In terms of the spatial taxonomy adopted here, the industrial "core" (OECD countries) has grown by adding members from the previous "rim" (Canada, Japan, Scandinavia, Austria, Switzerland, and Italy). The dominance of the (expanding) core in industrial and economic power is as large as ever. The OECD countries still account for 70% of the world's industrial output (cf. *Table 6.1*) and 75% of the world's merchandise trade (World Bank, 1992). Over 80% of the OECD's imports of manufactured goods are imported from other OECD members, and another 9% from the industrial rim (Eastern Europe and the four Tigers).[2] Only about 10% comes from the rest of the world ("periphery" in *Table 6.1*).

It is beyond the scope of this book to discuss the reasons, or possible remedies, for the widening disparities in industrial development. Much of the blame is often assigned to falling real prices for primary resources and the resulting deteriorating terms of trade. However, the constant change in the industrial structure of the core is also important, especially the decreasing material intensity of advanced industrialized countries. The prices of primary resource inputs in highly industrialized or even postindustrial economies matter less and less. Copper and bauxite prices, and even crude oil prices, hardly affect industries producing computers, software, or other high-value density products like aircraft. And it is precisely these industries that have shown the greatest growth over recent decades. At the same time, when raw material prices do increase, they affect the comparatively material-intensive economies of developing countries more severely.

Thus, deteriorating terms of trade partly explain why growth rates in the periphery are smaller than would be expected based on factor endowments. However, they do not fully explain the persistently higher growth rates in the

---

[2]The rapidly industrializing Asian economies of Hong Kong, Singapore, South Korea, and Taiwan are frequently referred to as the four Tigers.

core. The "success" of the core appears to derive more from its dynamics of industrial innovation and the resulting rise in industrial factor productivity, particularly the high growth of labor productivity (cf. Section 6.4). This raises the question of the importance of natural resource endowments for industrialization and economic development.

The availability of energy and mineral resources is often considered a *conditio sine qua non* for industrialization. We argue that resource availability is of secondary importance. This is especially true given the spatial division of primary activities (raw materials) and secondary activities (manufacturing) made possible by modern transportation systems, particularly since the 19th century. Domestic resource endowments are no longer a precondition for industrialization. Resources per se also matter less than technology. First, natural resources in and of themselves are useless without the technology to harvest them. Second, the availability of resources is itself a function of technology. Geological knowledge, exploration and production technologies, and so forth, all determine the quantity of resources available to humanity. Third, technology development provides for material substitutes such as synthetic fertilizers and rubber, alleviating possible resource constraints.

Thus, we view disparities in development and industrialization as "technology gaps" resulting from differences[3] in technology accumulation and innovativeness, rather than from differences in natural resource endowments. Innovative capacity – and the production, income, and growth that result – is thus "created" by an appropriate socioinstitutional framework and an expanding technology base and not determined by resource endowments. Industrialization is therefore socioeconomically and technologically "constructed" and not geologically "predetermined".

France, Scandinavia, Austria, Japan, and South Korea are all examples of successful industrialization with only modest national natural resource endowments. Indeed the abundance of resources appears to be a mixed blessing as it can lead to persistently higher intensity development trajectories (cf. David and Wright, 1996). This suggests that coal-rich China might well develop along the energy- and resource-intensive path of the USA, rather than the more material-efficient paths characteristic of France and Japan.

---

[3]Note here an important extension of the traditional definition of comparative advantage. The classical definition revolves around the comparative prices of factor inputs (resources, labor, capital, etc.) modeled via a production function approach. "Technology gaps" further incorporate differences in combinations and intensities of factor inputs that cannot be explained solely by their price differences. For a more detailed discussion see Dosi *et al.* (1990).

Table 6.2: Basic activity data for industry, 1990.

Economy	No. of people employed (10^6)	Value added (10^9 US$)	7 Major commodities produced (10^6 tons)[a]	Ton-km transport (10^{12})	Mtoe final energy consumed (without feedstocks)[b]	Carbon emissions (10^6 tons)[c]
Market	130	4,632	1,095	7	877	766
Reforming	80	975	515	5	640	584
Developing	300	1,068	895	5	841	733
World	510	6,675	2,505	17	2,358	2,083

[a]In decreasing order of tonnage: cement, steel, paper, fertilizer, glass, aluminum, copper.
[b]Mtoe, million tons oil equivalent (44.8 10^9 joules).
[c]This value includes the manufacture of cement, and carbon emissions from electricity production allocated to industry in proportion of industrial to total electricity consumption.
Source: Grübler (1995a:3).

It also cautions against rapid "convergence" scenarios for developing countries, as desirable as these may be from a human development perspective. While history provides examples of successful catch-up strategies, these take considerable time and must exploit limited windows of opportunity for development. Initial conditions are important, as are the education of the work force and the existence of a supportive socioinstitutional framework. A careful balance must be found between learning from past successes and investing in new niches and emerging technology clusters. Military generals must always avoid the temptation to plan for the last war. Similar caution is needed in planning industrial development strategies.

Table 6.2 summarizes selected macroindicators illustrating the present scale of global industrial activities. Transport is included in the table because everything that is shipped either originates as an industrial product or is processed at some stage by industry. Note that the 2,400 Mtoe of final energy used and its related 2 billion tons of carbon emissions rival the total tonnage of the seven top industrial commodities produced.

Table 6.3 shows how powerful industry is as an agent of global change. Industry accounts for about 20% of formal employment, and 40% of value added, final energy use, and industrial carbon emissions. The table also shows substantial variation in these values among countries with different degrees of industrialization and overall development. The dominance of industry in the material, energy, and smokestack-intensive economies of Central and Eastern Europe and the former USSR (labeled "reforming economies" in *Table 6.2* and *Table 6.3*) stand out clearly.

Table 6.3: Percent share of industry in anthropogenic activities, for selected indicators, 1990.

Economy	Employment	Value added	Final energy (without feedstocks)	Carbon emissions[a]
Market	34.0	33.0	31.5	31.9
Reforming	40.1	59.3	52.6	46.7
Developing	16.9	36.8	36.3	47.9
World	21.6	37.4	37.3	38.7

[a]Emissions from cement and for the generation of electricity purchased by industry are included.
Source: Grübler (1995a:4).

6.3.   Clusters

In Part I we introduced the concept of technology clusters. In this section we elaborate and present the history of technological change in industry as a succession of technology clusters. There is always some overlap between successive technology clusters as older technological and infrastructural vintages coexist with the new dominant cluster. In some cases older technologies are actively perpetuated, as illustrated by the former USSR's post-World War II industrial policy. Overlap also exists with the forthcoming cluster as its initial elements are developed within specialized applications and specific market niches. Eventually they will emerge, after extensive experimentation and cumulative improvements, as part of a new dominant technological mode. The process takes considerable time as cross-enhancing connections among isolated developments build slowly. A new cluster finally emerges after a period of crisis in the current cluster involving painful structural adjustments in both economic relations and social and institutional settings.

However, at any one time it is the dominant technology cluster that drives most industrial and economic growth. The dominant cluster is frequently associated with the most visible technological artifact or infrastructural system of the time. Such lead technologies are also the focus of economic historians using the leading sector hypothesis (e.g., the railways era or the "age of steel and electricity"; Freeman, 1989). We emphasize technology clusters because any dominant sector or infrastructural system studied under the leading sector hypothesis can only explain a fraction of the economic and industrial output growth of the time.[4] Only the combination of a whole host of innovations in many sectors and technological fields, i.e., in the form

[4]For case studies of coal, steel, and railways see, e.g., Fishlow (1965), Holtfrerich (1973), Fremdling (1975), von Tunzelmann (1982), O'Brien (1983), and Freeman (1989).

of a "cluster", can appropriately account for industrial and overall economic growth.

The following sections elaborate on the four historical technology clusters introduced in Part I in *Tables 4.1* and *4.2*. The four historical clusters can be labeled according to their most important industries or functioning principles. These are the "textile" cluster, extending to the 1820s; the "steam" cluster until about the 1870s; "heavy engineering", lasting until the eve of World War II; and finally, "mass production/consumption" until the present. We illustrate the clusters with key examples of both dominant and emerging technologies. The discussion necessarily remains eclectic, brief, and an over-simplification, as each cluster would merit a book on its own for an adequate historical account.

6.3.1. Textiles (ca. 1750–1820)

We know industrialization as a process of structural change in which increasing proportions of the national income and employment are generated by industry. By this definition industrialization began around the middle of the 18th century in England. Technological innovations transformed textile manufacturing in England and gave rise to what later became a new mode of production: the factory system. Important obstacles to industrialization, and the population concentration in cities that it depended upon, were overcome. Fuelwood and (wood-based) charcoal shortages were surmounted by coal and Darby's coke. These combined with the stationary steam engine, which was particularly important for pumping water from coal mines, provided power densities in any location where required that previously were found only in exceptional instances of abundant hydropower. Improvements in parish roads and turnpikes and the "canal mania" around 1800 made it possible to supply rapidly rising urban and industrial centers with food, energy, and raw materials. Charcoal and the puddling furnace produced the first industrial commodity and structural material: wrought iron. Innovations in spinning and later weaving made dramatic cost reductions and output increases possible, particularly in cotton textile manufacturing. The introduction of fine porcelain from China created an expanding chinaware (Wedgwood) industry.

The most spectacular output growth between 1780 and 1820 was in the cotton-spinning and weaving industry. This was to become the first industry to apply mechanization and the new factory system on a large scale. In Rostow's words, it was the "original leading sector in the first take-off of industrialization" (Rostow, 1960:53). Previously, textile manufacturing was a small-scale family business. The only division of labor was between spinning and weaving. However, it typically took three to four spinners to provide enough yarn for one weaver. This imbalance was accentuated by John Kays' invention in 1733 of the flying shuttle, which doubled the output per weaver-hour on a hand loom (Ayres, 1989a:16). Powerful incentives were thus in place to improve the productivity of spinning.

The spinning jenny, patented in 1770, was the technological response to this productivity challenge. The jenny basically imitated a spinning wheel, but enabled a single operator to control a number of spindles simultaneously. In today's computer jargon we would speak of "parallel processing". The original patent was for 16 spindles, but by 1800 the number of spindles had risen to about 100 and by the 1830s it had reached 1,200 (Ayres, 1989a:17). As a consequence the productivity of a single spinner was multiplied by a factor of about 1,000 within two generations. However, the cotton yarn produced by the spinning jenny was too weak to be used exclusively, and the warp on the hand loom continued to be made from hand-spun flax. The resulting "hybrid" textile (fustian) was neither easy to sew nor easy to wear. The eventual solution was the so-called water-frame, which was able to spin much stronger yarn and was the first spinning machine designed entirely for complete mechanization. Its original design was powered by hydropower, and its inventor Arkwright ironically first used a stationary steam (Newcomen) engine simply to pump water on a water wheel driving the new spinning machine (Ayres, 1989a:17).

This example presages a recurrent pattern in the introduction of new technologies. Considerable time is needed before the new technology can be applied in a different context or configuration where it can demonstrate its own merits rather than being just "plugged in" to an existing configuration. Below, for example, we discuss the first industrial applications of electricity where electric motors simply replaced centralized stationary steam engines, and all the pulleys and transmission belts remained. Only later was full advantage made of electricity's ability to provide power exactly when and where it was needed for decentralized applications.

Mechanization in weaving was slower and more complex. The first attempts to introduce a power loom came at the end of the 18th century, but workable models emerged only in the first decades of the 19th century, and even the diffusion of mechanical weaving in England extended well into the mid-19th century. Ayres (1989a:18–19) highlights a fundamental reason for the delay: the inadequacy of the original dominant structural material, wood. The use of iron machinery only became possible after high quality iron (through Cort's puddling process) became widely available, and that was made possible only through the innovation of coking coal in place of increasingly scarce charcoal from wood. Finally, metal-working machine tools had to be developed to work the iron.

The nexus of innovations involving cotton textiles, the coal and iron industries, and the introduction of steam power constitutes the heart of the Industrial Revolution in England. However, in order for these developments to take place, important preconditions had to develop. The first was a dramatic increase in agricultural productivity (cf. Chapter 5). More complex crop rotation patterns, the abandonment of fallow lands, field enclosures, new crops, and improved animal husbandry allowed increased agricultural productivity with less labor. Freed from agriculture, people sought urban residence and industrial employment. Another important precondition was institutional. The separation of political and economic power, new institutions for scientific research and the dissemination of its results, and the organization of market relations all mark the breakdown of feudal and medieval economic structures and their associated monopolies, guilds, tolls, and restrictions on trade. Rosenberg and Birdzell (1986, 1990) argue that the intellectual, institutional, and organizational changes were in fact the most important in enabling and encouraging changes in industrial technology, products, markets, and infrastructures. However, given the general "laissez-faire" attitude of the era, no provision was made to smooth the disruptions caused by structural changes in employment, urbanization, value generation, and the distribution of income. The result was violent manifestations of social and class conflict. Luddites attacked textile machinery, and the Captain Swing movement attacked the first mechanical threshing machines introduced on farms (cf. *Figure 2.20*). They provide early evidence of the painful social adjustments that are linked to industrialization.

6.3.2. Steam (ca. 1800–1870)

In this period, which lasted until the recession of the 1870s, industrialization spread from limited regions and sectors to become a pervasive principle of economic organization. England continued to dominate and reached its apogee as the world's leading industrial power in the 1870s, accounting for nearly one-quarter of global industrial output. Industrialization spread to the continent – specifically Belgium, the Lorraine in France, and the Ruhr in Germany – and to the eastern USA. Its spread followed essentially the lines of the successful English model – textiles, coal, and iron.

Coal was the principal energy source for industry except in the USA, where fuelwood continued to fill this role. Energy needs for transportation and households, however, continued to be met mostly by renewable energy sources – wood and animal feed. Particularly characteristic of the "steam" period is the emergence of mobile steam power (locomotives and ships), but transport infrastructures were still dominated by inland navigation and canals. These reached their maximum extent in the 1870s in England, France, and the USA. Mobile steam (locomotives and steam ships) were to fully displace traditional long-distance transport modes (coaches and clippers)[5] only near the end of the subsequent period of industrialization ("heavy engineering"), i.e., by the 1920s.

Parallel to their diversification into mobile applications, steam engines were continually improved in terms of power, reliability, and especially energy conversion efficiency. Advances in materials, an improved understanding of thermodynamics, and especially the pervasive diffusion of steam engines outside their initial applications in coal mines all contributed to steady improvements. Steam engines became a common stationary power source in mechanized industry, although their share among all stationary power sources, including hydropower, did not reach a maximum until about 50 years later, ca. 1920. By then steam engines supplied virtually all mechanical and motive power in the then industrialized countries. *Figure 6.3* illustrates the improvements in the thermal efficiency of steam engines and the associated steady decline in specific fuel consumption.

In addition to technological improvements in steam engines, the period also saw important innovations in the field of materials (Bessemer steel production), transport and communications (railways and telegraphs), and energy (city gas) and the systematic development of a coal-based chemical industry. These were later to be at the center of the next technology cluster, "heavy engineering", which ran from the second half of the 19th century until the Great Depression of the 1930s.

6.3.3. Heavy engineering (ca. 1850–1940)

Fueled by coal, the "heavy engineering" technology cluster was dominated by railways, steam, and steel. It constitutes the most smokestack-intensive period of industrialization. With primary commodities and capital equipment as the dominant outputs, the industrial infrastructure spread globally.

---

[5]The displacement of sailing ships by steam ships started in the 1820s and took about 100 years to complete. The process followed a quite regular pattern of technological substitution (cf. Nakićenović, 1984). The technology literature devotes considerable space to what Ward (1967:169) labeled the "sailing ship effect", i.e., the major technical improvements in clippers when challenged by competition from steam ships. Such improvements were substantial, and in fact all speed records set by sailing clippers were set in the late 19th century. But steam ships also improved, and sailing ships could not keep up. They subsequently disappeared as transport vessels and have found only a limited substitute market as recreational vessels.

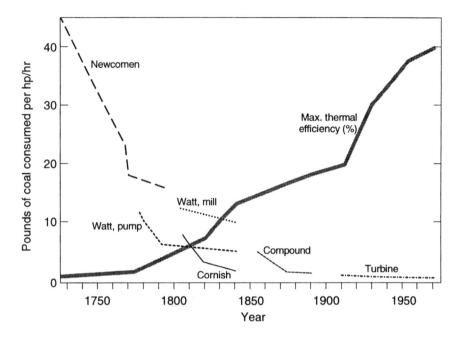

Figure 6.3: Evolution of the performance of steam engines, pounds of coal required per horsepower per hour and maximum thermal efficiency of steam engines (in %). Source: Ayres (1989a:16).

Indeed, expanding the industrial and infrastructural base became almost a self-fulfilling objective, driven by economies of scale at all levels of industrial production and organization. An essential characteristic of the "heavy engineering" technology cluster was the standardization of mass-produced components and structural materials, perhaps best symbolized by the Eiffel Tower and the skyscrapers of Manhattan.

Steel epitomizes the period. The steel "boom" began in the second half of the 19th century as a direct consequence of two important process innovations – the Bessemer process introduced in 1856, and the Siemens-Martin open-hearth process introduced some 10 years later. The technological breakthrough represented by the Bessemer process came from its simplicity and its exothermic reaction. Molten pig iron was transformed to a low carbon steel by blowing air through it. The exothermic heat from the rapid oxidation of carbon raised temperatures high enough to maintain the metal in a molten state for subsequent casting. A manganese-iron alloy (Spiegeleisen) was added to the molten metal to control excessive oxidation. Originally the process was restricted to low-phosphorus iron ores. The basic variant of the process (Thomas process) that was introduced in the 1870s extended applications to high-phosphorus ores including those then available in the Lorraine region (Grübler, 1990b:144). An important innovation of the 20th century was the basic oxygen, or LD (for the Austrian steel towns of Linz and Donawitz) process, in which oxygen was used as the oxidant instead of air. The Siemens-Martin open-hearth process was a slower process of iron decarbonization using an endothermic (heat absorbing) reaction, that required sophisticated waste heat conservation techniques, specifically the regenerative furnace. The two variants of the Siemens-Martin open-hearth process differed in their feed material: "pig (iron) and ore" for the Siemens process versus "pig and scrap" for the Martin process. The diffusion of both the Bessemer and the open-hearth processes (*Figure 6.4*) led to rapidly expanding steel production, from about 1 million tons worldwide in 1870 to some 100 million tons in the 1920s. Since then global steel production has continued to increase by nearly another factor of 10.

As is evident in *Figure 6.4*, different steel process technologies have very different diffusion dynamics. The diffusion of the Bessemer process and later the basic oxygen process was rapid. Each grew to market dominance in less than three decades. For the open-hearth process and electric arc steel production, diffusion was much slower. The time required to grow from a market share of 1% to 50% (Δt) was 80 years for the open-hearth process and 100 years for the electric arc process. The latter process offers advantages, such as, first, much flatter economies of scale, which counterbalances the trend toward ever larger steel converters (cf. *Figure 2.2*). Second, it became possible to use almost exclusively recycled steel scrap as raw material. Currently in OECD countries some 40% of steel is produced from recycled scrap (Wernick *et al.*, 1996:185).

The new steel processes, particularly the Bessemer process, led both to output increases and to drastically falling production costs. Much of these were passed on to consumers as price reductions. Twenty years after the introduction of Bessemer steel, real-term prices had fallen by about a factor of 10. In the USA, steel rails that had cost US$170 per ton in 1867 fell to US$15 per ton by 1898 (Ayres, 1989a:25). In Germany, real-term steel prices fell from some 1,000 DM/ton (in constant 1913 DM) in the early 1860s to less than 100 DM/ton by the 1880s and remained practically constant until World War I (Grübler, 1990b:127).

More important from an environmental perspective has been the continuous improvements in yield and efficiency, both from radical process technology changes and continuous incremental improvements. The combined effect has been dramatic. Since the 19th century the efficiencies of both

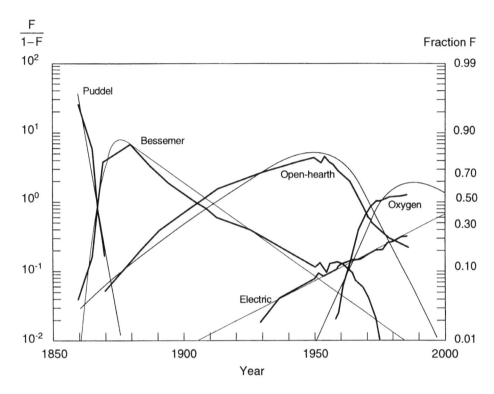

Figure 6.4: Substitution of process technologies in global steel manufacturing, in fractional shares (F) of total raw steel tonnage produced, logit transformation, i.e., log(F/1-F). Jagged lines represent historical data and smooth lines are model estimates based on a logistic substitution model. Straight lines indicate that the percentage shares of each process technology follow a logistic market share trajectory. Source: Grübler (1987). For the data of this graphic see the Appendix.

carbon (coal) use and energy use per ton of steel have improved by at least a factor of 10.[6]

The changing geography of steel production is shown in *Figure 6.5*. The scenario is similar to that for other dominant technologies of the "heavy engineering" technology cluster, such as railways. England, the innovation center, quickly loses market share first to other industrializing countries of Western Europe and then to the USA. This reflects the overall shift in industrial

---

[6]Relative to the start of industrialization improvements are even higher. Ayres (1989a:21) estimates that some 14 tons of carbon were required to manufacture 1 ton of iron prior to 1800. The current US average is about 500 kg carbon, nearly a factor of 30 lower. The theoretical minimum is 161 kg carbon for ore that is mostly hematite Fe_2O_3 (Elliott, 1991:380–381).

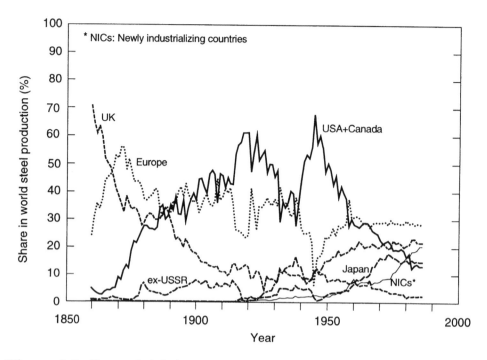

Figure 6.5: Share of global principal steel-producing regions in total crude steel output. Source: Grübler (1990b:124).

leadership. By the 1920s the USA emerged as the world's largest industrial power, accounting for 40% of global manufacturing output (Bairoch, 1982) and 60% of world steel production (Grübler, 1990b:123–125). Only after the 1950s do the former USSR, Japan, and the newly industrializing countries of Asia such as South Korea (NICs in *Figure 6.5*) expand their production and thus their share of global steel tonnage.

The overall diffusion of the "heavy engineering" technology cluster was not confined to process technologies, new materials, and steam power. It was above all a transport "revolution" (steam railways and ships) and a communications revolution, illustrated specifically by the telegraph. Railways and steam ships drew even the most remote continent into the vortex of international trade, dominated by the industrialized core countries. Facilitated by the universal adoption of the gold standard, free world trade grew exponentially. Its political counterpart was imperialism and colonialism. The industrial periphery provided expanding markets for the products of the industrialized core while supplying raw materials and food. Long-distance trade in food had become possible with the invention of canned food and refrigeration. Throughout, trade flows in industrial products remained dominated by trade among the industrialized core countries and with the industrializing "rim", i.e., Russia and Japan.

Over the course of the "heavy engineering" period the pace of technological change accelerated with the emergence of oil, petrochemicals, synthetics, radio, telephone and, above all, electricity. The institutional and regulatory picture was less progressive. Emerging industrial giants, monopolies, and oligopolies – best symbolized by Rockefeller's Standard Oil Company – drew the attention of government regulators. Social issues were only beginning to be tackled. Legislation was slowly introduced (and even more slowly implemented) to limit child labor, provide elementary health care, and shorten working days that could run up to 16 hours. Dissatisfaction with the prevailing ethos of capitalistic accumulation stimulated alternative theories, such as Marxism, and new social movements, such as the labor movement and trade unionization, that aimed at more equitable distributions of productivity gains.

High investments sustained the continuing diffusion of new production methods and technologies that in turn led to increasing returns to scale and significant productivity gains. Because the distribution of income favored entrepreneurs, demand generation was investment (profit) driven. Labor profited mainly in the form of increased employment, and, to a lesser extent, rising wages coupled with falling real prices for food and manufactured products.

There was a widening mismatch between industrial growth and the ability of institutions to provide an equitable distribution of the benefits from productivity gains. The resulting conflicts later began to be resolved by progressively internalizing labor and social welfare costs into the economics of industrial growth. This is characteristic of the social welfare state of the mass production/consumption cluster that emerged in the 1920s and was more fully developed after World War II. We suggest that much of the present discussion about internalizing environmental costs could find useful analogies in the institutional solutions that have been devised to take more fully into account the social externalities that have come with industrialization.

6.3.4. Mass production/consumption (ca. 1920–present)

The unprecedented post-World War II growth rates in industrial and overall economic output are based on a cluster of interrelated technical and managerial innovations that have produced productivity levels clearly superior to those of the heavy engineering period. In particular the extension of continuous flow concepts from the chemical industry to the mass production

Table 6.4: Growth of mass production/consumption, 1950–1990.

	1950	1990	Factor increase
Objects in use			
Merchant ships (10^6 dwt)	93	424	4.6
Motor vehicles (10^6)	70	439	6.3
Telephones (10^6)	70	526	7.5
Radios (10^6)	226	1,966	8.7
TVs (10^6)	45	826	18.4
Annual production			
New book titles (10^3)	<200	842	>4.0
Raw steel (10^6t)	188	773	4.1
Paper (10^6t)	50	270	5.4
Fertilizer (10^6t)	13	138	10.6
Activity per year			
International tourists (10^6)	28	456	16.3
Air passengers (10^6)	23	1,027	44.1

Abbreviations: dwt, deadweight tons; t, ton.
Source: UN *Statistical Yearbook* (various volumes, 1950–1995).

of identical units created real-term cost and price reductions that enabled mass consumption. Products typical of this technology cluster include the internal combustion engine and the automobile, petrochemicals and plastics, farm machinery and fertilizers, and consumer durables. Petroleum has played a particularly vital role both as a principal energy carrier and an important feedstock in the industrial, residential, and especially transport sectors (cf. Section 6.7).

Technological Change

The mass production/consumption technology cluster is characterized by three factors: (i) an unprecedented increase in the scale of production and consumption; (ii) an unprecedented number of new "designed" substances in the form of new materials (such as plastics), drugs, and chemicals; and (iii) unprecedented numbers of new artifacts (ranging from the useful to the superfluous) produced for consumers. As we discuss below, all of these generate wastes, and we remain largely ignorant of their long-term environmental implications. *Table 6.4* lists the growth of selected commodities and consumer products and services characteristic of this technology cluster.

   The number of new materials and substances introduced over the last 50 years is endless. Plastics, composite materials, pesticides, drugs and

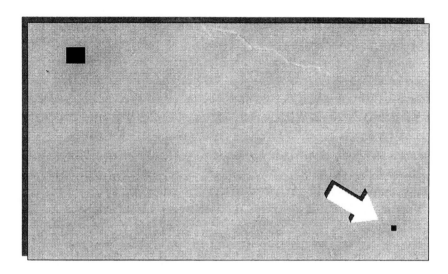

Figure 6.6: Relative proportions of chemicals known (5 million, total gray area) to those tested for carcinogenicity (7,000, black rectangle) and definitively related to human cancer (30, small black rectangle in right hand corner). Source: adapted from Adams (1995:46).

vaccines, and nuclear isotopes (and wastes) are just a few of the major ones. The properties, functions, and services these new products provide are indeed spectacular. Penicillin and antibiotics have basically wiped out a large number of infectious diseases and significantly increased life expectancy, particularly among the young. Isotopes, and later tomographics, have revolutionized medical diagnostics. Plastic containers are cheap and ubiquitous. New packaging has improved hygiene and food preservation. New metal alloys and ceramics that withstand unprecedented pressures and temperatures enable improvements ranging from ever higher energy efficiencies in turbines to space shuttle reentry.

Yet for all our progress, we remain frustratingly ignorant of both the positive and negative properties of all these new manufactured substances. Moreover, we seem to discover key properties "by accident" rather than through research and knowledge. Perhaps this is due to a systematic bias in (re)search procedures that always evaluate the new using yardsticks created for the old (Schelling, 1996). But it is also likely that we are simply introducing new substances faster than we can discover their properties, especially their long-term impacts on the environment and human health.

Consider the following example (*Figure 6.6*). The US National Research Council (1983) estimates that about five million known chemical substances exist whose safety theoretically falls under regulatory restrictions. Of these fewer than 30 have been directly linked to cancer in humans, 1,500 have been found to be carcinogenic in tests on animals, and about 7,000 have been tested for carcinogenicity. Adams (1995:45) calls this situation the "darkness of ignorance" *Figure 6.6*. The suggestion is not that millions of chemical substances will turn out to be carcinogenic, or that one can infer that the 1,500 substances that were carcinogenic in tests on animals will also prove to be so on humans. The calculation simply illustrates how little we know, or conversely, how many manufactured substances still remain outside the "assessment paradigm" of a systematic assessment with respect to their ultimate impacts on human health and the environment.

This veil of ignorance is lifted in most cases "accidentally", i.e., only *after* negative impacts become apparent. We mentioned the example of DDT in Chapter 5. A second example is that of quicksilver, or mercury, a heavy metal. The danger of mercury releases that subsequently accumulate in the food chain became clear only after numerous cases of mercury poisoning (e.g., the Minamata disease in Japan). In some instances negligence or straightforward misbehavior is to blame rather than ignorance. But in many instances the undesirable impacts of new substances are simply unknown or impossible to anticipate. It is only later with the advance of scientific knowledge that we really begin to understand them.

A good example is that of chlorofluorocarbons (CFCs), invented in the 1920s, and halons, introduced after World War II. Before CFCs, the principal substances used in refrigeration cycles were ammonia, methyl chloride, and sulfur dioxide. Butane gas was the main propellant in spray cans. All had serious drawbacks. The traditional refrigerants were all highly reactive, noxious and toxic, leading to corrosion, frequent leaks, and serious health risks. Butane is highly flammable and caused numerous accidents.

Researcher Thomas Midgley Jr. of General Motors developed the first CFC, dichlorofluoromethane (Freon 12) in 1930. (Midgley is equally well known for having invented the gasoline antiknock additive tetraethyllead in 1921, cf. Ayres and Ezekoye, 1991.) Freon 12 laid the foundation for the commercial success of General Motors' Frigidaire Division.[7] CFCs offered significant advantages over traditional refrigerants. They were chemically inert, nontoxic, and nonflammable. Their discovery created opportunities for new products with entirely new properties and applications. A principal new use was in fire extinguishers that provided a compact new product for combatting fire without water. Given their advantages, the use (and production) of these substances grew significantly (*Figure 6.7*). At their peak in the mid-1970s, about 750,000 tons of CFC-11 and CFC-12 were released

[7]See Friedlander (1989:168–169) for biographical information on Thomas Midgley. The name Frigidaire became synonymous with refrigerator in many countries.

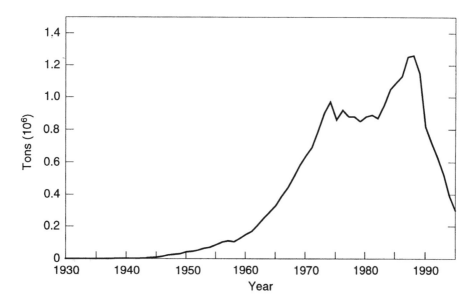

Figure 6.7: Trends in global emissions (million tons) of CFCs (CFC–11, CFC–12, CFC–113, 114, and 115). Data source: Boden *et al.* (1994: 474–476), and WRI (1997b: CFC data series).

into the atmosphere annually. Most releases were from applications using CFCs as propellants or as blowing agents in the manufacture of insulating foam materials. About 100,000 tons were released from leaking refrigerators.

It was only many decades after the successful introduction of CFCs that their disadvantages were discovered. Precisely the property that made them so attractive, their chemical inertness, made them very problematic. Long-lived CFCs changed chemical reactions in the stratosphere in a way that reduced stratospheric ozone concentrations, and even caused a seasonal "ozone hole" over Antarctica.[8]

The basic scientific understanding of stratospheric chemistry and the possibility that CFCs could destroy stratospheric ozone[9] emerged only in the early 1970s and was initially controversial. Lovelock's invention of the electron capture detector for gas chromatography in 1970 provided the first capability for measuring CFCs in the atmosphere in parts per trillion (10^{12})

---

[8]Note that stratospheric ozone depletion is not to be confused with increasing ozone concentrations at lower altitudes. Tropospheric ozone concentrations are increasing due to motorized traffic and other emissions. Policy debates therefore often distinguish between "good" (stratospheric) ozone and "bad" (tropospheric) ozone.

[9]For a concise overview see Crutzen and Graedel (1986); for a historical overview of the policy debate see Friedlander (1989). For a recent review of ozone protection in the USA see Cook (1996).

(Glas, 1989:137) and confirmed that CFCs were accumulating (Lovelock, 1971). Stratospheric ozone depletion was later confirmed through ground based and satellite measurements. These findings led to key international agreements, in particular the Vienna Convention for the Protection of the Ozone Layer (1985) and the ensuing Montreal Protocol (1987), that, together with the availability of substitutes, have resulted in traditional CFCs and halons being rapidly phased out (see the dramatic decline since the end of the 1980s in *Figure 6.7*).[10]

However, past emissions have left an environmental legacy that will remain in the stratosphere for many decades to come. This illustrates one of the difficulties of counteracting global environmental change: even if technological change can be implemented rapidly, environmental changes are often not quickly reversible. CFCs also illustrate a more fundamental dilemma of technological development. A new solution is proposed that offers numerous advantages. Possible negative impacts are simply not fully known (or not known at all). These are only revealed subsequently through advances in scientific knowledge and further technological developments, in this case measurement, instrumentation, and remote sensing technologies, specifically satellites.

Institutional/organizational Change

The "mass production/consumption" technology cluster of the last 50 years, and its tremendous expansion in production, is not simply the result of "more technology". The industrial system of these 50 years required substantial organizational and institutional changes. These took time to implement, but have been essential for the large productivity gains associated with the cluster's new technologies. The prototypical organizational change for this cluster is the Fordist type of assembly line, complemented by a separation of management from production along the principles of Taylor's scientific management. Additional economies of scale also came from increasing vertical integration of industrial activities and the emergence of multinationals operating on a global scale.

Infrastructural developments in transport and communication systems have also been critically important. Railways have been replaced by roads and the internal combustion engine, either in cars in the case of market

---

[10] Arguably, the limited number of manufacturers, and the fact that they had already developed substitutes (largely in anticipation of forthcoming regulations), helped speed the negotiation and the CFC phase-out. For an industry perspective on CFC and ozone depletion, see cf. Glas (1989:137–155). For an overview of ozone diplomacy see Victor and Salt (1995:18–22).

economies or in buses in former centrally planned economies and the developing world. Air transportation and global communication networks (the telephone, radio, TV, and more recently the internet) have not only effectively reduced physical distances but also enhanced cultural and informational exchange. Science has grown "big" (de Solla-Price, 1963) and has been systematically integrated into industrial activities, from industrial R&D to quality control and even consumer research.

There are many examples of the mutual reinforcement existing between social and institutional developments and the new technologies that were largely material intensive, energy intensive, and oil based. Keynesian policies generally subsumed under the concept of the welfare state led to various forms of demand management. Some were direct, through increasing public demands via infrastructural investments in roads and highways, in defense, and in public services, such as health care and higher education. Some were indirect, such as income redistribution policies that provided more disposable income for mass consumption. Other important socioinstitutional innovations included large-scale consumer credit, new forms of public relations and advertising, the development of mass communication, and the inclusion of labour unions and others in various forms of *Sozialpartnerschaft* for building a social consensus on economic objectives and policy. It was the combination of such institutional developments with new technologies of the period that made possible the explosive economic expansion after World War II. Neither the institutional developments nor the new technologies alone could have fueled such growth.

Today it appears that the broad social consensus represented by such institutions is weakening and a growing mismatch is developing between this socioinstitutional framework and newly encountered limits confronting market expansion, environmental impacts, and overall social acceptance (Perez, 1983).[11] Currently changing social values, new technologies and growth sectors, new ways of organizing production, shifting occupational profiles, and shifting international cost advantages all suggest a need for structural and institutional adjustments.

To blame everything on intensified, cut-throat international competition, as done frequently in the current "globalization" debate (e.g. Martin and Schumann, 1996) may be overly simplistic. In our view, the real challenge ahead is to find new institutional and organizational configurations to ease

---

[11]Boyer (1988a,b) argues that once the Fordist/Tayloristic paradigm has been adopted everywhere, it cannot contribute further to productivity growth. Today's productivity slowdown is thus because it is impossible to "deepen" the Fordist organizational model. New solutions, such as the Japanese total quality control (TQC) model, are not yet embedded within existing industrial relations structures.

the transition away from the saturating "mass production/consumption" cluster. The current focus on restructuring the former centrally planned economies should not blur the need for similar far-reaching social and institutional "perestroikas" in the industrialized core countries. Faced with continued unemployment and environmental constraints, and with the obvious limits of traditional "end-of-the-pipe" regulatory approaches, our human ingenuity is challenged to devise new technological, organizational, and institutional innovations for sustainable growth. Obviously, the increasing short-term focus of both industry (on quarterly profits) and governments (on short-term macroeconomic targets, e.g., such as those associated with the European monetary union) and the resulting increasing neglect of long-term strategic goals (R&D, innovation, institutional reforms) perpetuates rather than alleviates the current "crisis" of industrialization.

Required changes, however, will not happen overnight. To choose an example of a technological transition that was nearing completion at the beginning of the mass production/consumption technology cluster, consider the electrification[12] of the shopfloor (*Figure 6.8*). The first use of electricity in the USA for motive power in factories dates back to 1884 (Devine, 1983:349). However, it took 50 years for electric motors to replace steam engines, which earlier had taken 50 years to replace water power (Auer *et al.*, 1983:30). The shift from steam to electricity was quite different from that of water power to steam.

The shift from water to steam required no fundamental change in factory design or organization. Both were centralized power sources from which power was distributed through a system of shafts and transmission belts to each individual machine tool. Power transmission was controlled, though with great difficulty, through numerous clutches and pulleys that both determined the factory layout and created significant risks to workers on the shopfloor. Plant design was dominated by the need to organize power transmission around a centralized shaft. Machine groups were frequently "stacked" on top of each other in the form of multistoried factory buildings that were expensive to build and resulted in production being organized suboptimally. Design elements that simplified power distribution complicated the supply of raw materials, the transfer of intermediate products between machines, storage, and delivery.

In their first applications in factories, electric motors were simply used in place of the centralized steam engines, and the power distribution systems remained unchanged. Subsequently a "compromise" design emerged, in which several centralized electric motors were used instead of just one.

---

[12]For an excellent historical account of electrification see Hughes (1983).

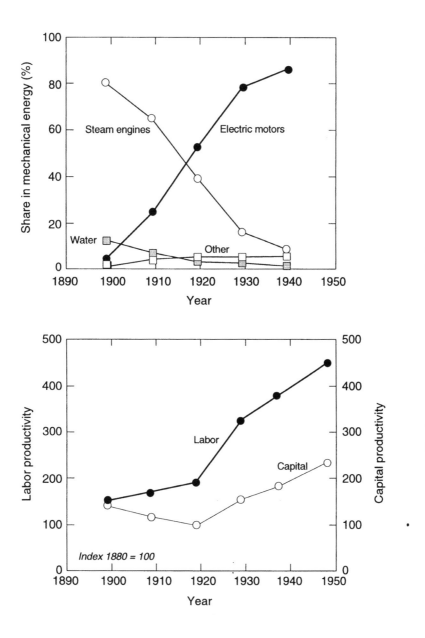

Figure 6.8: Diffusion of electrical drives (percent of installed horsepower, top) and labor and capital productivity (index 1880 = 100, bottom) in US manufacturing, 1890–1950. Source: adapted from Auer *et al.* (1983:30) and Devine (1983:349).

Neither approach required any drastic reorganization of the shopfloor. As a consequence, productivity increases were rather modest. The productivity turnaround finally came in the early 1920s (cf. *Figure 6.8*) when centralized power transmission was abandoned altogether in favor of the electric unit drive. Each machine now had its own electric motor and that meant, first, that it could be located so as to optimize production rather than simplify power transmission. Second, power was only used when it was needed. (For a more detailed discussion and quotations from contemporary engineers, see Devine, 1982, 1983.)

Devine (1982) estimates the overall energy efficiency of a steam engine coupled with mechanical power distribution to range from 3% to 8%. If the steam engine is simply replaced by electricity, and the mechanical power distribution system is unchanged, overall energy efficiency remains at 3–6%. But combining utility-generated electricity and decentralized unit drives raises the energy efficiency by a factor of up to three, bringing the overall system efficiency to 10–12%. All these efficiencies apply to the 1920s. Current overall system energy efficiency for industrial drives (including power generation, distribution, and motors) is on the order of 25–28%, i.e., twice as large as 70 years ago (Nakićenović *et al.*, 1990).

Reorganizing factory layout and production processes took much longer than the basic penetration of industrial electrification. This explains why noticeable productivity increases only became apparent in the 1920s, when electricity already supplied more than half the mechanical power in US manufacturing (*Figure 6.8*). The same was true for energy efficiency, which only increased significantly with electric unit drives. Overall industrial electrification took more than 50 years to fully diffuse, and noticeable impacts on labor, capital, and energy productivity came only three decades after the first introduction of electricity in industry, and after half of the industrial motive power was already supplied by electric motors.

David (1990) suggests a cautious analogy between electrification and current computerization. Despite recent heavy investments in computers and new information technologies, productivity has not noticeably increased. Indeed, after the mid-1970s, productivity growth slowed considerably. This "productivity paradox" is succinctly captured in Robert Solow's observation that "We see the computer everywhere but in the productivity statistics" (David, 1990:355). The expanding use of computers is much more subtle than the earlier expansion of industrial electrification, and it is very difficult to measure productivity increases in information processing. However, it is not unreasonable to argue that the measurable impact of computers on productivity may lag their original introduction by something on the order of the three decades it took productivity increases due to industrial electrification to show up. The reconfigurations that will eventually generate productivity increases go beyond "simple" technological or economic issues. They involve organizational and ultimately mental "reconfigurations" that take considerable time to evolve.

6.4.   Socioeconomic Impacts of Industrialization

Industrialization has far-reaching socioeconomic impacts. We have already mentioned its impact on employment patterns, on where people live, and on differential economic growth in different countries. Here we focus on three additional aspects: (i) labor productivity increases; (ii) income increases; and (iii) impacts on working time. Some of these will be revisited in Chapter 7's discussion of the rise of the service economy and the profound shift in time budgets in industrialized countries from formalized work contracts to informal work and free time.

Increased life expectancy, rising incomes, and reductions in working time are all social changes that result directly or indirectly from industrial output and productivity growth. Combined, they result both in direct and indirect global change impacts. For instance, higher labor productivity enables expansion of production and output, resulting in increases in resource use (direct global change impact). Higher productivity enables rising incomes and reductions in working time. In turn, higher incomes lead to more consumption (and wastes generated) and, combined with more free time, also induce more travel (and hence increased energy use and emissions). These latter type of impacts are examples of "indirect" global change impacts that went along with industrialization and its tremendous productivity gains.

Since the beginning of industrialization, growth in industrial labor productivity has outpaced the growth in overall industrial output (that, as discussed above, grew by about a factor of 100.) Data are uncertain, but the best evidence available today remains consistent with the picture presented in Colin Clark's classic work, *The Conditions of Economic Progress* (1940). Time series indicate growth in industrial labor productivity by a factor of at least 200 since the middle of the 18th century. Thus, an industrial worker in the USA today produces in one hour what took an English laborer two weeks of toiling 12 hours per day some 200 years ago.

Figure 6.9 estimates labor productivity growth in manufacturing for a number of industrialized countries. The international comparison of industrial and manufacturing labor productivity is one of the most complex tasks for comparative economic statistics. Differences in the industrial output mix, relative price structure, labor qualification, industrial relations,

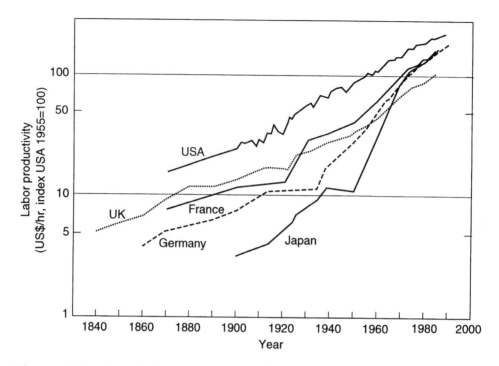

Figure 6.9: Growth in manufacturing labor productivity (US$ per hour worked), ratio scale. Source: Grübler (1995a:46).

hours worked, and so forth, still await definitive methodological and empirical resolutions. Therefore, *Figure 6.9* serves more to illustrate the evolution of labor productivity within a given country, rather than to provide an accurate comparison among countries.[13]

Figure 6.9 illustrates that for "older" industrial countries like the UK and later on the USA, labor productivity (measured in the dollar value generated per person hour worked in industry) has risen on average between 2.5% and 2.8% per year. Thus it has increased by a factor of 10 in less than 90 years. For "late" industrializers like Japan labor productivity growth has been even more spectacular. It has averaged close to 10% per year, thereby increasing by a factor of 10 in about a quarter of a century. Nevertheless,

---

[13]We have renormalized the individual country indexes to be roughly equivalent with the prevailing consensus on comparative international manufacturing productivity, e.g., the estimates of Dosi *et al.* (1990) for the year 1977/1978, and the overview of estimates by Broadberry and Crafts (1990) for the year 1985. From these we have adopted the median between industry of origin and expenditure-based estimates. For the early 1980s this yields roughly the ratios 2, 1.5, and 1 between manufacturing productivity per hour worked in the USA; Japan, Germany, France; and the UK, respectively. For a historical account of industrial labor productivity see Phelps Brown (1973).

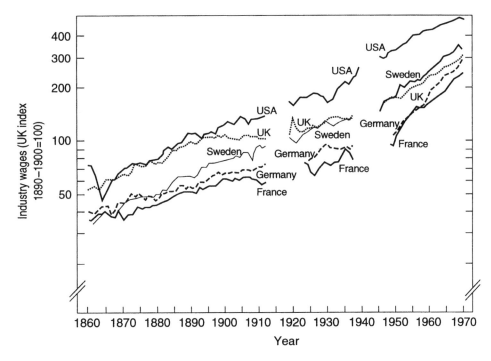

Figure 6.10: Long-term trend in real wages in industry (UK index 1890–1900 = 100). Source: Phelps Brown (1973:59).

there still remain persistent differences in manufacturing labor productivity even among the industrialized countries. Apparently, distinct national industrial systems and associated institutional settings have evolved, with different sectoral structures, technology bases, working time regulations, wage negotiation patterns, and so forth. These differences are cumulative and create distinct national paths of industrialization that persist despite intense international trade and competition.

Some of the historical differences relate to the relative availability of various factor inputs in industry. As we saw in the case of agriculture, labor was comparatively scarce for US industry. Consequently, industrial labor productivity was higher in the USA than in the UK even when the USA was still a newly industrializing country and the UK was the world's leading industrial power.

The benefits of the labor productivity gains shown in *Figure 6.9* have been distributed among rising wages and incomes (next to output growth and reductions in prices) on the one hand (*Figure 6.10*), and reduced working hours on the other (*Figure 6.11*). From the beginning of this century to the mid-1980s, for example, weekly work hours for an average manufacturing

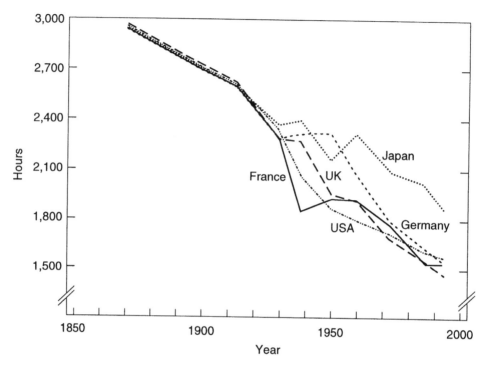

Figure 6.11: Average annual hours worked in selected countries. Source: Ausubel and Grübler (1995:202) based on Maddison (1991).

employee in the USA dropped from 53 to 41, hourly pay increased from US$1.70 to US$9.20 per hour, and the annual salary increased from less than US$5,000 to some US$20,000 (Starr, 1990:11).[14] Again national experiences are diverse, but the overall direction is consistent: rising wages and incomes, and reduced working hours.

Wage increases resulting from rising labor productivity are necessary to balance the equation of mass production with mass consumption. The implications of increased productivity for working time are less well documented (for a discussion of measurement issues, data sources, and implications see Ausubel and Grübler, 1995). During the last 100 years in all industrialized countries they are overlaid by significant growth in the population's available time budget as a result of life expectancy increases. These increased from 35 to 45 years in 1870 to about 70 years for males and more than 75 years for females today (Flora *et al.*, 1987:92–95), primarily as a result of reductions in mortality of infants and young (cf. Chapter 7). Over the same period, annual hours worked have, on average, dropped by half (*Figure 6.11*). This has

---

[14]US$ in this book refers to constant 1990 money and prices, unless otherwise stated.

resulted in more time available either for other socially obligatory activities like child care or for leisure activities.

The biggest gains in available time have accrued to children (many more survive now) and to the elderly, i.e., after retirement. Without the substantial productivity increases of the last 100 years, none of the social security and pension systems that are such important social achievements in the industrialized countries could have been financed. Their future financing will continue to depend even more on productivity increases in view of the rapidly aging populations in most OECD countries.

Overall, the social and economic gains from rising productivity are impressive and widespread, despite difficulties and disparities in their distribution. They are a compelling consequence of continued technological change. The result is a level of health and longevity, material well being, and leisure time in the industrialized countries that is beyond the imagination of even the most daring social utopias of the 19th century.

6.5. Environmental Impacts of Industrialization

Industrialization has had unprecedented environmental impacts stemming from effluents, new and old, coupled with the expanding scale of industrial activities. First, industrialization has intensified environmental impacts that were already concerns prior to industrialization, such as deforestation, land disturbances from mining, and local air pollution from burning coal.[15] It is important to stress that such impacts do not grow in direct proportion to the growth in industrial activity. Most environmental impacts are inherently nonlinear, and industrial productivity increases can provide a powerful check on rising environmental impacts. This latter point is discussed more fully in the sections below on "dematerialization" and "decarbonization".

Second, industrialization has also created entirely new environmental impacts by virtue of introducing new materials such as DDT and CFCs. These bring with them hitherto unknown impacts on the environment, such as the biotic accumulation of pesticides and changes in stratospheric chemistry.

However, while industrialization facilitated by technological change has had unprecedented adverse environmental impacts, it has also substantially improved selected environments owing to increased productivity and incomes. Industrial societies have a tremendous technological and economic

---

[15] Cf. George Perkins Marsh's 1864 classic *Man and Nature*; or Brimblecombe (1987). For a particularly grim perspective on the environmental future of the Victorian coal-based industrial economy see John Ruskin's (1884) *Storm Cloud of the 19th Century*.

capacity for environmental remedies. Urban air and water quality in industrialized countries is now much better than at the beginning of the industrial age. The result is longer life expectancy and the disappearance of infectious diseases like typhoid or cholera. Indeed, only an affluent society – elevated beyond the day-to-day struggle for survival, food, and shelter – can reasonably be expected to have the scientific and economic capabilities to understand its impact on the environment and be concerned about the well being of future generations, of "nature", and of "the planet" itself. Technology's rapid evolution argues for a departure from traditional anthropocentric "world views". As summarized by Meyer-Abich (1996:232), "we will need science and technology to treat problems that we would not have without science and technology". Note, however, the inherent difficulty of a circular argument for establishing appropriate *ceteris paribus* conditions. We simply would not be here, nearly 6 billion of us, without all the advances of science and technology, starting with the neolithic toolmaking revolution and continuing to the Industrial Revolution. Meyer-Abich (1996:232) argues further that "we need to diffuse a new understanding of nature, including our own nature, in order to drive our science". His arguments reinforce those of Lynn White (1967), who emphasized that western civilization's anthropocentric view of nature is deeply rooted in Christianity and the origins of western science and technology in the Middle Ages, i.e., significantly prior to the onset of the Industrial Revolution.

The central theme of this section is that industry has an inherent incentive structure to minimize factor inputs, and technological change provides mechanisms for doing so. Therefore, existing incentives and opportunities foster movement in the right direction, although historically this has not been fast enough to offset the continuing global expansion of industrial activities. Moving in "the right direction" means two things: (i) minimizing resource inputs per unit of economic activity, "dematerialization"; and (ii) improving the environmental compatibility of the materials used, processed, and delivered by industry. (For the energy sector, such improvement is summarized under the rubric "decarbonization" below.)

How best to make progress on both fronts is the subject of the rapidly evolving field of industrial ecology (Frosch, 1992) and of industrial "metabolism" (Ayres, 1989c).[16]

---

[16]The best introductory text on improving manufacturing's environmental impacts remains Frosch and Gallopoulos (1989), which is part of an eminently readable special issue of *Scientific American*. For a recent comprehensive overview see Socolow *et al.* (1994).

6.6. Industrial Metabolism and Dematerialization Strategies

6.6.1. Industrial metabolism

Worldwide, industrial activities mobilize a gigantic amount of materials. In 1990, for example, close to 10 billion tons of coal, oil, and gas were mined as fuel (BP, 1996:6–32); more than 5 billion tons of mineral ores were extracted (Ayres and Ayres, 1996:2); and over 5 billion tons of renewable materials were produced for food, fiber, fuel, and structural materials (FAO, 1991). Actual material flows were even larger, because all the materials included in the above estimates had to be extracted (overburden and waste rock must be removed), processed (with resulting wastes), transformed and upgraded (with inevitable conversion losses), converted to final goods (manufacturing wastes), and finally, disposed of after final usage (consumer wastes). Because of varying geology, conversion processes, and technological efficiencies, the ratio of the final material in a product to the total amount of material mobilized, handled, and processed varies enormously.

Consider the following examples. Refining crude oil into petroleum products (fuels, feedstocks, and lubricants) is perhaps one of the most material-efficient processes in industry. It has energetic and material conversion efficiencies well above 90%. Its materials mobilization ratio (MMR), defined as the ratio of primary to final material, is close to one. Alternatively, producing 1 ton of lignite in Germany's gigantic open cast mines, which is roughly equivalent to 0.3 tons of hard coal, requires the removal and disposal of more than 10 tons of overburden and 10 tons of water (Grenon, 1979:379–396). That is an MMR of more than 10 if water is excluded, and more than 20 if it is included.

In agriculture a measure that corresponds to the MMR is the feed production ratio (FPR), defined as the amount of feed required per unit of food produced. The FPR is particularly unfavorable for meat production, as much of the required feed supports the animals' basic metabolic functions rather than adds to net weight, i.e., meat. The FPR for beef is about 12. It is 4–5 for pork, poultry, and eggs, and between 0.7 and 1.2 for milk (Leach, 1995:29). This emphasizes the importance of diet in calculations of the calories (and land) needed for growing human populations. If everyone were a vegetarian, thereby eliminating conversion "losses" in meat products, even the most pessimistic estimates of future agricultural productivity coupled with a doubled global population would not lead to global food shortages. (This does not mean hunger will disappear, as hunger is less a consequence

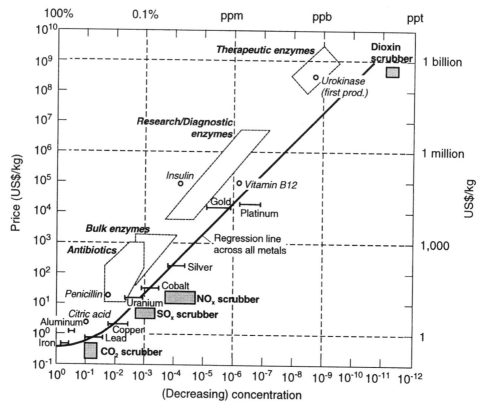

Figure 6.12: Price of selected materials (US$ per kg) as a function of the decreasing concentration of the final product in the initial raw material. Note in particular the double logarithmic scale extending over 12 orders of magnitude. Source: adapted from Dwyer (1984:957), CSST (1987:22), Holland and Petersen (1995:299), Akutsu (1996), and Tikovsky (1996).

of inadequate food production than a consequence of inadequate food *distribution* due to the inability of the poor to purchase food that is available.)

An extreme MMR example is the thoroughly researched case of gold. Ayres and Ayres (1996:2) estimate that for each ton of gold, some 150,000 tons of ore have to be mined and the wastes disposed. For even more precious materials like drugs and medicines, the MMR can approach 1 billion (CSST, 1987:22).

The MMR is also a good proxy for the amount of effort, energy requirements, other resource use,[17] and, consequently, the costs associated with materials production. *Figure 6.12* shows the price of different materials as a

---

[17]For a methodology of resource use accounting see Grenon and Lapillonne (1976).

function of the decreasing concentration of the end-product in the initial raw material, which ranges in the figure from mineral ores to biological broths and polluted stack gases. Also shown in the figure is a regression analysis by Holland and Petersen (1995:299) for current metal prices as a function of the metal concentration in the original mineral ore. The main conclusion is the upward trend for both prices and waste materials as less and less concentrated mineral ores are mined. The same conclusion holds for biotechnology materials. In *Figure 6.12* their costs are consistently above those for metals because of their purity requirements and enormous product differentiation. The figure also shows the costs of extracting various pollutants from stack gases. These also rise exponentially with increasing dilution.[18] Thus, it is much more economical to capture effluents as close to the source as possible before they become diluted in the environment.

All these examples illustrate how the total material flows associated with industrial production, especially minerals, are much larger than just the amount shown in industrial output statistics. Globally, metal production generates 13 billion tons per year of waste materials (Argawal, 1991:389) in the form of waste rock, overburden, and processing wastes. Similar estimates are unavailable for energy materials, but are likely to be at least one order of magnitude larger. Most of these materials do not pose environmental problems. The overburden, waste rock, or water that have to be "mined" and subsequently disposed of are generally not toxic or hazardous. But they do significantly disturb the land, cause infrastructures and settlements to be relocated, and substantially alter surface water and groundwater flows. Their long-term impacts can be remedied through land reclamation and water management, but *ceteris paribus* the extent of environmental impacts remains a function of the MMR.

Metals and hydrocarbons (see Section 6.7) are abundant in the earth's crust. Accessibility and concentration determine if a particular deposit is potentially minable. Neither is a fixed quantity. Both are rather functions of available technologies and prices (see also *Box 6.5*).

Prospecting and exploration efforts determine the amount of deposits identified as recoverable reserves. Prices and available technologies in turn determine what concentration levels can technically and economically be

---

[18]Data for biologicals are from Dwyer (1984:957) and CSST (1987:22), and apply to the early 1980s. Since then, advances in transgenic biotechnology have lowered dilution rates and the costs of the enzyme urokinase by as much as six orders of magnitude (BioPharm, 1994:14). Representative data for the scrubbing of carbon, sulfur, and nitrogen oxides from stack gases of electric power plants are based on an analysis of Japanese power plants (Akutsu, 1996). The estimates for dioxin apply to the hazardous waste incineration facility in Vienna, where investments of US$50 million were required to remove about one gram per year from the flue gases of the incinerator (Tikovsky, 1996).

exploited. Both change over time, and historically, advances in mining and recovery technology have been sufficiently large to increase reserves and compensate for the decreasing concentrations of minerals that are mined, keeping prices low (Barnett and Morse, 1967). Should similar advances not materialize in the future, costs and prices will rise and most likely trigger material conservation, recycling, and substitution efforts.

Thus resource availability is not simply a geological issue despite the ultimate finiteness of Earth's resources. It depends on the effort ("price" in the language of economics) that society is prepared to incur for its material supply. Even in the long term (which is well beyond 100 years given currently available reserves and resources), geology may not be the most important constraint. As we turn to progressively less accessible and less concentrated raw materials, environmental impacts from mammoth material handling requirements (and/or stiffening environmental regulation) may prove the ultimate constraining factor rather than geological availability.

Potential developments in biotechnology, such as bacterial leaching of mineral deposits, may break the pattern of an increasingly "heavy" industrial metabolism, but such breakthroughs are not in the immediate future. In the meantime, it is important to better understand long-term trends in the material intensiveness of economic activities. The best place to start is the USA. Excellent data are available, and by virtue of the size of its economy, the USA is one of the heaviest material users in the world.

Despite the proliferation of studies of "industrial metabolism" and a number of valuable case studies of individual materials, few studies are available that deal with all materials comprehensively. The best data for the USA (Wernick and Ausubel, 1995) are summarized in *Figure 6.13*. As shown in the figure, the total annual material input to the US economy is an astounding 6 billion tons (Gt) per year, or about 50 kg per day per person. Thus an average American requires about 20 tons of material inputs per year, or about 1,600 tons over an 80-year lifetime. The energy and construction industries are the largest suppliers/users at some 2 Gt per year each, followed by forestry and agriculture with close to 1 Gt combined. The remaining 1 Gt comprises 0.6 Gt of material imports (about two-thirds are crude oil and petroleum products) and less than 0.4 Gt of industrial minerals and metals. Slightly more than 0.2 Gt of materials are recycled annually. In addition, more than 15 Gt of extractive wastes and some 130 Gt of consumptive water use have to be added as "by-products" that are generated even before the major finished material flows enter use. These 6 Gt of materials and 15 Gt of extractive wastes are in agreement with a similar recent estimate of WRI (1997a:59) of 22 Gt total material requirements for the USA in 1994, corresponding to 84 tons per capita per year. Similar estimates for

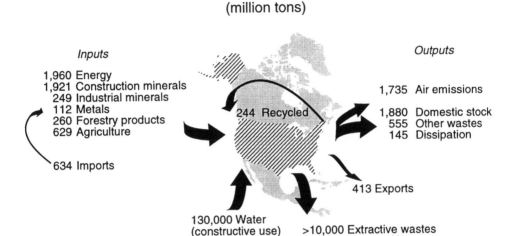

US Annual material flows
(million tons)

Inputs

1,960 Energy
1,921 Construction minerals
  249 Industrial minerals
  112 Metals
  260 Forestry products
  629 Agriculture

  634 Imports

244 Recycled

Outputs

1,735 Air emissions

1,880 Domestic stock
  555 Other wastes
  145 Dissipation

413 Exports

130,000 Water
(constructive use)

>10,000 Extractive wastes
(mostly waste rock)

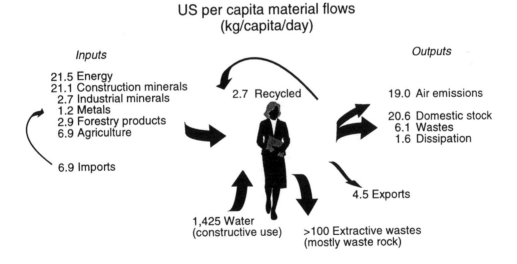

US per capita material flows
(kg/capita/day)

Inputs

21.5 Energy
21.1 Construction minerals
  2.7 Industrial minerals
  1.2 Metals
  2.9 Forestry products
  6.9 Agriculture

  6.9 Imports

2.7 Recycled

Outputs

19.0 Air emissions

20.6 Domestic stock
  6.1 Wastes
  1.6 Dissipation

4.5 Exports

1,425 Water
(constructive use)

>100 Extractive wastes
(mostly waste rock)

Figure 6.13: Major US annual material flows ca. 1990, total (million tons, top) and per capita per day (kilograms, bottom). Source: Wernick and Ausubel (1995:486–487).

other selected industrialized countries indicate comparable orders of magnitude: 76 tons/capita in Germany, 67 tons/capita in the Netherlands, and 45 tons/capita in Japan (WRI, 1997a:23). These numbers also include direct and indirect material requirements associated with imports, which are particularly important for economies outside the USA.

The mole composition of the total materials used in the USA is made up mostly of hydrocarbons (some 87%) and silicon dioxide (9%). Metals, nitrogen, sulfur, etc. account for less than 4% of total materials use (Wernick *et al.*, 1996:174). The preponderance of hydrocarbons in the mix explains the dominance of airborne emissions over wastes and other forms of materials dissipation. The oxidation of carbon to carbon dioxide (CO_2) and of hydrogen to water account for 1.6 Gt of an estimated total of 1.7 Gt airborne emissions. A comparable amount of materials (1.9 Gt annually) are added to the domestic stock of materials in the US economy in the form of long-lived structures (houses, bridges, etc.) or consumer durables such as refrigerators and automobiles. These in turn add to the waste generated in subsequent years. Such postconsumer wastes are estimated at 0.3 Gt per year, to which processing wastes, water and water sludge, and other wastes are added for a total waste stream of some 0.6 Gt per year (excluding the airborne pollutants discussed previously).

Finally, an additional 0.15 Gt of materials are dissipated each year into the environment. These are through applications where no recovery of the material is practical. These materials are lost for potential recycling and thus "consumed" (even though strictly speaking there is no materials consumption as all materials used are transformed, dissipated, or added to the materials stock of a society). Typical examples are fertilizers and road salt. (All the estimates cited are from Wernick and Ausubel, 1995:474–475, which also describe data sources and quality of these numbers.)

Such quantitative descriptions do not give a complete picture of environmental impacts because they leave out important qualitative characteristics of different wastes, most prominently toxicity. Total US dioxin and furan emissions, for example, amount only to one metric ton per year (Wernick and Ausubel, 1995:485), but that ton causes considerable environmental concern. Heavy metal emissions are another example. Global emissions of arsenic were estimated at 78,000 tons in 1980 (Ayres and Ayres, 1996:4), a factor of four higher than releases from natural sources. The enormous expansion of metal production worldwide has led to emissions for copper, lead, and zinc (*Figure 6.14*) that approach or have already surpassed natural fluxes for these metals (cf. Holland and Petersen, 1995:384).

Note that while heavy metal pollution is now at an unprecedented scale, such pollution is not exclusive to the industrial age. In particular, ice cores

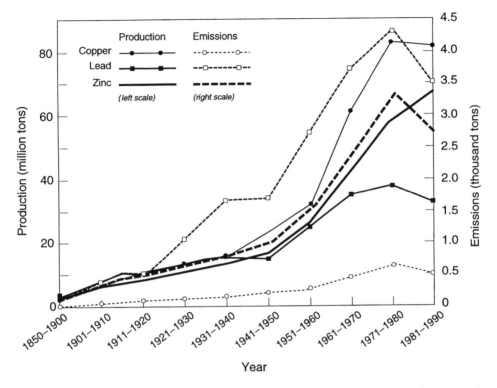

Figure 6.14: Worldwide production (million tons) and emissions (thousand tons) of copper, lead and zinc since 1850. Source: adapted from Nriagu (1996:223).

from Greenland have identified lead deposition from atmospheric pollution exceeding four times natural background levels in the period from 500 B.C. to A.D. 300. The likely cause was widespread pollution of the northern hemisphere from Roman mines and smelters. With the collapse of the Roman empire, lead deposition fell to background levels. Eventually it rose again, reaching Roman levels around 1700. In this century deposition levels in Arctic ice increased to some 100 times natural background levels, but since the 1970s have dropped by a factor of seven due to the phase out of leaded gasoline in North America and Europe (Nriagu, 1996:223–224).

Although in the aggregate heavy metal emissions have increased roughly in line with metal production – at least until the 1970s – regional and sectoral patterns are much more diverse. In highly industrialized areas emissions have already been on the decline for several decades (see *Box 6.1*). Recent studies also suggest that heavy metal contamination is a localized "hot spot" phenomenon (see *Box 6.2*) even in regions where transboundary pollution

Box 6.1: The Heavy Metal Pollution Trend-Shift in Western Europe: The Rhine Example*

During the last few decades the trend of increasing heavy metal pollution has been broken in most OECD countries. The Rhine Basin illustrates this pollution trend-shift, with cadmium as an example. In this region, air and water emissions of cadmium decreased between the mid-1960s and the end of the 1980s by approximately 90%. Emissions of cadmium and most other heavy metals have been dominated by coal-fired power plants, iron and steel and nonferrous metals refining. Reductions of air and water emissions have primarily been related to such industrial point-sources and caused by gradual development of pollution control, more energy and materials efficient processes, changed and sometimes reduced production, and abolishment of bad waste practices, e.g., dumping of sludges into the river. These decreases were connected to a redirection of waste flows from air and water to solid waste. The improved pollution control has been closely linked with the development of legislation on a national and, sometimes, regional level. The EC legislation has had an effect on the cadmium-using industry. It is, however, impossible to find any direct influence from the International Commission for the Protection of the Rhine, which is often praised for its achievements. In competitive industries, such as the chemical industry, improvements have been more important. New and larger plants with modern technology and emissions control have replaced numerous old inefficient, heavily polluting, plants. Despite stable or reduced material volume of production, the production value has increased, indicating a development toward final products of higher price and quality. Less competitive state-owned industries, such as coal-related activities, have not shown such a dynamic development.

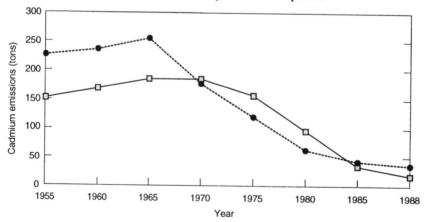

Cadmium emissions to air (circles) and water (boxes) in the Rhine Basin (in tons per year). Based on Anderberg and Stigliani (1994) and Stigliani *et al.* (1994).

*Based on Anderberg (1998).

Stefan Anderberg
Institute of Geography, University of Copenhagen
Copenhagen, Denmark

Box 6.2: Is Atmospheric Heavy Metal Pollution a Large-Scale Transboundary Problem?

During the last few years heavy metals have been launched as a transboundary air pollution problem. The convention on long-range transboundary atmospheric pollutants has recently included the heavy metals cadmium, lead and mercury to its agenda and reduction plans for atmospheric emissions of these elements have been made. This is linked with the concern of large-scale degradation of soils by heavy metal pollution and associated health risks from consumption of food cultivated on heavily contaminated soils.

Since the beginning of industrialization, large amounts of heavy metals have been released into the environment by mining, fossil fuel combustion and various industrial activities. Fertilizers and sewage sludge have added to the heavy metal load to agricultural soils. Alarmingly high concentrations of lead and cadmium have been measured in both air, soil and human blood in major industrial and mining regions. An IIASA study (Prieler and Anderberg, 1996) analyzed the long-term development of heavy metal pollution of agricultural soils in Europe in general and in Central Europe in particular. According to the estimates, 1,000–3,000 tons of cadmium have been emitted annually and deposited in Europe between 1955 and 1994. In the project study area, including large parts of the Czech Republic, Poland, and the former German Democratic Republic, which has been one of the areas of Europe with the highest atmospheric deposition during this period, the average cumulative cadmium load (70–90% from atmospheric deposition) on agricultural soils has been 400 g/ha and the maximum load 900 g/ha. But the resulting accumulation over the 40-year period is relatively moderate: 0.08 mg/kg on an average agricultural field and a maximum in Upper Silesia of 0.2–0.3 mg/kg, compared to background concentrations for unpolluted soils (0.35 mg/kg in Poland). This period must be regarded as quite extreme, and current deposition is already less than one-third of that in the 1960s. Emissions in the future will definitely be lower and not lead to any important large-scale heavy metal accumulation in soils. The most pessimistic scenario results in a maximum increase of 0.1 mg/kg between 1995 and 2050 in extreme areas with intensive agriculture, high pH-values, and high organic matter content.

The IIASA study also concluded that it is impossible to explain the extreme deposition and soil concentrations in "hot-spot" areas such as the Katowice voivodeship (average 3.2 mg Cd/kg) through long-range atmospheric deposition. Emission sources there have primarily been local and the contributions from diffuse emissions connected to mining and industrial activities and coal burning seem to have been badly estimated. Heavy metal pollution in such areas should definitely be of concern because of immediate health risks, but compared to local pollution sources, transboundary atmospheric pollution of heavy metals is not significant.

Stefan Anderberg
Institute of Geography, University of Copenhagen
Copenhagen, Denmark

was previously thought to impact large areas, such as the heavily polluted "black triangle" in Central and Eastern Europe.

For different industries, environmentally harmful wastes are various proportions of total material output. For instance, in 1990 the US chemical industry produced some 90 million tons each of organic and inorganic chemicals, but generated 350 million tons (wet basis) of hazardous wastes, which works out to be more than half the US total (Wernick and Ausubel, 1995:488). Depending on the definition of "hazardous" (itself a formidable challenge), waste estimates for the USA vary from 100 million tons (UNEP, 1993:349), to 180 million tons (OECD, 1993:147), to the 350 million tons mentioned above. A detailed toxic release inventory (TRI) covering 329 chemicals and 23,638 US industrial facilities prepared by the US Environmental Protection Agency (EPA, 1992) indicates total releases of 2.2 million tons of toxic chemicals per year, including close to 1 million tons from chemical industries and some 400,000 tons from metal industries. Proposals to enlarge the inventory by 313 chemicals and to extend the list of reporting facilities beyond manufacturing (to waste treatment companies, laundries, etc.) are being considered (UNEP, 1993:334). Such figures illustrate how difficult it will be to fully understand the environmental impacts of the volume and variety of industrial wastes.

There are several generic strategies for responding to environmental impacts without waiting for a full understanding of all the details. In addition, each aims to prepare for surprises comparable to the surprise that arose from using CFCs. Four such strategies will be discussed below: (i) dematerialization; (ii) material substitution; (iii) recycling; and (iv) "waste mining" (cf. Ayres and Ayres, 1996:15–16).

6.6.2. Dematerialization

Dematerialization is a decrease in the materials used per unit of output. Input can be measured in terms of a specific material used for a particular purpose (e.g., aluminum for beverage cans), or as an aggregate for a particular economic sector, or for the entire economy. Similarly output can be either a specific product, or output aggregates for a specific sector or the whole economy. For the whole economy, useful aggregates are industry value added and GDP.

At the level of individual products perhaps the best example of "dematerialization" is miniaturization in the electronic industries from vacuum tubes to transistors and finally integrated circuits. The first electronic computers filled several rooms. Today their functions can be easily fulfilled by a light and cheap pocket calculator. In construction a similar evolution can be seen in the materials (and mass) required per unit volume as one moves from stone and brick structures, to skyscrapers, to Buckminster Fuller's geodesic domes, which provide for the highest inside volume per unit of material required. In these examples dematerialization is achieved both by radical design changes and technological change, typically in the form of new structural materials. But products can also become lighter without changes in basic design. A good example is the ubiquitous aluminum beverage can. Improvements in aluminum rolling and forming have enabled the production of much thinner cans. As a result, the mass of an aluminum can has dropped 25% since 1973 (Wernick, 1996:117–118).

However, the fact that individual products may become lighter does not necessarily mean an overall reduction in material requirements. After all, as the material requirements for each unit drop, so do costs, and falling prices can lead to increased demand including totally new uses. For example, in the case of building structures, material requirements per unit volume have fallen drastically, and new materials and designs have made possible entirely new sorts of buildings. Indoor artificial ski slopes and ocean beaches have been built in Japan. Thus in addition to looking at dematerialization for individual products, it is important to look at larger aggregates and absolute levels of materials usage.

Figure 6.15 shows economy-wide materials use per unit of GDP, and *Figure 6.16* shows physical structure materials use in terms of absolute amounts per capita. The picture is mixed, both over time and from one material to the next. Two broad general patterns emerge. First is the distinction between traditional "bulk" materials and modern materials. Traditional bulk materials such as timber, steel, and copper show persistent declines in material intensity over the last 50 years, i.e., "dematerialization". Conversely, "modern" materials such as aluminum and plastics show increasing material intensities. Their use has grown faster than overall economic activity. Second, material intensities for some materials, most notably steel, increase initially and then decline, especially after World War II. The reversal reflects the different phases of industrialization discussed earlier. The "heavy engineering" phase is characterized by the build up of the material-intensive infrastructure of the steam age in which steel use grows faster than GDP. In the subsequent period of "mass production/consumption" steel use continues to grow, but more slowly than the overall economy. The result is the decline in material intensity shown in *Figure 6.15*.

The most important observation from *Figure 6.15* is that there is no instance where material intensity falls faster than GDP grows. Even in the case of timber, material intensity improvements have fallen just short of economic output growth. As a result, absolute timber use has grown by

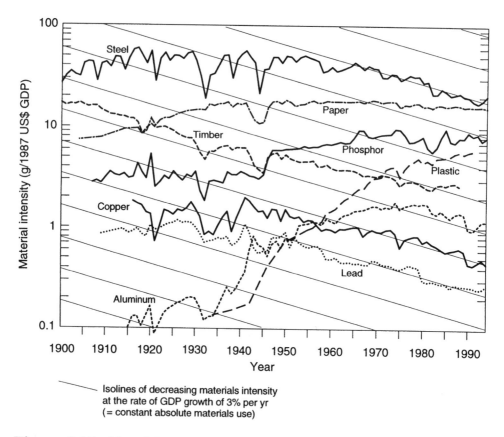

Figure 6.15: Material intensity of the US economy (grams per constant 1987 US$ GDP) since 1900. Isolines show the declines in materials intensity that would be required to keep total materials use constant with GDP growing at the average historical rate of 3% per year. Source: adapted from Wernick (1996:114).

approximately a factor of two since 1900 (from 7 to 16 billion cubic feet). Thus, "dematerialization" does not mean absolute declines in material use. Although the productivity of material use has improved, economic output has grown still faster. At best, "dematerialization" has stabilized material use at high levels for some bulk materials. This is precisely the case for US steel and lead use since 1970. For copper, usage has even slightly decreased: by 10% since 1970.

Shorter-term (1975–1994) data for other industrialized countries confirm this conclusion from the analysis of *Figure 6.15*. Total materials requirements per unit of (constant) GDP have declined between 1.3% per year in Germany, 2% per year in Japan, and 2.5% per year in the Netherlands (WRI,

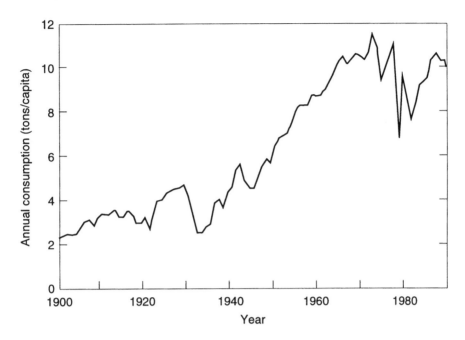

Figure 6.16: Annual per capita use of physical structure materials (i.e., excluding food and energy) in the USA since 1900 (tons per capita). Source: Wernick (1996:114).[1]

1997a:14). Material intensities per unit GDP have declined at a slightly slower pace than GDP growth. The result is a continued, albeit very slow growth in absolute material requirements, despite "dematerialization".

Overall material use has therefore increased in absolute terms. Whereas the average American demanded about two tons of physical structure materials[19] in 1900, current individual use amounts to 10 tons (*Figure 6.16*). However, growth in per capita use of physical structure materials stalled in the early 1970s, suggesting saturation of the materials intensity characteristic of the "mass production/consumption" technology cluster.

Decreases during the last few decades in the material intensity of bulk materials indicate a progressive decoupling of growth in materials use from growth in economic output, at least in the most advanced industrialized countries. There are several explanations. First are changes in the structure of the economy. Growth is no longer centered in the traditional

---

[19]Physical structure materials are construction materials, industrial minerals, metals, and forestry and animal products excluding meat. Current use of these materials in the USA is about 10 tons/capita/year. Total US material use is 20 tons/capita (excluding extraction wastes and water), the difference consisting of energy (8 tons) and food (2 tons).

material-intensive sectors, but has moved toward more "immaterial" economic activities, particularly in the service sector, such as information services and software. Ironically the "information revolution" has had no discernible impact on paper use. Since the mass diffusion of computers, paper use has continued to grow in tandem with economic growth. Apparently, more information on computers means more information needs to be stored and circulated on paper. Perhaps no myth of technology has been so clearly debunked by statistics and everyday experience than that of the "paperless office" that was envisioned to result from the computer revolution. A second cause of the decoupling of material use and economic growth is the substitution of new light-weight materials for older heavier materials, and, of course, increased material recycling. We address each in turn below.

6.6.3. Material substitution

Material substitution is a core phenomenon of industrialization. Indeed we can associate a key substitution with each of the four technology clusters or periods with which we have described industrialization. For the textiles period the key substitution is the replacement of traditional textiles by mechanically spun and woven cotton. For the steam period it is the replacement of fuelwood and charcoal by coal and coke. For the heavy engineering period it is the replacement of iron by steel, and for the mass production/consumption period it is the displacement of coal by oil and gas and the displacement of natural materials by synthetic fibers, rubber, plastics, and fertilizers.

Material substitution serves three main purposes. First, it can overcome resource constraints and diversify key supplies. Second, it can introduce materials with new properties that allow entirely different applications (e.g., the use of cast iron and subsequently steel as structural materials in bridges and buildings). Third, sometimes new materials simply provide the same functional characteristics better (e.g., with less material input), more cheaply, or both. An example is the replacement of copper cables in electricity transmission by aluminum and in telephone lines by optical fibers.

The environmental implications of material substitution are two-fold. First, environmentally harmful substances can be replaced by environmentally less harmful ones. The example of HFC-134a as a substitute for CFC-12 has already been given. Note that HFC-134a is not entirely benign. While it represents an improvement concerning stratospheric ozone depletion, it is a potent greenhouse gas (1,300 times more potent than CO_2 over a 100-year period, cf. IPCC, 1996b:26). Thus it qualifies as a temporary rather than a definitive solution. A second example of environmentally driven material substitution is the ban on tetraethyllead as an antiknock gasoline additive

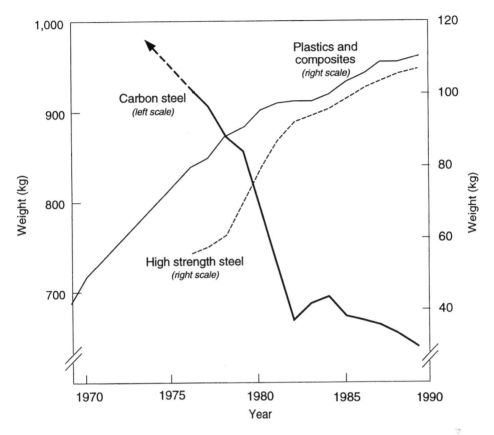

Figure 6.17: Changing weight (in kg) of major materials in the average US automobile. Source: Wernick (1996:117).

and its replacement by aromatics and alcohols. A third example is the development of biodegradable packaging materials based on starch to replace plastics. In addition to being more environmentally benign, some new materials have the added advantage of replacing a nonrenewable raw material (e.g. petroleum) by renewable ones.

The second environmental implication is the overall reduction in material use through the introduction of substitutes. *Figure 6.17* gives an example from the US automobile sector, where traditional heavy carbon steel has been replaced by materials that are either lighter or have improved structural characteristics. Since the early 1970s the weight of the average US car has fallen by some 300 kg, or by 35%. An increase of about 100 kg in plastics and composites, and of high-strength steel, helped to "downmass" the automobile, while preserving its structural integrity. The main motivation for "lightening" the automobile has been to reduce fuel consumption rather than to conserve materials. Fuel consumption is roughly proportional to car weight. The new materials substitute for carbon steel at a ratio of one to three, i.e., 1 kg of new material replaces 3 kg of traditional materials. In its shift to lighter materials the automobile industry has essentially followed a path started earlier by the aircraft industry, where weight reduction has always been a central objective.

Unfortunately, the use of new materials can often introduce disadvantages when it comes to recycling. New materials, such as composites, often mean artifacts have a more complex material composition that increases the difficulty of disassembly and recycling. For cars, substantial design and engineering, plus legislative support, are still required before a car is available in which major components can be replaced and recycled with minimal effort (Wernick *et al.*, 1996:182).

6.6.4. Recycling and waste mining

Recycling and waste mining are discussed together because of their fundamental similarity. Both use wastes as raw material either from discarded artifacts (recycling) or from processing of manufacturing wastes (waste mining). Both are intimately related to technology. Recycling requires technologies for material separation as well as technologies for material reprocessing. The technologies required to manufacture products from scrap are generally different from the technologies required to manufacture the same products from virgin raw materials.

As is the case for recycling, waste mining also depends on technology, either for separation of different scrap fractions, wastes, etc., or in the case of mining and processing tails because the materials being "mined" are generally orders of magnitude less concentrated than they are in raw material deposits.

Recycling is far from new. Since antiquity people have collected and smelted scrap metals. Paper has been manufactured from collected rags since the Middle Ages. Even today, the paper with the highest density and quality is made from textile fibers. It is also quite costly and therefore used only for the arts and for special occasions, like wedding invitations. Wood fibers, on the other hand, are relatively abundant and cheap. They have made the mass production of paper possible and therefore qualify as perhaps the most important technological innovation contributing to the widespread dissemination of literacy and knowledge.

Table 6.5 summarizes current OECD recycling rates for selected materials. For some materials and countries, recycling rates reach two-thirds of total materials use. For rare and expensive materials, recycling rates can

Table 6.5: Recycling rates (percent of use) of selected materials.

	OECD Europe	OECD N. America	OECD Pacific[a]	OECD Average	Other countries	World
Aluminum	26	34	29	30	n.a.	–
Copper	52	63	48	55	n.a.	–
Glass[b]	39	20	55	33	n.a.	–
Lead	59	65	13	55	37	49
Paper	39	28	48	35	n.a.	–
Steel	54	63	47	55	36	45
Zinc[c]	17	31	26	23	n.a.	–

[a] Japan, Australia, New Zealand.
[b] Glass bottles and containers only.
[c] Minimum estimate.
Sources: Metallstatistik (1993:13–44), OECD (1993:149), IISI (1995:141–168), and UN *Statistical Yearbook* (1995:587).

even be higher. Nearly 100% of the platinum used in catalytic converters and of the rare metals used as catalysts in petroleum refining is recycled. What is surprising, at least from an economic perspective, is the relatively high recycling rates of 20–55% for low-cost materials such as glass and paper, for which there are abundant (and renewable) raw material supplies. High recycling rates for these materials demonstrate the importance of environmental movements and newly emerging consumer ethics that have pushed recycling rates, in some cases, far beyond where purely technological and economic considerations would have led. Moreover, it is not true that more recycling is always environmentally better. In the case of paper, too much recycling can be bad for the environment, as shown by a careful cradle-to-grave analysis (Virtanen and Nilsson, 1993) comparing environmental impacts from recycling paper to those from using virgin raw materials[20] (see *Box 6.3*).

Another approach, in addition to recycling, is to mine wastes. This includes coprocessing or recovering secondary materials along with a main "parent" material. It includes material recovery from liquid and airborne waste streams, and it includes material recovery from the wastes of previous material production.

---

[20] One of the main reasons for this is as follows. Paper manufacturing from virgin wood fibers is energy autosufficient, i.e., the industry supplies all its energy needs from waste materials (bark, refuse fibers, pulping liquor, etc.). If paper recycling rates are pushed up to high levels (above approximately 60%), the paper industry increasingly becomes dependent on external energy inputs (fossil fuels) that entail additional airborne emissions. Of course this conclusion depends on numerous specific factors such as pulping technologies used and the structure of energy supply (with coal use entailing particularly high emissions). These are discussed in detail in Virtanen and Nilsson (1993).

Box 6.3: Recycling of Paper

The total world annual consumption of paper is currently in the magnitude of 270 million tons. Over the last 30 years, global paper consumption has increased by a factor of three. Demand for paper is estimated to continue to grow over the long term, particularly as the economies of many developing and newly industrialized countries expand.

Landfill is still the predominant destination for waste paper in the developed countries, even though recycling has increased in many countries in recent years. In order to reduce the amount of wastepaper in landfills, most European countries have enforced legislation (IIED, 1996) or set up targets for recovery and recycling of paper and fiber-based packaging waste (legislation or targets request recycling of 50–80% depending on product). Hence, legislation on recycling is claimed to be driven by environmental concerns.

Quite a few life cycle environmental comparisons have been carried out (cf. Virtanen and Nilsson, 1993; Nilsson, 1994) which indicate that paper recycling is preferable to landfills mainly due to high methane emissions from the decomposition of organic waste. Most of the studies carried out show that incineration can have environmental advantages over recycling depending on the pulping process, technological level, and structure of energy supply in the area. Public attitudes tend, however, to be strongly opposed to incineration. Recycling of waste paper from household waste has especially incurred financial problems due to the high costs of collection and sorting and low sales income.

It can be concluded that the regulations introduced on recycling of paper have driven the paper cycle system into chaos, where the market forces seem to be bypassed. A system in chaos can move in any direction depending on how some of the marginal factors of the system are tackled. The complexity of recycling illustrates that laws and regulations introduced and driven purely by political processes and based on regulations following a flat rate of recycling are bound to be counterproductive in one way or another from an environmental standpoint. Policy measures that encourage recycling of waste paper may be justified in terms of specific environmental or social impacts. These policies should aim to correct market failures in waste collection and disposal and not in other stages of the paper cycle. Policies that stimulate recycling may reduce the amount of virgin fibers used, but they do very little to improve the quality of forest management in social and environmental terms.

Thus, there is a strong need to employ very broad boundaries in the analyses of environmental impacts of recycling, otherwise, it is highly probable that inappropriate conclusions will be made in a broader sustainability context.

Sten Nilsson
International Institute for Applied Systems Analysis
Laxenburg, Austria

Coprocessing and secondary material recovery, i.e., by-products "recycling", have a long history in connection with certain metals like cobalt, silver, and gold where the secondary material is a minor constituent of the ore of another metal. In some cases, such as copper mining, "by-products" can account for a significant fraction (even for a majority) of revenues. Secondary material recovery can also be motivated by environmental considerations beyond economics. Arsenic recovery from copper mining and smelting and cadmium recovery from zinc mining and smelting are prominent examples. Although secondary recovery can help to avoid uncontrolled emissions of arsenic and cadmium, their toxicity limits their ultimate use. The motivation for recovering arsenic and cadmium is thus environmental and not economic. Without recovery these toxic materials would be dispersed and diffused into the human and the natural environment. Once dispersed, they would be much harder to control than directly during the mining and smelting process.

An environmentally and economically more positive example of "waste mining" is the recovery of sulfur from petroleum refining and from coal and oil-fired power plants. Not only are sulfur emissions reduced but new sources of elemental sulfur and gypsum (a product of limestone flue gas desulfurization scrubbers) are made available that can replace sulfur and gypsum that would otherwise come from new mining. Currently the market is better for recovered sulfur, which is already an important input for the chemical industry. However, the use of gypsum from power plant scrubbers is not yet widespread. In Germany, the marketing of gypsum from power plant scrubbers received a setback when traditional gypsum producers began to market their product as "bio" gypsum, i.e., as "natural" mined material.

Other examples of possible waste mining are the use of coal ash in the cement industry and the recovery of fluorsilicic acid from phosphate rock processing to replace mined fluorspar (Ayres and Ayres, 1996:16). In the long term, mining much more diluted wastes may become economic as new technologies are developed (for instance low-cost bacteriological leaching), and as large quantities of metals accumulate in processing wastes (tailings) over the next decades.

All things considered, waste mining has an important and expanding role to play in controlling materials use. Like recycling, however, that does not necessarily mean it can keep up with the huge and increasing flow of waste materials that we anticipate.

Waste mining also faces an ultimate limit when the waste materials generated are in fact so large as to surpass any possible economic uses. The most evident case is that of carbon emissions from the burning of fossil fuels, currently amounting to over 6 billion tons (Gt) elemental carbon. This quantity is more than two times larger than the total weight of the seven most important manufactured materials (refer to *Table 6.2*) taken together (2.5 Gt). Evidently, even if it were possible to scrub all of the carbon from fossil fuel use (e.g., as sulfur is scrubbed from flue gases of power plants), only a small fraction could be used for various purposes. The remainder would have to be disposed of either in depleted oil and gas fields, or in the deep ocean.

6.7. Energy

Of all industrial activities, energy industries best epitomize what is implied by "global change". First, energy industries operate in literally every environment and in almost every corner of the planet, from the arctic circle to the tropics. Energy companies and energy markets operate on a global scale. Oil companies such as Rockefeller's Standard Oil, British Petroleum, and Shell were among the first of the multinational companies that now dominate global industrial activities. Concerns over their market power and the globalization of activities far beyond the jurisdiction of nation states have been with the industry since its inception. Antitrust regulation broke up Standard Oil's monopoly, although a number of the companies created by the break up are among the "seven sisters" (the seven major multinational oil companies) that dominate the global oil scene.

Second, environmental impacts from the production, conversion, and end use of energy are ubiquitous. Some, like air pollution from dense motorized traffic, are local impacts that are nonetheless common to cities around the world. Some are regional impacts such as transboundary air pollution ("acid rain") caused mainly by energy sector emissions. And some are truly global, such as possible climate changes due to energy-related greenhouse gas emissions, most notably carbon dioxide (CO_2).

Finally, energy is the lifeblood of modern industrial societies and affects nearly every aspect of daily life for all inhabitants of the planet – from the subsistence farmer in Bangladesh cooking a warm meal to the business woman boarding a jet aircraft in Los Angeles. Energy use is pervasive – spatially, socially, and environmentally – and the energy industry cannot be considered separately from its billions of customers, the consumers. In fact the traditional separation between energy supply (industry) and end use (consumers) has created both analytic and policy impasses. The need to overcome this rather arbitrary distinction is becoming increasingly recognized by industry. As stated by the World Energy Council, the oldest and most global association of all energy industries:

> On a more fundamental level, however, the period post-2020, if not before, must see the implementation of new concepts of the energy "demand" and "supply" process. Indeed, the energy community is the captive of its own terminology in continuing to use these distinctive terms in ways which fail to recognize overtly or covertly the interdependence of procurement, provision, processing, transformation, transportation, distribution and utilization as elements in a system which should be driven not by the exigencies of primary energy supply, trade or the energy market but by the end-point services which energy is the means of providing." [WEC, 1993a:246]

In this section we describe the historical evolution of the global energy system. We start with "grand patterns" and continue with three major examples of the energy system's many environmental impacts: urban air pollution, regional acidification, and global CO_2 emissions. We conclude the chapter by focusing on the diversity of energy end use patterns and the role of technological change in lessening the energy system's environmental burdens.

6.7.1. Two grand transitions

Prior to the Industrial Revolution, the energy system relied on harnessing natural energy flows, and on animate and human power, to provide energy services in the form of heat, light, and work. Power densities and availability were constrained by site-specific factors, with mechanical energy sources limited to draft animals, water, and windmills. The only form of energy conversion was from chemical energy to heat and light, through burning fuelwood, for example, or tallow candles. Energy use typically did not exceed 0.5 toe[21] per capita per year (Smil, 1994).

Two "grand transitions" have since shaped structural changes in the energy system at all levels. The first was initiated with a radical technological end use innovation: the steam engine powered by coal. The steam cycle represented the first conversion of fossil energy resources into work rather than simple heat. It allowed the provision of energy services to be site independent, since coal could be transported and stored as needed. As noted earlier, it permitted high power densities previously only possible in exceptional locations of abundant hydropower. Stationary steam engines were first introduced for lifting water from coal mines, thereby facilitating increased coal production. Later, they provided stationary power for what was to become an entirely new way of organizing production: the factory system. Mobile steam engines on locomotives and steam ships drove the first transport revolution as railway networks were extended to increasingly remote locations and ships converted from sail to steam. Characteristic energy use levels during the "steam age" were about 2 toe per capita per year. By the turn of the 20th century coal had replaced traditional nonfossil energy sources and supplied virtually all the primary energy needs of industrialized countries.

---

[21]One ton of oil equivalent (toe) equals about 1.5 tons coal equivalent (tce) and 44.8 10^9 joules (J). The principal energy unit used in this section is Gtoe, i.e., gigatons (10^9 tons) of oil (equivalent), equaling 44.8 EJ (10^{18} joules). See *Box 6.4* for definitions and an illustration of energy units.

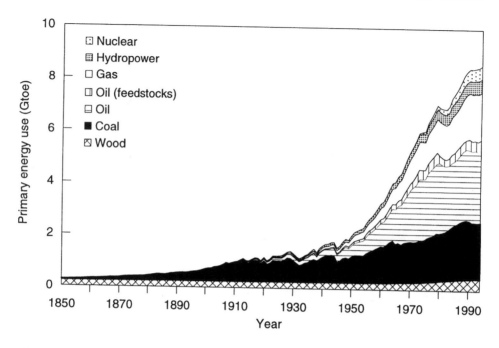

Figure 6.18: Global primary energy use in Gtoe. Source: adapted from Nakićenović *et al.* (1996) in IPCC (1996a:82–83, Working Group II). For the data of this graphic see the Appendix.

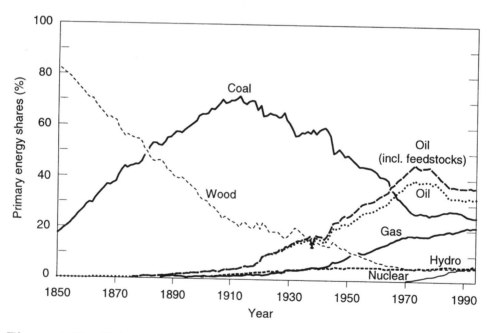

Figure 6.19: Global primary energy shares (in %). Source: *Figure 6.18*.

The second grand transition was the greatly increased diversification of both energy end use technologies and energy supply sources. Perhaps the most important single innovation was the introduction of electricity as the first energy carrier that could be easily converted to light, heat, or work at the point of end use. A second key innovation was the internal combustion engine, which revolutionized individual and collective mobility through the use of cars, buses, and aircraft. Like the transition triggered by the steam engine, this "diversification transition" was led by technological innovations in energy end use, such as the electric light bulb, the electric motor, the internal combustion engine, and aircraft. However, changes in energy supply have been equally far reaching. In particular, oil emerged from being an expensive curiosity at the end of the 19th century to occupying the dominant global position for the last 30 years (reflected in our use of ton oil equivalent as the main unit in this section; see *Box 6.4* on energy units).

Overall, global energy production and use during these two "grand transitions" increased from 0.2 Gtoe in 1850 to 9 Gtoe in 1990, i.e., by a factor of about 40 (*Figure 6.18*). This tremendous increase is intimately linked to the structural changes in energy supply and end use just described. On the supply side, the structural changes can be seen most clearly by plotting the market shares of different primary energy sources over time, as shown in *Figure 6.19*.

Figure 6.19 shows many things, such as the following: (i) the long transition away from traditional renewable energy sources (fuelwood) toward fossil fuels; (ii) the subsequent dominance of coal, supplying close to two-thirds of global energy needs by the eve of World War I; (iii) the introduction of oil and later natural gas, first as a by-product of oil production and more recently as an energy carrier in its own right; (iv) the peak in oil's market share in the 1970s; and (v) a reduction in the dynamics of change in the primary energy supply structure during the last two decades.

The two grand energy transitions have been an essential part of the industrialization transformations described earlier, with their far-reaching structural changes in employment, the spatial division of labor, and international trade. Within the energy sector, particularly important changes include the following.

Commercial energy.   There has been a transition from noncommercial to commercial energy forms reflecting the structural economic shift from agriculture to industry, the related monetarization of the economy, and increased urbanization.

Box 6.4: Energy Units and Scales

Energy is defined as the capacity to do work and is measured in joules (J), where 1 joule is the work done when a force of 1 newton (1 N = 1 kg m/s^2) is applied through a distance of 1 meter. Power is the rate at which energy is transferred and is commonly measured in Watts (W), where 1 Watt is 1 joule per second. Newton, joule, and Watt are defined as units in the International System of Units (SI). Other units used to measure energy are toe (ton oil equivalent; 1 toe = 41.87 × 10^9 J), used by the oil industry; tce (ton coal equivalent; 1 tce = 29.31 × 10^9 J), used by the coal industry; and kWh (kilowatt-hours; 1 kWh = 3.6 × 10^6 J), used to measure electricity. Frequently used multiples of these units are Gtoe/gigaton, i.e., 10^9 tons, oil equivalent, Gtce, and TWh (Terawatt-hours, i.e., 10^{12} Wh, or 10^9 kWh). The figure below shows some of the commonly used units of energy and a few examples of energy consumption levels, along with Greek names and symbols for factors to the power of ten (e.g., exa equals 10^{18} and is abbreviated as E). In 1990 global primary energy use was about 9 Gtoe (376 EJ), out of which 7 Gtoe (293) were accounted for by fossil fuels. Final energy use in 1990 was above 6 Gtoe (268 EJ).

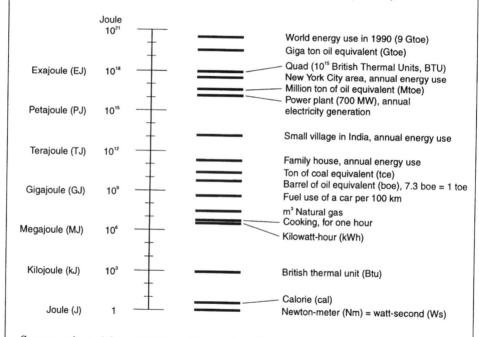

Source: adapted from Nakićenović *et al.* (1996) in IPCC (1996a, Working Group II).

Increasing energy "quality".   There has also been a transition from "batch," solid energy forms, such as traditional biomass and coal, to liquids and grid-dependent energy forms that are more flexible, more convenient, and cleaner. This is a function of three major underlying trends in the "quality ladder" of different energy currencies:

- Industrial processes and technologies are becoming ever more complex requiring energy in forms that are easier to handle, easier to store, more continuous, and are more flexibly available (cf. the case history of electrification in manufacturing given above).
- More convenient energy forms are demanded with rising levels of affluence, and there is a willingness to consider a "convenience premium" in addition to price in fuel choices in residential and commercial end uses.
- Both of these trends lead to higher "form value" (quality) of the energy currencies that are at the interface between energy supply and demand, favoring flexible, clean energy forms such as electricity, gas, and ultimately hydrogen.

Decreasing energy intensity. Although per capita energy needs have increased with economic development (cf. discussion below), the specific energy needs per unit of economic activity (energy intensity) have decreased.

Four major patterns characterize *Figure 6.20* and also explain part of the persistent differences in the overall energy intensities of different economies.

Aggregate energy intensities, including noncommercial energy, generally improve over time, and this is true in all countries. A unit of GDP in the USA, for example, now requires less than one-fifth of the primary energy needed 200 years ago. This corresponds to an average annual decrease in energy intensity of roughly 1% per year. The process is not always smooth, as data from other countries illustrate. Periods of rapid improvements are interlaced with periods of stagnation. Energy intensities may even rise in the early take-off stages of industrialization, when an energy- and materials-intensive industrial and infrastructure base needs to be developed.

While aggregate energy intensities generally improve over time, *commercial energy* intensities follow a different path. They first increase, reach a maximum, and then decrease (see dashed lines in *Figure 6.20*). The initial increase is due to commercial energy carriers substituting for traditional energy forms and technologies. Due to the energy inefficiency of traditional fuel use, increasing use of modern, commercial energy forms improves total system efficiencies (and thus energy intensities) significantly. This is shown in the persistent decline of aggregate total energy intensity over the substitution period (full lines in *Figure 6.20*). Once that process is largely complete, commercial energy intensities decrease in line with the pattern found for aggregate energy intensities. Because most statistics document only modern, commercial energy use, this "hill of energy intensity" has been frequently discussed in the literature (e.g., Reddy and Goldemberg, 1990). Its existence in the case of commercial energy intensities, however, does not diminish the power of the result for aggregate energy intensities – there is a

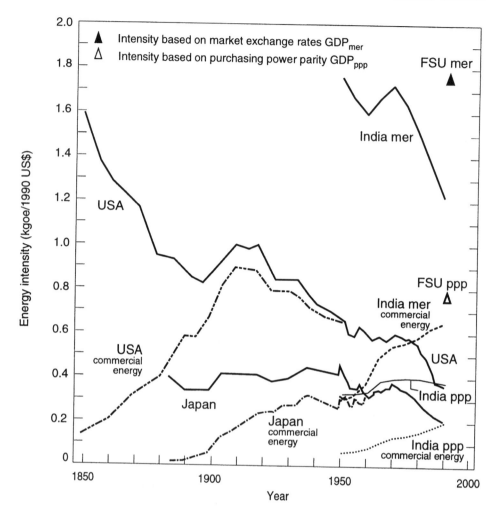

Figure 6.20: Primary energy intensity per GDP (calculated at market exchange rates, mer) for selected countries (kgoe/1990 US$), total and commercial energy only. For India and the former USSR (FSU), intensities are also shown per unit of GDP calculated at purchasing power parities (ppp). Source: IIASA–WEC (1995:31).

decisive, consistent long-term trend toward improved energy intensity across a wide array of national experiences.

History is important. While the trend is one of conditional convergence across countries, the patterns of energy intensity improvements in different countries reflect their different situations and development histories. Economic development is a cumulative process, leading to different consumption lifestyles, different settlement patterns and transport requirements, different industrial structures, etc. Thus the evolution of national energy intensities is *path dependent*. In *Figure 6.20*, for example, there is an evident distinction between the "high intensity" trajectory of the USA, and the "high efficiency" trajectory of Japan.[22] Despite improvements, aggregate energy intensities in developing countries and those in transition from centrally planned to free markets remain consistently higher than in industrialized countries. Although comparisons are complicated by measurement problems, this conclusion holds whether economic output is compared using market exchange rates or purchasing power parity exchange rates.[23] There is a consistent pattern in the decrease of energy intensity with increasing economic development, whether measured using market exchange rates or purchasing power parity. Moreover, energy intensities in developing countries – as well as their energy use structures – are *ceteris paribus* comparable to those of industrialized countries past similar levels of economic development (GDP per capita) (IIASA–WEC, 1995:32).

Constant technological change since the onset of the Industrial Revolution has thus been the principal driver of continuous structural change in energy systems and, as a result, continuous energy productivity increases. Resource scarcity, a consideration that might also be expected to be a major driver of energy system restructuring and efficiency improvements, has proved less important. Although fuelwood in 18th century England was becoming increasingly scarce and expensive and wood imports from Scandinavia soared, this was not the main cause for the rise of the coal industry. It was rather coal's comparative advantages in terms of much higher energy densities and new applications owing to new end-use technologies, specifically stationary and mobile steam engines. Similarly, although the progressive extinction of the whale population made illuminants more expensive, and kerosene from petroleum a very welcome substitute, resource scarcity cannot explain the spectacular rise of oil from a minor curiosity to

---

[22]We have already introduced the concept of path dependency in Chapters 2 and 3. Path-dependent processes of productivity increases have also been illustrated in Chapter 5 for agricultural labor and land productivity. The example of energy productivity (intensity) discussed here, or that of different mobility levels discussed in Chapter 7 reconfirms the importance and pervasiveness of path-dependent processes and history arising largely out of the cumulativeness of technological change.

[23]The purchasing power parity exchange rate is calculated by comparing the prices of a standardized basket of consumer goods and services in two countries. Because most developing countries have larger informal sectors, lower labor costs, and limited trade, their prices for food, services, etc. are substantially lower than those in industrialized countries. The resulting purchasing power-based exchange rate is thus much more favorable than the official market exchange rate. For instance, while the per capita GDP of India is only some US$300 when calculated at prevailing market exchange rates, it is approximately US$1,200 per capita based on purchasing power parity, nearly four times larger. The resulting energy intensity is correspondingly smaller (as shown in *Figure 6.20*).

the dominant fuel of today. Oil was always much more scarce than coal, but nevertheless replaced coal as the dominant fuel of industrialization. The story of oil's advantages over coal is similar to that of coal's advantages over fuelwood. As a liquid rather than a solid, oil was much easier to transport and to store. New technologies dependent on oil as a fuel or raw material (cars and aircraft, petrochemical processes) drove the petroleum industry's ascent even while oil was much more expensive than coal.

If history provides one lesson, it is that much of the current trade of energy modeling should seriously question its persistent attention given to quantities and prices, while entirely ignoring *quality*. Analyzed on the basis of quantity and price, oil should never have made it to market during the first 50 years of its history. Only after World War II, and the beginning of large-scale oil production in the Middle East and the development of cheap transport technologies in the form of supertankers, did oil prices start to challenge those of coal, at least in Western Europe with its expensive deep-mined coal. Coal continued to be cheaper than oil in the USA, but nevertheless lost all of its major markets to oil (and gas) except for electricity generation and some captive industries, most notably steel.

Fears of resource scarcity have surfaced periodically since the 19th century. Although coal is now recognized as the most abundant of conventional hydrocarbons (excluding methane hydrates),[24] the first influential publication on resource depletion focused precisely on coal in the midst of the steam age. William S. Jevons' *The Coal Question: An Enquiry Concerning the Progress of the Nation and the Probable Exhaustion of Our Coal-mines* was published in 1865 by Macmillan, London. Similarly, fears of oil resource scarcity have resurfaced ever since 1919,[25] particularly in the USA. "In 1920 the Geological Survey predicted that all oil reserves would be depleted

---

[24]The first inventory of global coal resources dates back to the International Geological Congress in 1913, where they were assessed at 10,000 Gt, a figure that has not changed since (for a discussion see Fettweis, 1979). It is estimated that the largest occurrence of hydrocarbons is in the form of methane clathrates (hydrates), i.e., methane (natural gas) molecules trapped in the crystalline structure of ice. Clathrates exist in large quantities in permafrost areas and offshore continental shelves (e.g., the Gulf of Mexico). Due to limited exploration, the quantities reported (cf. MacDonald, 1990) are necessarily speculative and may be overestimated by perhaps as much as a factor of three (cf. Holbrook *et al.*, 1996). Even with this uncertainty, methane clathrates remain the most abundant form of hydrocarbons in Earth's crust (cf. *Box 6.5*).

[25]In a saying attributed to Winston Churchill, the allied forces of World War I "swam to victory on a wave of oil", largely coming from US oil fields. Consequently, in 1919 the supply of oil products ran short in the USA, prompting rationing and imminent resource scarcity fears. Churchill himself was instrumental in changing the British Navy's main fuel from coal to oil. Oil provided for higher energy density and thus reduced refueling requirements for battleships. It also had the added security benefit that smoke plumes from oil combustion were much less visible than those of coal steamers.

in 14 years. Yet, by 1960, for every barrel believed available in 1920, eight had been extracted and an additional five had been proven to exist" (Starr, 1996:245). Similar resource depletion concerns were voiced at the beginning of the 1970s, leading to costly and massive programs for synfuel development. Yet throughout the 20th century, oil reserves have continued to rise, maintaining a reserve-to-production ratio of between 30 and 40 years. Oil reserves are now at an all-time high. Reserves still constitute only a small fraction of total petroleum occurrences in the form of discovered and undiscovered, conventional and unconventional resources (see *Box 6.5*). The resource potential for natural gas is larger still.

The real issue concerning the availability of fossil fuel resources is therefore not whether they are ultimately finite – they are. The real issue is the time horizon over which resource availability would begin to constitute a constraint. Both historical experience with scarcity concerns and the current quantitative evidence suggests it is very improbable that this would happen before the end of the 21st century.

In the meantime a more immediate constraint on the further expansion of fossil fuel use seems likely to be the environment. The conventional fossil fuel resource base tabulated in *Box 6.5* of some 5,000 Gtoe corresponds to a near equal quantity of carbon (4,700 GtC). All fossil fuel occurrences tabulated in *Box 6.5* represent some 29,000 Gtoe, or some 20,000 GtC. This compares with a current global use of fossil fuels of some 7 Gtoe, resulting in 6 GtC of annual emissions (IIASA–WEC, 1995), and a current atmospheric carbon content of 770 Gt (IPCC, 1996a). The latter therefore may be of more concern than the former, suggesting that for future energy systems the binding constraint is not geology, or what we can dig out of Mother Earth, but rather the environment, i.e., how much we can impose upon it.

6.7.2. Environmental burdens and impacts: Local, regional, and global

The essence of the energy system is a simple chemical reaction: combustion. Air pollution has been a consistent result, ranging from local to global levels, and from short episodes to century-scale changes in atmospheric composition.

Local environmental impacts. By the Middle Ages Europe had long forgotten the technological sophistication of Roman hot air central heating systems (hypocausts). The cold European climate left few alternatives: either freeze, or choke on polluted indoor air. For the frugal Cistercian monks freezing was the main solution. Originally only one room in a monastery (the calefactory) was heated and monks were allowed to warm up for limited

Box 6.5: On Energy Reserves and Resources

In 1950, global oil reserves amounted to 10.4 Gtoe with annual oil production running at 0.52 Gtoe. The ratio of remaining reserves over annual production, the so-called reserve-to-production ratio, for 1950 says that, on a pure calculatory basis, global oil reserves would last another 20 years at constant 1950 production levels. Since then cumulative 95 Gtoe of oil were produced by early 1996. Adding to this past production today's identified reserves of some 150 Gtoe results in a total of 235 Gtoe discovered since 1950 – more than 20 times the reserves of 1950. It appears that efforts to estimate "reserves" at any given point have had a poor track record.

Generally, reserves are those occurrences of a natural resource that are geologically identified and known to be technically recoverable under present market conditions. Therefore, estimates of future reserves are influenced by the current and future states of geological knowledge, production technologies, economics and expected levels of production. Knowledge and technology are routinely improving while economics and demand may change substantially over time. Although these factors do not evolve independently from each other, historically the net result of advances in knowledge and technology is a continuous replenishment of reserves from occurrences that previously did not qualify as reserves.

Because of technology advances, reserves estimates of the day do not provide a useful basis for analyses involving horizons extending over several decades or centuries. What is required is a comprehensive account of fossil occurrences irrespective of short-term technological or economic considerations. In addition to reserves of conventional hydrocarbons, i.e., the oil, natural gas and coal we are consuming today, the account should also include nonconventional occurrences of oil (shale oil, tar sands, heavy oil) and natural gas [coal-bed methane, tight formation gas, geopressured gas, gas from fractured shales, ultradeep gas, gas hydrates (clathrates)]. These occurrences are likely to come on-stream in the future as their respective production technologies continue to improve and market conditions change.

Aggregation of fossil energy occurrences, in Gtoe.

	Consumption				Resource	Additional
	1860–1995	1995	Reserves	Resources[a]	base[b]	occurrences
Oil						
Conventional	106	3.25	150	145	295	
Unconventional	6	0.16	183	336	519	1,800
Natural gas[c]						
Conventional	50	1.91	141	279	420	
Unconventional			192	258	450	400
Clathrates						18,800
Coal	136	2.23	1,003	2,397	3,400	2,900
Total fossil occurrences	299	7.55	1,669	3,415	5,084	23,900

[a] Reserves to be discovered or resources to be developed as reserves.
[b] Resource base is the sum of reserves and resources.
[c] Includes natural gas liquids.

Aggregation of fossil energy occurrences, in Gt carbon.

	Consumption				Resource	Additional
	1860–1995	1995	Reserves	Resources[a]	base[b]	occurrences
Oil						
Conventional	87	2.7	124	119	243	–
Unconventional	5	0.1	151	276	427	1,500
Natural gas[c]						
Conventional	29	1.1	81	161	243	–
Unconventional	–	–	111	149	260	200
Clathrates	–	–	–	–	–	10,800
Coal	140	2.3	1,034	2,470	3,505	3,000
Total fossil occurrences	262	6.2	1,501	3,175	4,678	15,500

–, Negligible volumes.
[a] Reserves to be discovered or resources to be developed as reserves.
[b] Resource base is the sum of reserves and resources.
[c] Includes natural gas liquids.

The first table summarizes such an account (based on Rogner, 1996) for coal, oil and natural gas occurrences broken down in reserves and resources (which together form the resource base) plus additional occurrences. Here, resources are occurrences with current uncertain geological assurance or lack of technical recoverability or doubtful economics. But if historically observed rates of technical progress in the hydrocarbon upstream sectors are applied, that share of occurrences labeled resources may become available to replenish reserves throughout the 21st century.

To date, some 300 Gtoe fossil fuels have been produced, with current reserves amounting to 1,670 Gtoe. The fossil fuel resource base exceeds 5,000 Gtoe, with additional fossil occurrences mostly in the form of methane hydrates (clathrates) adding another 24,000 Gtoe. At face value, geological occurrences are therefore unlikely to constrain fossil fuel use within the next 100 years. Such constraints are likely to emerge rather from environmental impacts.

The second table shows that to date some 260 Gt of carbon have been oxidized and released to the atmosphere since 1860. Out of a total global fossil energy carbon endowment of 20,000 Gt C, some 1,500 Gt C of fossil energy (about twice the current carbon content of Earth's atmosphere) have been identified to eventually reach the market place at costs not significantly higher than today's market prices (Rogner, 1996). If the past 260 Gt of carbon emissions are already considered responsible for a "discernible human influence on global climate" (IPCC, 1996b), then potential adverse environmental impacts rather than physical fossil resource availability will limit the ultimate utilization of fossil resources by future generations.

Hans-Holger Rogner
Institute for Integrated Energy Systems
University of Victoria, Victoria, BC, Canada

periods of time. Before the diffusion of chimneys (which was a long process lasting well into the 18th century in Europe), indoor air was heavily polluted from smoke from open fires,[26] much like the indoor air pollution in many developing countries today (cf. Chapter 5). Chimneys improved indoor air quality, but at the expense of energy efficiency; over 90% of a fire's heat did not warm the room, but escaped through the chimney. However, as long as the smoke and air pollution also left through the chimney, the result was an improvement, albeit a chilly one. Unfortunately, what was previously indoor air pollution now became outdoor air pollution with the rise of large urban agglomerations and the widespread diffusion of open coal fires. London in particular can claim to be the "innovation center" of a new form of air pollution: smog.[27]

London smog incidents became frequent with the increasing use of coal and in the 19th century assumed a new qualitative dimension in the form of "killer smogs". Smog incidents in 1873, 1880, 1891, 1892, and especially the killer smog of 1952, were all characterized by exceedingly high air pollution levels. Soot deposition in the 1891 smog was close to 10 grams of soot per m^2. Particulate and sulfur dioxide concentrations reached recurrent daily concentration levels of up to 4,000 micrograms (4 milligrams) per cubic meter. This exceeds current World Health Organization exposure guidelines for particulate matter and sulfur dioxide by a factor of 25 (WHO and UNEP, 1993:222–223). Excess mortality reached 1,000 deaths per smog episode, with 4,000 lives claimed during the 1952 killer smog (Brimblecombe, 1987:124). *Figure 6.21* illustrates the dark days of December 1952.

The Clean Air Act of 1956, prompted by the 1952 smog disaster, attempted for the first time to control domestic sources of pollution as well as those of industry. Among other pollution control measures, coal burning in open fireplaces was banned in "smokeless zones", which were designated under the Act. These zones covered over 90% of London. As a result, smoke air pollution has been reduced today by 80% compared to the 1950s (Brimblecombe, 1987:171). Although the Clean Air Act controlled only smoke (particulate matter), there have also been notable reductions in sulfur dioxide emissions from domestic sources as a result of people switching from using coal to either gas, electricity, or low sulfur oil. Moreover, emissions

---

[26]In some regions of Austria, "smokehouses" (so named because they lacked a chimney, and smoke from the open fire would rise to the ceiling and escape from there through small wall openings, usually above the entrance door) disappeared only in the 1950s. Retrofitted with chimneys or central heating systems, such wooden "smokehouses" are now much researched aesthetic secondary residences of the urban wealthy.

[27]The term, a combination of smoke and fog, seems to have been first proposed in 1905 to describe the mixed smoke and fog that settled commonly over London in Victorian times (Brimblecombe, 1987:165).

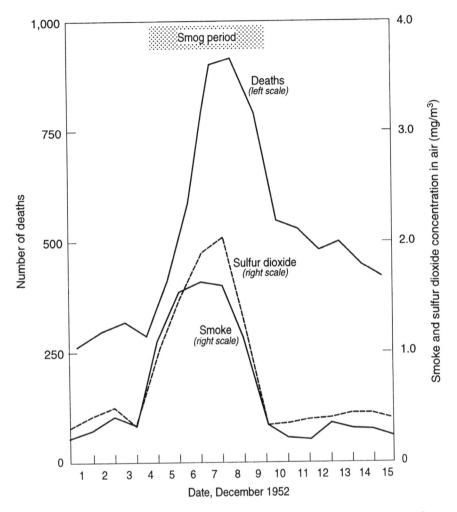

Figure 6.21: Deaths and pollutant concentrations (milligrams per m^3) during the London smog of 1952. Source: Brimblecombe (1987:168).

from large point sources have been reduced using dust filters (by electro-static precipitators) and dispersed using tall stacks resulting in additional improvements in local air quality. However, dispersion via tall stacks has not eliminated pollution problems entirely, but rather shifted them from the local to the regional level. At the regional level they are complicated by the addition of politically sensitive transboundary issues, as will be discussed below.

Nevertheless, in London and other cities in industrialized countries, air quality has improved significantly. As discussed earlier (cf. *Figure 5.20*) pollution in these cities from traditional sources is largely under control. New local air pollution problems have arisen from "modern" sources, particularly motorized traffic (cf. Chapter 7), but even there progress has been made with the introduction of unleaded gasoline and the diffusion of catalytic converters in cars.

In the rapidly growing cities of developing countries, however, local air pollution from both traditional and modern pollutants remains serious. Traditional local air pollution from smoke and sulfur dioxide in cities like Beijing resembles the situation in London many decades earlier. In Beijing particulate concentrations exceeding 1,000 micrograms per m^3 during the winter months are common (WHO and UNEP, 1993:16), and sulfur dioxide concentrations exceed WHO guidelines by a factor of two or more. As was the situation in London, large quantities of low-grade coal and biomass burned in millions of individual (and energy inefficient) stoves are the principal culprits. Hopefully London's policies to curb air pollution will also be imitated in developing countries. Ideally they will be improved upon, given our additional experience, particularly our understanding that taller stacks are not the definitive answer.

Regional environmental impacts. Compared to our relatively good understanding of local air pollution and its impacts on human health, vegetation, and materials (such as crumbling historical sandstone buildings) (cf. e.g., Crutzen and Graedel, 1986:213–250), regional and transboundary air pollution impacts are still fraught with considerable uncertainties. Since the early 1970s researchers have been aware of acidification of surface water bodies in Scandinavia and Canada due to "acid rain" (the presence of the strong inorganic acids H_2SO_4 and HNO_3 from SO_2 and NO_x emissions) and its deleterious effects on fish. It has been more difficult to relate acidic precipitation directly to impacts on vegetation. Comparatively modest levels of sulfur and nitrogen deposition seem to act as fertilizers, particularly for agricultural crops (Fischer and Rosenzweig, 1996). Beyond certain threshold levels and depending on the widely varying buffering capacity of soils, however, negative impacts on vegetation can occur. Nonetheless, direct cause-and-effect relationships for ecosystem impacts are scientifically poorly understood and inherently difficult to prove. That is, it is difficult to tie specific impacts to a particular group of pollutants.

Forest die-back in the Erzgebirge, the border region between Poland, the Czech Republic, and the former German Democratic Republic, seems to stem from extremely high sulfur deposition levels (reaching 15 grams per m^2 per year) from coal burning in the area. However, beyond areas of extremely high pollution, it is very difficult to link forest damage with particular deposition levels of pollutants. Damage results from a combination of factors – acidic precipitation, photooxidant smog, "overfertilization" from nitrogen

Table 6.6: European sulfur emissions, 1965, 1980, and 1994 (in million tons elemental sulfur).

	1965	1980	1994
Western Europe			
"Old" industrialized countries[a]	9.8	6.9	3.6
"New" industrialized countries[b]	3.5	5.5	1.7
Central and Eastern Europe	4.0	6.3	4.0
European part of former USSR	7.7	5.9	2.5
Total	25.0	24.6	11.8

[a]UK, Germany (including German Democratic Republic), France, Benelux countries (Belgium, Netherlands, Luxembourg).
[b]Remainder of Western Europe and Turkey.
Data sources: 1965: Mylona (1993, Appendix B) and Mylona (1996). 1980 and 1994: official country submissions to EMEP (1996).

deposition, and increased vulnerability due to monocultures – and it is very difficult to disentangle their separate effects. One approach to connecting overall vulnerability to specific contributory factors makes use of the new concept of "critical loads" (see *Box 6.6*).

Regional air pollution issues also illustrate a relatively new dimension of environmental policy-making: the involvement of many governments and actors with very different circumstances. Different countries have different emission levels, different types of emission sources, and different economic abilities to reduce their emissions. They must then negotiate agreed environmental targets in the midst of scientific uncertainties. Initial European agreements on reducing sulfur dioxide emissions (and to a lesser extent nitrogen oxide emissions) have been nonetheless quite successful and have resulted in dramatic emission reductions. These have been accomplished using a whole range of policy measures, including "end of pipe" technologies (e.g., scrubbers at power plants and desulfurization of oil products) and structural changes in energy systems (e.g., greater use of natural gas, cogeneration of electricity, and district heat). In addition, total energy demand growth in Europe has remained flat over the last 20 years, and has even declined precipitously in Central and Eastern Europe since 1990. As a result, the agreed policies have led to substantial absolute reductions in emissions as shown in *Table 6.6*. Yet further reductions are required (and indeed have been agreed upon) to lower remaining high deposition levels, especially in Central Europe (cf. *Figure 6.22*). The international sulfur reduction negotiations have also benefited from a new "technology", i.e., new integrated models that can simulate the effects of emission reductions in one country on deposition levels and resulting ecological impacts in another country.

Box 6.6: Critical Loads and Emission Reductions in Europe

The concept of critical loads emerged in the mid-1980s, and in a series of workshops (starting in 1986, most of them sponsored by the Nordic Council of Ministers) the concept has been further refined. A critical load has been defined as the deposition "below which significant harmful effects on specified sensitive elements of the environment do not occur according to present knowledge". Over the past decade methodologies for computing critical loads for sulfur and nitrogen have been elaborated and compiled by the Task Force on Mapping (TFM) under the Working Group on Effects (WGE), which operates under the United Nations Economic Commission for Europe's (UN/ECE) Convention on Long-Range Transboundary Air Pollution (LRTAP). Critical load values for a variety of ecosystems (forest soils, lakes, seminatural vegetation) are compiled on a national level and submitted to the Coordination Center for Effects (CCE), located at the Dutch National Institute for Public Health and the Environment (RIVM). The CCE contributes to the development of the critical loads methodology in Europe, and collates and merges national data into European maps and data bases. These maps are then approved by the TFM and the WGE before being used in emission reduction negotiations under the LRTAP Convention.

Critical loads of sulfur have been used in negotiating control strategies for transboundary air pollution in Europe, as evidenced by the signing of the Second Sulphur Protocol in Oslo in June 1994. This protocol is the first international agreement on emission reductions taking explicitly into account environmental vulnerability, in addition to technological and economic considerations. Earlier protocols on sulfur (1985), nitrogen (1988) and volatile organic compounds (VOCs) (1991) – although aiming at protecting the environment – did not consider the spatial variation in the vulnerability of ecosystems. In fact, the savings from taking into account critical loads in the Second Sulphur Protocol are estimated at five billion DM compared to flat rate reductions.

Acidification, however, is caused by the deposition of both sulfur and nitrogen, and both compounds "compete" for the counteracting (neutralizing) base cations, which are mostly provided by deposition and weathering. And, in contrast to sulfur, for nitrogen there are additional natural (sources and) sinks such as uptake by vegetation, immobilization and denitrification. Consequently, it is not possible to define a single critical load, as was the case when looking at sulfur alone, but a function, called critical load function. This function defines pairs of S and N deposition for which there is no risk of damage to the ecosystem under consideration. In addition to acidification, nitrogen deposition also acts as a nutrient for ecosystems. Consequently, in order to avoid eutrophication, critical loads for nutrient nitrogen have been defined and mapped for various ecosystems.

The negotiations for the revision of the nitrogen protocol are aiming at a so-called multipollutant multieffects protocol, taking into account both the acidifying and eutrophying aspects of nitrogen and, in addition, the role of nitrogen oxides (together with VOCs) as precursors to the formation of tropospheric ozone. This ambitious goal requires sophisticated yet transparent integrated assessment models, which allow an easy evaluation of (cost-optimal) emission reduction scenarios by linking submodels of the energy system, the long-range transport of the different pollutants and the sensitivity of ecosystems to these pollutants. Such a model, RAINS, has been under development at IIASA for several years (cf. Alcamo *et al.*, 1990). The concept of critical loads has turned out to be simple enough to allow an application on a pan-European scale and scientifically sound enough to be widely accepted as a means for supporting negotiations on emission reduction strategies.

Maximilian Posch and Jean-Paul Hettelingh
Coordination Center for Effects, RIVM, Bilthoven, Netherlands

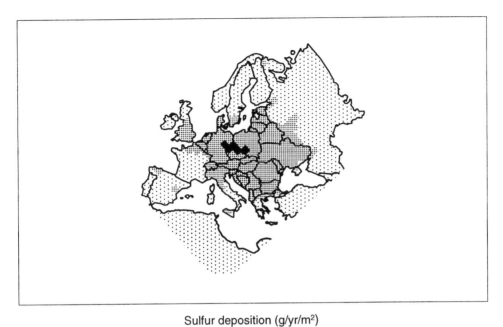

Sulfur deposition (g/yr/m²)

0 – 1   1 – 5   > 5

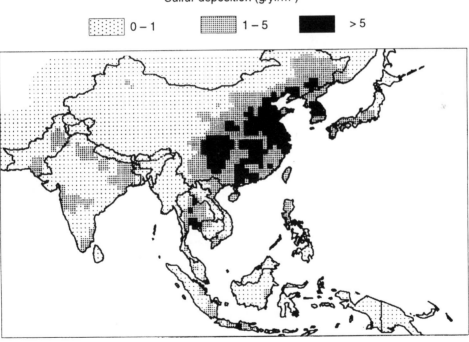

Figure 6.22: Current sulfur deposition in Europe (top) and projections for a high-growth scenario for Asia in 2020 (bottom), in grams sulfur (S) per m². Source: based on Amann *et al.* (1995); graphic courtesy of Transboundary Air Pollution Project, IIASA.

The progress in Europe (and North America) has been impressive, and it will be important to replicate it in other parts of the world. Transboundary air pollution is a growing problem for, in particular, the rapidly growing, coal-intensive economies of Asia. *Figure 6.22* contrasts current European sulfur deposition levels with those calculated for Asia in 2020, assuming continuing vigorous growth of economies and energy demand, and hence coal use.

Global environmental impacts. Sulfur emissions, which began as a local environmental concern and developed into a regional concern, have recently been determined to be also a key factor in the principal *global* environmental impact of the energy sector – the greenhouse effect. Let us turn therefore to the greenhouse effect, beginning with the basics before returning to the recently identified role of sulfur aerosols.

The fact that Earth is habitable at all is due to the existence of greenhouse gases, principally water vapor but also other trace gases, of which the most prominent is CO_2. As in a greenhouse, infrared radiation is unable to escape, thereby raising the average temperature about $30°C$ above what it would be in the absence of greenhouse gases. Comparing Earth with its neighbors, Mars and Venus, illustrates the delicacy of the atmosphere's radiative balance that makes life possible. Mars, with its thin atmosphere and no greenhouse effect, is far too cold to support life as we know it. Conversely, Venus with its thick cloud cover and dense CO_2 atmosphere illustrates a "run away" greenhouse effect with temperatures around $500°C$.

Since the onset of the Industrial Revolution, the concentrations of natural greenhouse gases (most prominently carbon dioxide, methane, and nitrous oxide) have increased, and new synthetic greenhouse gases have been added. As noted earlier these include CFCs and their halon substitutes. Atmospheric CO_2 concentrations have risen by some 30% (see *Figure 6.24*). Taking into account all greenhouse gases except water vapor, the total increased effect is equivalent to an increase in carbon dioxide by almost 50% (Bolin, 1995). *Figure 6.23* shows the estimated effect of changed greenhouse gas concentrations on changes in the heat balance of the planet, the so-called radiative forcing (in W/m^2).

By far the largest single source of changes in radiative forcing is CO_2, mostly from fossil fuel burning but also from land-use changes such as deforestation. Smaller contributions have come from methane (CH_4) and halocarbons (CFCs). Their aggregated effect is estimated at some $2.5 \ W/m^2$. This is partly counterbalanced by the cooling effects from aerosols in the form of fossil fuel soot and sulfur, and from biomass burning (fuelwood, deforestation, and natural wildfires). Sulfur emissions, in addition to their local and regional impacts, therefore also have global implications.

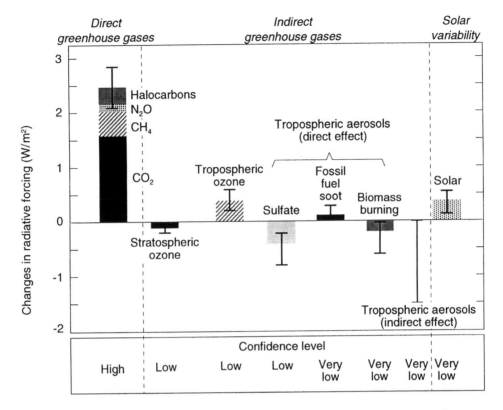

Figure 6.23: Estimated relative contribution of different factors on changes in radiative forcing (in W/m²). Contributing factors include human-induced changes in atmospheric concentrations of direct and indirect greenhouse gases between 1750 and 1992, and estimated changes in solar output from 1850 to the present. Positive values indicate possible global warming; negative values represent cooling from aerosols. Source: adapted from IPCC (1996b:19).

Excluding aerosols, model calculations indicate an expected base increase in global mean temperature between 0.8 and 2.2°C for the estimated change in radiative forcing of 2.5 W/m². However, the inertia of the climate system acts to delay the resulting climate change by 30–50%, and the cooling effect of aerosols, principally from sulfate aerosols, diminishes the impact by another 20–40%. Thus, the currently observed climate change of 0.3–0.6°C is within the lower bound of theoretical expectations (Bolin, 1995; IPCC, 1996a, Working Group I). It thus remains difficult to disentangle the human-induced climate change effect from naturally occurring climate variability.

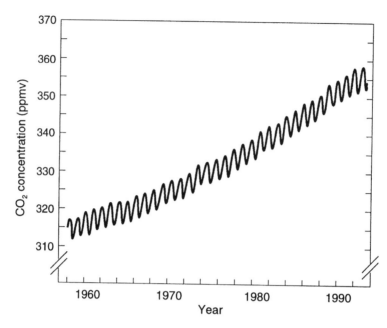

Figure 6.24: Atmospheric concentration of CO_2 (parts per million by volume, ppmv) as measured at Mauna Loa, Hawaii. Note in particular the rising trend and annual variations resulting from seasonal variations in vegetation growth. Source: IPCC (1995:43, Working Group I).[2]

Despite the substantial uncertainty and controversy as to whether a significant human-induced climate change signal can already be discerned, there is no doubt that atmospheric concentrations of greenhouse gases have increased ever since the mid-18th century. The first systematic measurements were initiated by C. David Keeling in the late 1950s, and the resulting time series from the Mauna Loa observatory on the island of Hawaii (*Figure 6.24*) has become one of the most frequently cited graphical "icons" of the global change literature.

CO_2 concentrations prior to the Mauna Loa data, and indeed the Industrial Revolution, have been estimated by analyzing air bubbles trapped in ice core drillings (IPCC, 1996b:18–19). Both types of measurements illustrate the critical importance of technology as an observational and diagnostic tool. The measurements confirm that anthropogenic greenhouse gas emissions have had a truly global impact on atmospheric concentrations. Today, industrial activities release some 6 ($\pm$0.5) Gt carbon per year, almost exclusively through burning fossil fuels. Deforestation releases another 1.6 ($\pm$1) Gt (IPCC, 1996b:20). From 1850 to 1994, fossil fuel use released up to 300 Gt of carbon into the atmosphere (Grübler and Nakićenović, 1996:101), and land-use changes including deforestation added another 120 ($\pm$40) Gt of carbon (IPCC, 1995:52). Burning fuelwood may have released some 44 Gt of carbon since the mid-19th century, a figure that is usually included in the land-use change and deforestation figures of historical carbon emission inventories. Among fossil fuels the largest contributor has been coal, followed by oil (see *Box 6.5* above). The contribution of natural gas has been relatively modest because gas use has grown only comparatively recently and it is also the fossil fuel with the lowest carbon emissions per unit of energy (cf. discussion on "decarbonization" in Section 6.7.3).

While the overall correlation between increasing carbon emissions and rising CO_2 concentrations is clear, there remains significant uncertainty about the carbon cycle. A comparison of emissions and concentrations indicates that only about half of CO_2 emissions remain "airborne". The rest is absorbed by natural sinks, principally the oceans. Currently the oceans are estimated to absorb 2 Gt of carbon per year (IPCC, 1996b:20); however, we do not know where the rest of the carbon that does not enter the atmosphere goes. Its unknown destination is referred to as the "missing carbon sink". Seasonal variations in atmospheric CO_2 concentrations (see *Figure 6.24*) illustrate the influence of seasonal vegetation growth cycles in the northern hemisphere. Consequently the "missing carbon sink" is speculated to involve increased CO_2 uptake by vegetation in northern latitudes stimulated by "fertilization" caused by increased CO_2 concentrations in the atmosphere.

Additional uncertainties abound. Carbon dioxide is only one of several greenhouse gases, and important feedback mechanisms are poorly understood. These include changes in atmospheric water vapor content, cloud formation and resulting albedo changes (i.e., the reflection of incoming sunlight back into space), the role of aerosols, and how regional changes in temperature and precipitation patterns are related to the global averages calculated by models. Three factors in particular exacerbate uncertainties. First, little is known about the response of ecosystems to transient, and possibly abrupt, climate changes.[28] Second, we know very little about possible impacts in developing countries where few studies are available, but potential vulnerability to climate change is high. Third, it is nearly impossible to aggregate impacts across different cultures, economic systems, and ecosystems.

---

[28] Most impact studies are based on simulations assuming CO_2 concentrations twice the current or preindustrial levels. Impacts are determined by comparing such a scenario to the present climate. Questions of adjustment time and transient responses are rarely addressed.

The current state of knowledge[29] may be summarized as follows. First, unmanaged natural ecosystems are relatively more vulnerable than managed ecosystems such as agriculture. This is simply because, as the term "managed" implies, technology can be applied to adapt to a changing climate and to mitigate adverse impacts. Drought-resistant crops can be developed and planted, farming practices changed, and coastal areas protected by dikes. Conversely, natural ecosystems generally have slow rates of adaptation and are therefore more vulnerable to changes in climate.

Second, poorer societies are more vulnerable than affluent ones. This is because larger shares of their economies depend on climate-sensitive activities like agriculture, and also because poorer societies have limited resources to adapt to climate change. Such societies also face numerous competing and pressing short-term policy needs, particularly alleviation of poverty. Conversely, affluent societies have less to fear. Smaller proportions of affluent economies involve climate-sensitive activities (agriculture typically accounts for only a few percent of their economies). Affluent economies are richer in resources and technology which enables them to adapt to climate change. Indeed, for some industrialized countries climate change may prove a net benefit, providing it does not come too suddenly. Such calculations, however, are complicated by the fact that many of the potential impacts (e.g., damage to coastal wetlands or the loss of human life due to heatstrokes) cannot be quantified in economic terms. Yet according to current knowledge, they dominate the possible impacts of a changing climate.

The debate in policy circles is even more complicated than the scientific debate. In addition to scientific uncertainties, the policy debate must deal with the different values assigned by different countries and interest groups to the various potential impacts of climate change and to the costs of alternative policies to reduce their likely magnitude. The principal forum for scientific debate is the Intergovernmental Panel on Climate Change (IPCC), although some of the policy debate has spilled over into the latest IPCC evaluation of the science of climate change, the Second Assessment Report of the IPCC (1996a). The principal forum for the policy debate are the negotiations within the United Nations Framework Convention on Climate Change (FCCC) aiming to arrive at binding protocols to curtail emissions growth (see *Box 6.7*).

Neither the policy debate nor the scientific uncertainties are likely to be resolved easily. It is noteworthy that while this book was being written, an important anniversary took place. In April 1896, Svante Arrhenius published

---

[29]For an overview see the IPCC (1996a) reports of Working Groups II and III.

Box 6.7: The Framework Convention on Climate Change

The Framework Convention on Climate Change (FCCC) was signed at the "Earth Summit" in Rio in 1992. Negotiated in response to the fear of global warming, its stated objective is to achieve "stabilization of greenhouse gas concentrations in the atmosphere at a level that would prevent dangerous anthropogenic interference with the climate system." However, because the impacts of climate change are uncertain and vary by region, science is currently unable to provide clear guidance on what concentration levels would be dangerous.

In December 1997 governments strengthened the Convention by adopting the "Kyoto Protocol." Industrialized countries agreed to cut their emissions of six greenhouse gases on average 5% below 1990 levels by around 2010. Commitments were "differentiated" in an attempt to reflect the varied conditions of different countries. Marginal costs vary, as does public pressure to slow global warming. Countries where willingness to pay the cost of cutting emissions was high (e.g., in the European Union) generally agreed in Kyoto to cut more than those whose governments were less eager to abate (e.g., Australia and Iceland). The Kyoto Protocol also allows industrialized countries to create a system of tradable emission permits, which could lower the cost of meeting the Kyoto targets for some countries, although the crucial administrative details remain to be resolved.

The most thorny political problem is whether and how developing countries should commit to slow global warming. Emissions from developing countries are rising rapidly (although they will remain lower in per capita terms for the foreseeable future). Favoring economic development over potentially costly climate control, developing countries ensured that neither the FCCC nor the Kyoto Protocol regulated their emissions. Developing countries may benefit from incentives such as "joint implementation (JI)," which could allow industrialized countries to satisfy their regulatory commitments by investing in emissions controls in developing countries where abatement may be less costly. Appealing in principle, schemes for verification and credit tracking have yet to be created. As JI expands, so will the scope of worldwide activities that are subjected to the climate change regulatory regime. As Korea, Mexico, and other developing countries prosper, pressure will mount to explicitly subject their emissions to regulation.

Ultimately, the pace of strengthening the FCCC's controls depends both on public pressure to address the climate problem and on technology. The FCCC was concluded in 1992 at a high point in public environmental concern; the topic of global warming was especially a public concern in industrialized countries. The Kyoto Protocol was adopted when concern was lower and public pressure weaker, but governments still felt the need to show that they were doing something. Over the very long term, technology probably matters most. Decarbonization of the energy system will make it less costly–or even costless–to regulate carbon, which will make it easier for governments to adopt stricter treaties. Perhaps international law will also help speed the pace of decarbonization, although little is known about how international treaties spur (or dampen) technological innovation and diffusion.

David Victor
International Institute for Applied Systems Analysis
Laxenburg, Austria;
Council on Foreign Relations
New York, USA

an article "On the influence of carbonic acid in the air upon the temperature on the ground". It was the first comprehensive study to analyze the
"greenhouse effect" and assess the implications of elevated atmospheric CO_2
concentrations (carbonic acid in the terminology of Arrhenius) on temperature. Thus science has known about potential climate change due to CO_2
emissions from fossil fuel burning for 100 years.[30] However, for nearly eight
decades that knowledge went largely unnoticed apart from a few specialized
scientists. (The author recalls vividly the perplexity of numerous scientific
audiences when the "greenhouse issue" was raised by IIASA scientists in
the mid-1970s.) Today, mass media, chief executive officers, policy makers,
and even the public at large in some countries, take an active interest in
detailed and intricate scientific issues concerning possible climate change.
Harvey Brooks (1996) describes this as an "attention management" problem. There is simply an enormous difference between scientific knowledge,
and the diffusion of that knowledge at the right time to the right people.
More science and new technological options can create "more knowledge",
but that knowledge ultimately needs to be diffused, evaluated, and translated into concrete actions by society at large to be truly useful. This is a
far more complex process than is sometimes assumed, particularly in light
of the wide range of diverse interests, initial conditions, anticipated impacts,
economic capacities, technological capabilities, and policy levers associated
with the different governments, industries, interest groups, and individuals
involved.

6.7.3. Diverse patterns and generic environmental strategies

Disparities in Current and Past Energy Use and Environmental Burdens

Levels of economic development, standards of living, and access to energy services are distributed around the world extremely unevenly (*Figure 6.25*). Disparities are evident even at high levels of regional aggregation,
e.g., between the industrialized countries of the "North" and the developing countries of the "South", and become accentuated as we consider more

---

[30]Arrhenius calculated a possible temperature increase of 5–6°C for a doubled CO_2
concentration (600 ppmv; Arrhenius, 1896:266). For comparison, the most recent IPCC
assessment estimates an increase between 1.5°C and 4.5°C for a 600 ppmv scenario (IPCC,
1996b:39). Fossil fuel use in Arrhenius' time was almost entirely coal. Oil was an unimpor-
tant curiosity, and gas was unknown as an energy source. Nuclear and solar photovoltaic
electricity generation were not even subjects of speculation. The fact that today's energy system, and the sources of CO_2 emissions, are so different from those analyzed by
Arrhenius illustrates both the dramatic impacts of technological change and the inherent
difficulty in anticipating similar changes 100 years in the future.

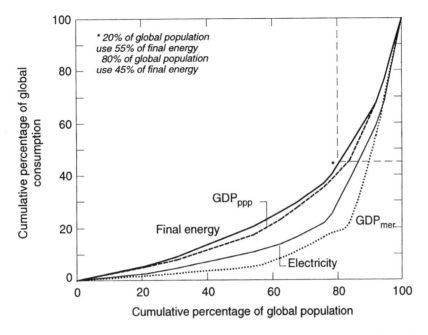

Figure 6.25: Disparities in economic activity and energy use. Cumulative percentage of the global population versus cumulative percentage of global production and use of GDP (at market exchange rates and at purchasing power parities), final energy, and electricity. Source: IIASA–WEC (1995:12).

disaggregated regions, individual countries, and eventually different social strata within countries.

Comparisons based on Gross Domestic Product (GDP) show the richest 20% of the world's population producing and consuming 80% of the value of all goods and services globally (*Figure 6.25*). The poorest 20% dispose of only 1% of total world GDP. Between the countries of the Pacific OECD (Japan, Australia, and New Zealand) and the countries of the Indian subcontinent, GDP per capita varies by a factor of 70, from US$22,800 per capita to US$330 per capita.[31] Economic disparities are somewhat reduced if we use GDP calculated on the basis of purchasing power parities (ppp) rather than market exchange rates (mer). Nonetheless, they remain significant. The richest 20% dispose of 60% of global GDP_{ppp}, while the poorest 20% dispose of only 5%. The relative per capita income ranking of regions

---

[31] Disparities among individual countries and among different social strata are even more pronounced. The poorest 20% in Bangladesh, for example, have a per capita GDP of less than US$90. That is a factor of 700 lower than the US$60,000 annual per capita income of the top 20% in Switzerland.

remains quite stable whether the comparison is based on per capita GDP_{mer} or per capita GDP_{ppp}.

Figure 6.26 illustrates these economic disparities in a novel way. The size of different world regions are renormalized to be proportional to their 1990 GDP (at market exchange rates).[32] The 1990 economic map of the world (lower left corner of *Figure 6.26*) looks highly distorted. Most developing regions where the majority of the world population resides are barely discernible compared to Japan, Western Europe or North America, that became the three largest economies through their successful industrialization.

Equally shown in *Figure 6.26* is the same economic map, but for a medium economic growth scenario that spans out to the year 2050 and 2100. Regions become larger, as their economies grow and disparities become reduced, bringing the economic map in line with the geographic map with which we are all so familiar. This, however, will be a long-term process likely to span the entirety of the 21st century. A range of plausible scenarios (for a review cf. Alcamo *et al.*, 1995) indicates that some 100 years from now the world economy could be anywhere between 10 and 25 times larger than it is today. Without continued technological change and resulting productivity improvements such growth would not be feasible, both economically[33] as well as environmentally.

Given this future outlook, the challenge for technology is two-fold. First, rising levels of economic activity entail (require and cause) rising energy services for light, power, mobility, and comfort (heating and cooling). The extent of the increase of required energy services and corresponding energy use for a given increase in economic activity is mediated to a large extent by technology that can help to progressively decouple energy demand growth from economic growth. Second, technology is also a key element influencing the quality of energy use, both in terms of energy quality as well as environmental quality (i.e., emissions). Higher energy efficiency and cleaner energy supply are therefore two central objectives for improved technology. Historical improvements have been largely unplanned side effects of technological change and have fallen short of the extent by which economic activity has expanded. As a result both energy use and emissions have grown. The technological potential for vastly improved efficiency and zero-emission energy systems exists (cf. Häfele *et al.*, 1986; Nakićenović *et al.*, 1990). But in the race between growing economic activities and energy use and emissions,

---

[32]Only comparisons between regions are made in *Figure 6.26*. Countries within regions are therefore projected according to their relative geographical area (except in the case of Australia and Japan).

[33]Recall here the discussion of the sources of long-term productivity advances and economic growth from Chapter 2.

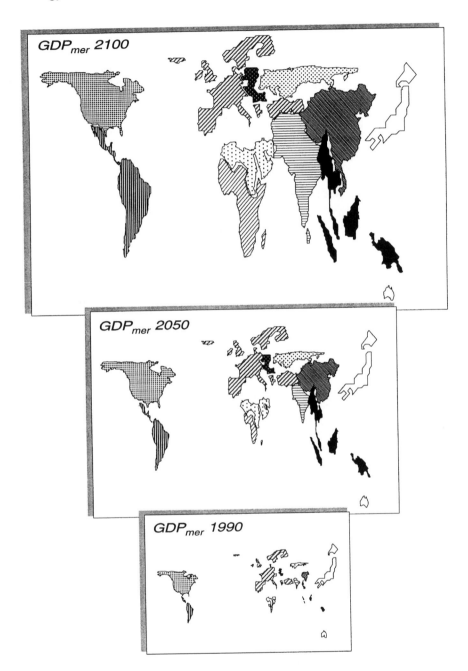

Figure 6.26: Economic map of the world. Size of 11 world regions proportional to their 1990 GDP (expressed at market exchange rates) for the year 1990 and for a medium economic growth scenario for 2050 and 2100. Source: IIASA–WEC (1995:96).

continued, dedicated efforts (R&D, technology demonstration and diffusion, and the appropriate economic signals to induce them) will be required to realize these technology potentials to the greatest extent possible.

Current disparities in energy availability both in terms of quantity and quality mirror the economic disparities among regions as illustrated above. The richest 20% of the world's population use 55% of final and primary energy, while the poorest 20% use only 5%. Per capita use of final energy varies by a factor of 18 between the Indian subcontinent (0.3 toe per capita) and North America (5.3 toe per capita). Of all energy carriers, the disparities are largest for high-quality energy forms like electricity. The richest 20% use three-quarters of all electricity, while the poorest 20% use less than 3%.

Current per capita commercial energy use varies by more than a factor of 20 between North America and South Asia, the highest and the lowest energy-using regions of the world. Total primary energy use in Bangladesh is about 0.4 toe per capita, 10 times lower than per capita use in Western Europe and the Pacific OECD and about 20 times lower than in North America. These disparities are even higher in the use of modern, commercial energy forms (i.e., excluding the use of traditional, noncommercial energy forms such as biomass). For example, commercial energy use is only 0.1 toe per capita in Bangladesh. This is nearly 100 times smaller than Canada's 9.4 toe per capita. In the most extreme cases between individual countries, commercial energy use per capita can differ by factors as high as 500.

Western Europe and Japan have much lower per capita energy use than North America but about the same level of affluence. This indicates a substantial degree of diversity in energy use patterns even among the industrialized countries. Disparities are also large among developing countries. Nonetheless there is a visible and statistically significant relationship between per capita energy use and per capita economic output across individual countries and regions and over time. This relationship is shown in *Figure 6.27* and confirms the need to exploit technologies to lower the energy use and environmental implications of increases in per capita incomes.

At one extreme are the low-income countries with the lowest per capita energy use. They include sub-Saharan Africa (AFR) and South Asia (SAS). As incomes rise so does energy use. At intermediate levels of per capita economic output and energy use are the economies of North Africa and the Middle East (MEA), Pacific Asia (PAS), and Latin America (LAM). Current per capita primary energy use in some of the higher income economies of Asia already exceeds that of some OECD economies. For example, Hong Kong, South Korea, and Taiwan use more energy per person than Turkey, Spain, Portugal or Greece. Per capita energy use in Singapore is about the same as in the UK (IEA, 1994).

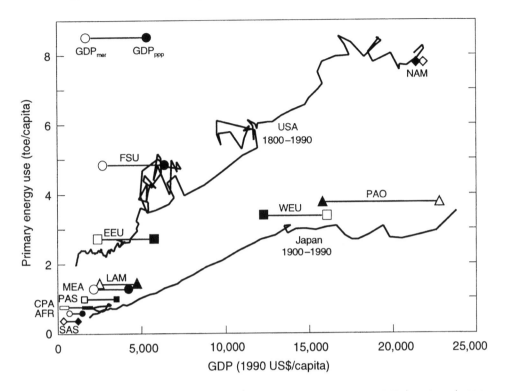

Figure 6.27: Primary energy use (toe/capita) versus GDP/capita (1990 US\$ at market exchange rates and purchasing power parities) for selected world regions (see *Figure 6.28*) and two historical trajectories for the USA (1800–1990) and Japan (1900–1990). Source: IEA (1993, 1994); World Bank (1995). Historical data are based on Nakićenović (1984); Maddison (1995).

Although this pattern of growing energy use with economic development is pervasive, there is no unique and universal "law" that specifies an exact relationship between economic growth and energy use. The relationship depends on many specific and individual factors prevailing in a given country or region. It depends on the historical development paths, natural resource endowments, settlement patterns, transport requirements, structure of the economy, policy and tax structures, and other geographic, climatic, economic, social and institutional factors. And it also depends a great deal on levels and types of technologies used, both at the consumer's end and within the energy sector.

Differences in such factors also explain the persistent differences in energy-use patterns among countries and regions even at comparable levels of income. Different development pathways span all the extremes from "high energy intensity" to "high energy efficiency". This is illustrated in

Figure 6.27 by the development paths of the USA and Japan, which show no apparent signs of convergence. Throughout the whole period of industrialization and at all levels of income, per capita energy use is lower in Japan than in the USA.

Figure 6.28 shows variations in the per capita carbon emissions from different regions that reflect both variations in per capita energy use as discussed, and variations in the structure of economic activities, energy systems, and technologies. The height of the individual bars in *Figure 6.28* is proportional to per capita carbon emissions disaggregated into fossil energy uses (coal, oil, and natural gas) and emissions related to land-use changes. The angled tops of the bars reflect uncertainties in estimated carbon emissions from tropical deforestation. The width of the bars is proportional to population, hence the area of each bar is proportional to total carbon emissions.

The figure emphasizes two characteristics of the history of industrialization. First, industrialized countries, including the formerly centrally planned economies of Europe, have persistently higher energy use and carbon emissions per capita than developing countries. Second, the structure of emissions is different. In developing countries most emissions stem from land-use changes, such as deforestation and unsustainable biomass use for energy purposes, and coal burning. In industrialized countries emissions are almost exclusively from fossil fuel use.[34]

Moreover, the structure of fossil fuel use is more diversified than in developing countries, having shifted away from coal use toward energy forms with lower carbon content (oil and natural gas) or energy sources that are altogether carbon free (not shown in *Figure 6.28*). This primarily reflects technology differences at the levels of energy end use and supply: cooking with fuelwood or coal in a traditional stove in India, versus an electric oven supplied by hydroelectricity in Canada; coal fired steam trains in Africa versus (nuclear) electric high speed trains in France; heating with coal ovens in Northern China versus district heat cogenerated with electricity from natural gas in Russia. Such structural changes in energy systems ("decarbonization") are both an important historical trend and an important generic response strategy to reduce the environmental impacts of energy production and use.

---

[34] Currently carbon emissions from land-use changes in industrialized countries are small or even negative (i.e., carbon is sequestered by forest regrowth which has been made possible by increases in agricultural productivity and surplus production). Historically, however, deforestation was also important in these countries. It is estimated that about half of the cumulative carbon emissions from land-use changes originated in the currently industrialized countries (Grübler and Nakićenović, 1994).

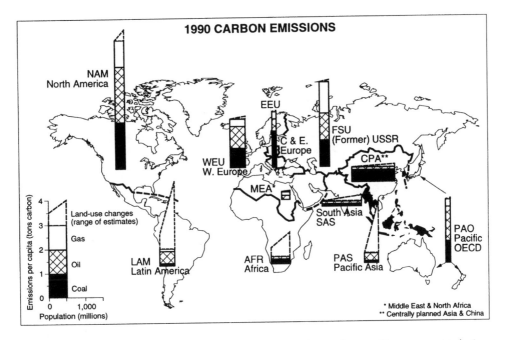

Figure 6.28: 1990 per capita carbon emissions (tons C per capita) by source, versus population (millions) for selected world regions. Source: Nakićenović *et al.* (1996) in IPCC (1996a:86, Working Group II).

The disparities apparent in *Figure 6.28* indicate the direction of future changes. With future population growth concentrated in the developing countries their emission bars in the figure will widen. At the same time, emissions per capita in these regions will rise with economic development and industrialization, and their emission bars will grow taller. Overall, the share of emissions from developing countries will increase. Two caveats are in order. First, growth in energy use (and emissions) in the developing countries should be seen more as an important *precondition* for development, rather than as a regrettable *consequence* of development. Second, it will still be a long time before the "South" becomes as big a contributor as the "North" to excess CO_2 concentrations in the atmosphere. Eighty-five percent of industrial carbon emissions (70% including deforestation) have been emitted since the onset of the Industrial Revolution by the currently industrialized countries (Grübler and Nakićenović, 1994). The major environmental responsibility (as well as the benefits) of industrialization therefore lie with the industrialized and affluent societies of the "North". This disparity is recognized in the FCCC that refers to the "common, but differentiated responsibility" for increases in CO_2 emissions and concentrations.

Given the greater economic resources and technology of the industrialized countries, and their greater historic responsibility for increased CO_2 concentrations, both practical considerations and ethics argue that they should take the lead in addressing the problem. No immediate "technological fix" is in sight, so the focus must be on generic, long-term strategies for lowering environmental impacts of energy use. Ideally such generic strategies should also lead to substantial benefits beyond limiting possible climate change, particularly in case future research shows our initial concerns about climate change to be overstated. But such strategies should also provide the basis for more aggressive policy actions should climate change prove a more serious concern than it is considered today.

Generic Energy–Environment Strategies

The first generic strategy for reducing emissions is to use energy more efficiently and therefore more sparingly. This has been the historical trend (see *Figure 6.20*), and there remains substantial room for continuing improvement. Today in the industrialized countries a unit of GNP requires just one-fifth the energy needed 200 years ago (Nakićenović, 1984). Detailed thermodynamic assessments indicate at least a similar order of magnitude in the potential for further improvement. Analysis based on the second law of thermodynamics (so-called exergy analysis) has shown that as little as 5% of energy inputs may end up as useful service, indicating a theoretical factor of 20 for possible future improvements (Ayres, 1989b; Nakićenović *et al.*, 1990; IPCC, 1996a:79–82, Working Group II).

The second generic strategy for reducing emissions is to use cleaner forms of energy. The ultimate objective would be an energy system that has shifted entirely away from the carbon and sulfur atoms, for instance via methane as a transitional fuel, to an economy based on hydrogen and electricity Ausubel *et al.*, 1988). Again, historical trends have been in this direction, and there remains substantial room for further improvements.

Figure 6.18 showed the growth in global energy use since the mid-19th century. The structural change away from (unsustainable) fuelwood use, to coal, oil, and more recently natural gas and carbon-free energy sources, has led to a gradual "decarbonization" of the global energy system. Decarbonization means a decrease in the specific amount of carbon (or CO_2) emitted per unit of energy used. Carbon can be used generally as a proxy for other emissions like sulfur dioxide or carbon monoxide. These also improve with energy forms that have lower carbon content.

Structural changes lead to decarbonization because the emission factors of fuels vary – from 1.25 tons of elemental carbon (tC) per ton oil equivalent

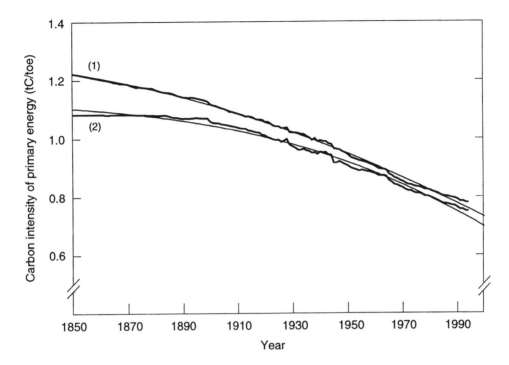

Figure 6.29: Carbon intensity of global primary energy use (in tC/toe). (1) "Gross" intensity includes all forms of energy use and emissions; (2) "net" intensity excludes fuelwood use and nonenergy feedstock uses of fossil fuels. Source: Grübler and Nakićenović (1996:101). For data used to calculate the indices see the Appendix.

(toe) for fuelwood, to 1.08 tC/toe for coal, 0.84 for oil, and 0.64 for natural gas (IPCC, 1996a:80, Working Group II). Combining these emission factors with data on global energy use from *Figure 6.18* provides a clear picture of past decarbonization (*Figure 6.29*).

Figure 6.29 provides two measures of decarbonization. First, it is difficult to determine how much fuelwood use leads to a net accumulation of carbon in the atmosphere.[35] Second, it is difficult to estimate how much of the carbon contained in petrochemical feedstocks (nonenergy uses) is eventually oxidized. We therefore include in the figure one index that includes

---

[35] If the amount of fuelwood burned is smaller than forest regrowth (and carbon uptake by trees), then carbon fluxes are "neutral", i.e., do not lead to a net accumulation in the atmosphere. This is generally the case in industrialized countries today, where fuelwood use is comparatively small. However, in many regions of the developing world that rely to a large extent on traditional fuelwood use, deforestation rates and land-use changes are also high (cf. *Figure 6.28*). The situation in the industrialized countries in the 19th century was also similar. In such cases burning of fuelwood leads to "net" emissions.

both these factors ("gross" intensity), and one that excludes them ("net" intensity).

The key observation from *Figure 6.29* is that the carbon intensity of primary energy use today is some 30–40% lower than in the mid-19th century. "Decarbonization" of the global energy system has proven to be persistent and continuous, although slow at an average rate of 0.3% per year. Despite fundamental changes in both energy supply and end use, as discussed in the following section, decarbonization has been surprisingly regular and nonlinear. Such patterns are common to many dynamic, self-organizing systems.

The historical data in *Figure 6.29* can be closely approximated by a three-parameter logistic curve. The result emphasizes that decarbonization of the global energy system is extremely slow (and seems to have slowed even further, beginning in the 1970s). The logistic approximation has a "half-time" (Δt) of nearly 300 years. Thus, if historical trends continue in this direction, the fossil fuel age may be only half completed as of today. It would draw to a close only late in the 22nd century.

Before turning to the driving forces behind decarbonization, let us briefly summarize some major implications of the overall trends. These are moving in the right direction and justify cautious optimism that development and economic growth can be reconciled with a precautionary policy of avoiding large-scale human interference with the radiative balance of the atmosphere. The task of controlling energy-related carbon emissions appears less daunting when understood as a need to accelerate existing historical long-term structural change trends rather than as a requirement to depart in an entirely new direction.

However, it is also clear that it will not be enough to rely on "autonomous" decarbonization. At 0.3% per year, decarbonization is dwarfed by both historical and anticipated future growth rates in economic output and energy use. Decarbonization will have to be substantially accelerated, and that will require ambitious technological and policy changes. To be successful, such changes must rest on a solid understanding of the forces that have driven decarbonization in the past and how these may evolve in the future.

Decarbonization has been driven by technological change both in energy supply and in end use. Trends in the case of energy supply – primary energy – have been discussed above: the emergence of steam (and coal), later on of internal combustion (oil), the yet more recent growth of natural gas use, and of course electrification.

Turning to end use, *Figure 6.30* shows how carbon intensities have decreased in different regions as economic development has progressed. As for *Figure 6.29*, carbon intensities for developing countries are calculated on a

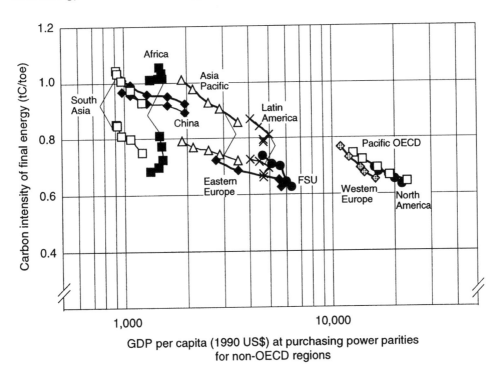

Figure 6.30: Carbon intensities of final energy versus degree of economic development (GDP per capita; for non-OECD regions GDP is expressed at purchasing power parities), 1971–1991. The range for developing countries refers to "gross" and "net" carbon intensities, respectively. Source: Grübler and Nakićenović (1996:105).

"gross" and "net" basis (i.e., including/excluding fuelwood), respectively. Four observations from *Figure 6.30* are noteworthy. First is the considerable variation among regions. Second is the persistent trend toward lower carbon intensities with rising per capita income. Third is how much lower carbon intensities of final energy are (below 0.7 tC/toe) than those of primary energy.[36] Fourth, decarbonization at the level of final energy proceeds faster than for primary energy. The data in *Figure 6.30* translate into an

---

[36] Primary energy carbon intensities are higher primarily because they include the emissions arising from the conversion of different primary energy to final energy forms. They are also higher because high carbon fuels such as coal that are of limited attractiveness for the final consumer have retained an important market niche in electricity generation. In this case emissions accrue at the level of primary energy (conversion of coal to electricity), but not at the level of final energy (use of electricity). For a discussion of different carbon intensities at the level of final and primary energy and of energy conversion (i.e., the difference between primary and final energy) see Nakićenović (1996a).

annual improvement rate of 0.6% per year in the case of the USA or India. This is twice as fast as carbon intensity improvements in the case of primary energy. In some instances improvements are even faster. In France, with its ambitious program of nuclear electrification, the carbon intensity of final energy has declined since 1960 by an average of slightly more than 1% per year.

Lower carbon intensities for final energy are the result of shifts cited earlier reflecting both technological requirements (one cannot run a computer with fuelwood) as well as consumer preferences for higher quality energy forms like electricity, district heat, gas, and liquids. Therefore, these shifts are particularly pronounced in high-income economies. Conversely, final energy use in low-income, developing countries is dominated by solids with a high carbon content, fuelwood and coal. Such high-carbon solids have virtually disappeared as end use fuels in high income countries, with the notable exception of some coal use in the metallurgical industry.

In addition to consumer preferences for cleaner, more convenient energy forms, decarbonization is more rapid for final energy because of additional constraints on primary energy development. Some constraints are techno-logical. For example, electricity as an end use fuel is carbon-free, but carbon emissions are usually produced at the point of electricity generation. Some constraints are imposed by government policy. In the USA and the European Union, for example, it is only recently that restrictions on using clean natural gas for electricity generation have been lifted.

All things considered, however, the long-term persistent trend is one of decarbonization as a result of continuous technological change and the quest for ever higher quality energy in terms of flexibility, convenience, and cleanliness. Together with improvements in energy efficiencies and aggregate energy intensities (energy needs per unit of economic output), this trend promises to help lighten environmental burdens in the future, other things being equal. As an illustration of this promise, the final section of this chapter describes how such trends in the past have worked together to lessen relative environmental burdens in the USA, where historical data allow us to go back as far as 1800.

6.7.4. Efficiency improvements and decarbonization

Figure 6.31 presents the growth in primary energy use and economic activity in the USA since 1800. The data include not just commercial energy forms, but also the fuelwood and water power (water mills) that dominated US energy use in the 19th century. For instance, coal only surpassed fuelwood use by the 1880s. Reliable long time-series data of energy consumption represented by feed for working animals and riding horses are unavailable. One estimate (Nakićenović, 1984) indicates that feed energy peaked at some 60 Mtoe around 1910, roughly representing 10% of all energy use for that year. Although renewable energy sources are today frequently assumed to be environmentally benign, historically they had substantial environmental impacts. By the 1920s, growing feed for working animals and riding horses required vast tracts of land that were then unavailable for food production. At the height of the horse era, some 40 million ha were required in the USA alone (cf. Chapter 5). Intense demand for fuelwood for domestic uses, industry (charcoal for the iron industry), and also transportation (railways) resulted in significant deforestation and even fears of a "timber famine". Throughout the 19th century, railroads in the USA were heavy consumers of wood. Locomotives used wood for fuel, railroad cars were made from wood, and railroad cross-ties were wood. Without preservation, cross-ties needed to be replaced regularly. The result was a tremendous demand for timber. To quote from a speech by President Theodore Roosevelt to the American Forest Congress:

> Unless the vast forests of the USA can be made ready to meet the vast demands which this [economic] growth will inevitably bring, commercial disaster, that means disaster to the whole country, is inevitable. The railroads must have ties .... If the present rate of forest destruction is allowed to continue, with nothing to offset it, a timber famine in the future is inevitable. [Quoted in Ausubel, 1989:72]

Progressive deforestation and the timber famine threatened by the railroads were environmentally unsustainable and contributed, among other things, to the rise of the conservation movement (cf. Hays, 1959) that led to the beautiful US National Parks system. Two technological developments, however, helped avert the worst: the replacement of wood by coal for locomotive fuel, and the development of chemical preservatives (creosote, cf. Ausubel, 1989:73). As summarized by Ausubel (1989:72): "... in the railroad timber story, new technologies are both cause and cure of environmental problems. The new transportation system [railroads] placed intense demand on natural resources, and innovations in turn alleviated the demand to the extent that today the issue is obscure and forgotten". Somewhat ironically, present-day fuelwood use in the USA is estimated at 75 Mtoe (EIA, 1997), which is an all-time historical high, but without provoking fears of resource overexploitation.

Richards (1990:164) estimates that between 1700 and 1950 some 77 million ha of forests were cleared in North America. Houghton and Skole (1990:400) estimate the resulting carbon flux at close to 40 Gt of carbon, equal to the cumulative carbon emissions reached by fossil fuel use in the

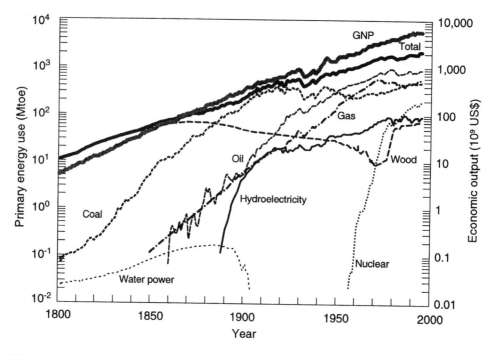

Figure 6.31: Primary energy use, by source (in Mtoe, left scale) and economic output (GNP, billions of constant 1990 US$, right scale) in the USA since 1800. Source: updated (US DOC, various years) from Nakićenović (1984:213–216). For the data of this graphic see the Appendix.

USA only at the end of the 1960s (cf. Grübler and Nakićenović, 1994). Given these numbers, carbon emissions from fuelwood are incorporated in all the calculations presented here, and no separate carbon intensities excluding fuelwood are calculated.

Applying the carbon emission factors introduced earlier to the data in *Figure 6.31*, we can calculate the overall carbon intensity per unit GNP for the USA since 1800. The result is striking. A unit of GNP in the USA today is produced with only one-tenth of the carbon releases of 200 years ago. By 1800, 2.5 kg of carbon was released from energy use per (1990) US$ GNP, compared to less than one-quarter of a kilogram of carbon today. In 1800 the releases came almost exclusively from fuelwood. Today they are almost exclusively from fossil fuels.

The improvement by a factor of 10 in the carbon intensity of GNP translates into an average annual improvement rate of 1.3% per year. *Figure 6.32* shows the relative contributions of energy efficiency improvements on the one hand, and reductions in the carbon intensity of primary energy on the other.

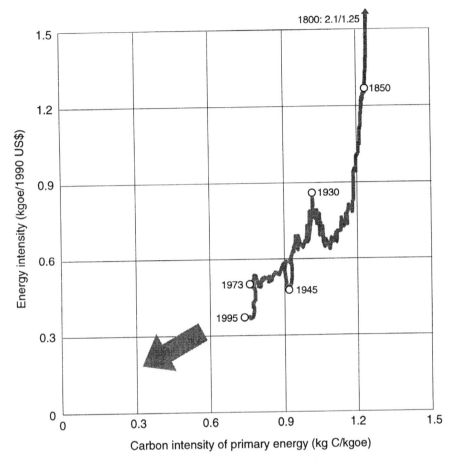

Figure 6.32: Declines in the primary energy intensity of GNP (kgoe/1990 US$) and in the carbon intensity of primary energy (kg C/kgoe) in the USA since 1800. Source: calculated from data given in *Figure 6.31* (see also the Appendix).

Three-quarters of the overall improvement come from using energy more sparingly, i.e., an energy-intensity decline averaging 1% per year. The other one-quarter comes from reduced carbon emissions per unit energy used, i.e., a carbon intensity decline averaging 0.3% per year, the same as the global decarbonization trend given in *Figure 6.29*. Despite these tremendous improvements in the "carbon productivity" of the US economy, however, they fell significantly short of the growth rate in economic output, approximately 3% per year. The difference of 1.7% per year translates into a corresponding absolute increase in US carbon emissions from energy use. By 1993 these had reached 1.4 Gt annually (EPA, 1995:Executive Summary, p. 3).

As this example illustrates, even substantial productivity improvements in resource use and environmental impacts per unit of economic output can be overtaken if the economy grows fast enough. There is a continual race between productivity improvements and growth in economic activity, both driven by technological change. Those who are skeptical should simply try to imagine the US economy operating at its current size with the technologies, and the energy and carbon productivity levels, of 200 years ago.

We have deliberately chosen carbon emissions as an environmental indicator of productivity improvements because other factor inputs such as labor are well documented, and their tremendous improvements since the onset of the Industrial Revolution are beyond debate (and were presented above). All productivity increases are driven by the same basic process of technological change in the form of new production methods, new products, and new forms of organization – in short, the process of "industrial mutation that incessantly revolutionizes the economic structure from within" to return to the quote in Part I from Joseph A. Schumpeter. Thus both productivity increases and economic growth are closely related and are the result of continuous technological change that arises from "within" the economic system. That is, it is an endogenous process that cannot be treated as an "externality" either in economic theory or in environmental policy.

An analogy can be drawn between environmental productivity increases as illustrated by carbon intensity improvements, and more traditional productivity improvements, e.g., in costs or labor. Both evolve along a classical learning or experience curve (introduced in Part I) as shown in *Figure 6.33*. Just as manufacturing costs decline and productivity increases as experience is accumulated (technological learning), the same process appears to occur for environmental productivity. For each doubling of cumulative output (measured by GNP), the US economy has used some 20% less carbon per dollar of GNP. The "learning" rate is surprisingly persistent even across successive technology clusters, from a fuelwood economy, to a coal economy, and to one based on oil, gas, and electricity. It is tempting to relate the persistence of this overall technological trajectory of "decarbonization" only to technology push factors and economic efficiency. Historically, however, initial technology choices were neither governed by *ex ante* calculations of economic efficiency, nor did they come in the form of discrete technology choices of a limited number of decision agents. Rather, they were gradually acquired through cumulative technology improvements and learning over successive technology clusters, including literally hundreds of individual technologies.

The overall regularity and persistence of this process over such a long historical time span and across so many technological generations (clusters) is simply astounding. We agree with Kates' (1996:51) interpretation of such

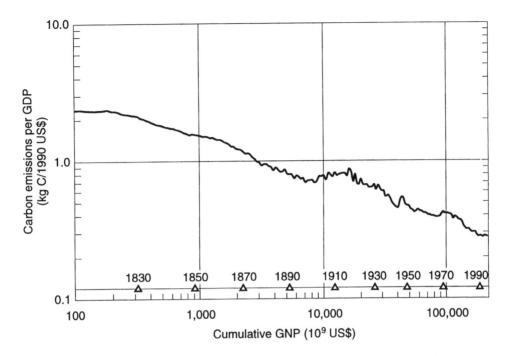

Figure 6.33: Learning (experience) curve of US carbon emissions per unit GNP since 1800. Specific carbon emissions (kg C per 1990 US$ GNP) versus cumulative output (accumulated US GNP, in billions of 1990 US$ GNP), double logarithmic representation. Source: calculated from data given in *Figure 6.31* (see also the Appendix).

trajectories as the result of a "complex coevolutionary process of technology selection within the totality of the human environment".

Nonetheless, even if the driving forces of these coevolutionary processes need further detailed exploration, *Figure 6.33* provides strong evidence that improvements in environmental productivity cannot be separated from other productivity improvements within a growing economy driven by a constantly changing technological base. Identical evidence at the sectoral level confirms this macroeconomic pattern (e.g., for the carbon intensity of US steel manufacturing, cf. Grübler, 1995a:52). Both environmental and other productivity improvements are driven by continuous technological change, and this resulting shift in energy systems toward a leaner carbon diet (i.e., increased energy "quality") was first initiated by the great transformations subsumed here under the "Industrial Revolution."

However, whatever technology has "given" in the form of increased environmental productivity, it has more than "taken back" through concomitant increases in output. This has been enabled by new technologies, products, forms of organization, and their impact on productivity. This is the fundamental environmental dilemma of technological change. Escaping this dilemma is *the* challenge for the 21st century. The task is to substantially accelerate historical trends toward dematerialization and decarbonization. Fortunately, trends are promising and powerful winds seem to be blowing in the right direction. The challenge for society at large is to install an additional motor in the sailing ship of technological change and social choice in order to more speedily reach the shores of long-term environmental compatibility.

Copyright acknowledgments

Chapter 7

Services

Synopsis

For the service sector the most important impacts of technological change are changes in how individuals use their time – their "time budgets" – and changes in consumer expenditures. Longer life expectancies, shorter working hours, and vastly rising incomes have changed time budgets and expenditure patterns in ways that have significant environmental impacts. A principal example is increased personal mobility – a consumer demand that appears far from satiated. Increased demands for ever more personal mobility have been largely met by motorized vehicles. Thus emissions from transportation, along with a whole variety of other environmental impacts, have grown substantially. Fortunately, projecting future transportation growth from historical innovation diffusion patterns indicates lower environmental impacts than are suggested by traditional linear extrapolations, assuming business-as-usual. Yet, the growth of the service economy and the consumer society is such that these could soon rival agriculture and industry as major sources of global change. Thus individual lifestyle decisions, particularly decisions about *which* artifacts are used and *how*, become ever more important in determining the type and scale of environmental impacts. One important example described in more detail is that of food. With rising incomes food demands become increasingly saturated. In the industrialized countries, further agricultural productivity increases from biological and mechanical innovations can then be translated into actual absolute reductions in agricultural land use, even while production and exports continue to increase.

7.1. Introduction: From Work to Pleasure

Of the three major economic sectors – agricultural, industrial, and services – the service sector is the most varied and, with respect to technological change, the least studied empirically. Therefore our discussion is necessarily highly stylized focusing on "grand" patterns of structural change. Productivity increases in the primary (agriculture) and secondary (industry) sectors have been so substantial that there has been a significant shift away from these sectors toward services. This move toward services occurred both in the amount of hours worked and money generated (and spent), particularly in connection with numerous newly emerging free-time activities. This over-all transition from "work" to "nonwork", or straightforward "pleasure", is the central theme of this chapter. It is environmentally significant because the incentives that drive consumptive behavior are radically different from those in the traditional "productive" sectors, and they are inherently more difficult to influence.

In advanced industrialized societies, services now account for typically two-thirds of economic output and jobs. In the USA and Switzerland, for example, the service sector constitutes 75% and 66% of GDP, respectively. It provides 72% of US and 64% of Swiss employment (ILO, 1995:198–241). Some of the shift to services is due to new types of economic activities (e.g., winter tourism) and disproportionate growth in traditional services, e.g., the increasing commercialization of traditional domestic chores such as food preservation and preparation or cleaning. However, much of the shift results from changes in the organization of economic activities. Just as numerous activities previously performed directly on farms are now the do-main of industry (manufacturing seeds, fertilizers, tractors, food processing, etc.), many industrial activities are now performed within the service sec-tor, ranging from design and advertising to distribution and retail business. Previously *vertically* integrated industrial activities (i.e., where an indus-try performed all activities from raw material production and acquisition to design and manufacturing, to distribution and retail) are becoming increas-ingly *horizontally* integrated. The shift also partly explains the fact that services, traditionally considered mostly "low-tech" activities, have become one of the largest consumers of new technologies, particularly information and communication technologies. In the USA, out of an estimated 1991 total of US$153 billion in nongovernmental investments in information technology hardware (i.e., excluding software), the service sector accounted for US$127 billion (over 80%) (NRC, 1994:2).

Although technology changes have been important in some areas of the service sector – most notably in transportation, communication, financial services, trade, and retail services – technology "hardware" has not been the central driver of change in services. Most radical innovations in the service sector can be characterized instead as "software", i.e., new forms of *orga-nizing* services. In financial services, new "technologies" have included new financial instruments as well as new communication technologies (telegraphs in the 19th century, and the telephone, computer networks, and electronic trading in the 20th). In retail services, the introduction of large centralized department stores in France and catalogue shopping in the USA[1] revolutionized shopping in the 19th century, as did hypermarkets (another French innovation now known as supermarkets) in the 20th century.

In terms of hardware, the service sector has remained largely a technology "taker", using new technologies that originated primarily in the industrial sector. Only in the second half of this century have there been a significant number of new service sector technological artifacts. Some have transformed typical traditional service activities, such as household chores. Others have made possible genuinely new types of service activities, such as entertainment (radio, TV, cinema, records, CDs, VCRs, etc.), recreation (e.g., alpine skiing), or social interaction (e.g., the rapid growth of e-mail and the internet).

The concentration of technological change in the service sector in the "mass production/consumption" technology cluster during the second half of the 20th century is striking. Unfortunately, quantitative data and analyses comparable to those discussed in the last two chapters for agriculture and industry are much harder to come by for services. In general, the service sector suffers from a lack of attention from technology historians and literature alike.[2] Consequently evidence presented in this chapter will necessarily be more anecdotal.

7.2. Measurement: Time Budgets and Consumer Expenditures

7.2.1. Spending time

A day cannot be made longer than 24 hours. However, life expectancies can increase, and with them the ultimate time budgets of individuals (cf. *Figure 7.1*). In 1870 in Europe life expectancy at birth was typically around 40 years for both women and men. Approximately 100 years later life expectancy had grown to 77 years for women and 70 years for men. Disparities

---

[1] The most prominent example is the Sears Roebuck Company catalogue. Among other things this is an invaluable source of data on consumer products for social science research. The catalogue existed for nearly a century and has been the basis for some of the most innovative insights into consumer products research. Examples are the stability of the distribution of relative prices of consumer items (Montroll, 1981) and the significant improvements in consumer product quality (Payson, 1994). Much to the detriment of research the Sears Roebuck catalogue no longer exists.

[2] Notable exceptions are publications from SPRU (Smith, 1986) and the US National Academy institutions (NAE, 1987; NRC, 1994) that discuss current technology trends and impacts in services. Other than Jonathan Gershuny's work (e.g., 1978, 1983, 1989) for the UK, comparable historical accounts have not yet been written.

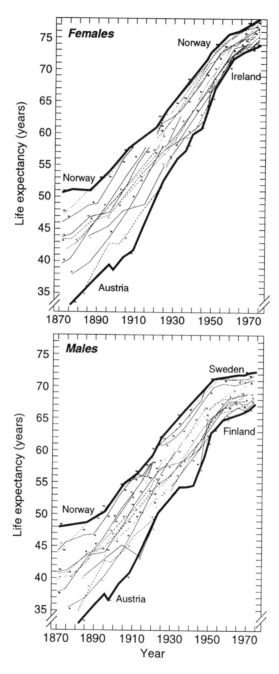

Figure 7.1: Life expectancy (years) at birth for females (top) and males (bottom), from 15 industrialized countries, 1870–1975. Source: adapted from Flora *et al.* (1987, II:92–95).

among countries have also narrowed significantly. In 1960 there were only 12 industrialized countries where the average life expectancy exceeded 70 years. Now it exceeds 70 years in all industrialized countries. In developing countries average life expectancy has increased by one-third over the last 30 years, and 23 developing countries now have average life expectancies above 70 years (UNDP, 1993:12–13). Over the last 100 years, life expectancies and available lifetime time budgets have roughly doubled in industrialized countries. At the same time, the number of hours worked (cf. *Figure 6.11* and discussion below) has fallen to about half its level in 1870. In summary, we are "working less and living longer" (Ausubel and Grübler, 1995).

It is tempting to attribute both trends to technological change, either directly or indirectly. However, in the case of life expectancy increases, at least, the argument is not straightforward. Life expectancy started to increase as early as the 19th century when modern medical technologies and *techniques* in the form of vaccines, antibiotics, and public health care services (that have been important contributors to declining mortality from infectious diseases in the 20th century) were not yet available on a large scale. Life expectancy increases in the 19th century also appear at odds with the abominable living conditions of the rural and urban poor (i.e., the majority of the population), as so vividly described in Friedrich Engels' writings on the working class in Victorian England (Engels, 1845). They appear equally at odds with deteriorating local environmental conditions, especially indoor and outdoor air quality, characteristic of the beginnings of the steam and coal age.

Instead, initial increases of life expectancy are perhaps better explained by less frequent and less severe epidemics of traditional infectious diseases such as typhoid, measles, or scarlet fever (Ponting, 1991:233–234). This in turn can be partly explained by increasing immunity, by gradually improving diets (cf. Chapter 5 on agricultural productivity increases), and by improved basic hygiene due to "reinventing" the old Roman technology of urban sewers and clean water supply systems. None of these changes is directly related to the quickening pace of technological change of the early 19th century. With the exception of smallpox immunization through cowpox vaccination, introduced around 1800, modern medical technologies appear not to have played a major role in the initial increases in life expectancy. The death rate from tuberculosis, for example, a major infectious "killer" of the 19th century, dropped by one-half in England between 1838 and 1882, at a time when the bacillus causing TB had not yet been identified (Ponting, 1991:234). It is only much later, in the 20th century, that one can clearly link the reduction in mortality rates from infectious diseases to technology developments in hygiene (e.g., milk pasteurization, food preservation, and refrigeration) and medical technologies (vaccines and antibiotics). Today, aquatic and airborne diseases, the main killers of especially children, are largely eradicated; however, some fear their possible resurgence as the resistance of diseases to antibiotics grows, with the increasingly widespread (and sometimes careless) use of antibiotics in human medicine and industrial animal husbandry.

In addition to the overall decline in mortality over the last 100 years, there has been a major transition in the causes of death – from infectious diseases 100 years ago, to cardiovascular (e.g., heart attacks) and malignant diseases (cancer) today. In 1900 in the USA infectious diseases claimed 660 lives per 100,000 people. Cardiovascular and malignant diseases claimed another 200 lives. Since then the death rate for cardiovascular and malignant diseases has climbed to 625 per 100,000 people, comparable to the level for infectious diseases some 80 years ago. The death rate for infectious diseases has decreased from 660 to less than 30 per 100,000, i.e., by a factor of more than 20. The aggregate mortality rate has decreased as reflected in increased life expectancy. Average US death rates declined from 860 per 100,000 in 1900 to 650 per 100,000 in 1980 (all data from Quinn, 1987:142).

This "mortality transition" holds important policy and lifestyle implications. Revolutionary medical progress has limited promise as long as simple and clear causes (and cures) for today's principal causes of death (e.g., numerous forms of cancer) remain an enigma to biological and medical sciences. Moreover, traditional external risk factors such as exposure to polluted water and air, contaminated food, and the lack of medical care, have been increasingly replaced by self-determined risk factors, like diet, smoking, and physical fitness.

We now turn to the second element of the time budget equation, the reduction in formal working time. Here the influence of technological change is more obvious. Technological change has led to tremendous increases in productivity, which in turn have been distributed partly via increased incomes, and partly via reduced working time. However, while productivity increases through technological change created the *potential* to reduce working time, actual realized reductions depended on social and political processes and institutions.

Reduced working hours have been a goal of laborers since time immemorial. The first big break came in agriculture where mechanization and external energy inputs, culminating in the cheap and dependable tractor, made the shift from farms to cities possible. Prior to these developments, 80% of the population was needed for farming, afterward, only 20%. Initially, however, city jobs proved to demand even more time, on an annual basis, than farm jobs.

In agriculture, the work schedule was built around the growing season with the peak load usually coming at harvest time. Industrialization brought a qualitative and quantitative transformation. Working time during early industrialization increased dramatically, up to 14–16 hours per day (Nowotny, 1989). This represented an extension of peak agricultural working hours, such as at harvest time, to a year-round norm. Annual industrial working times in excess of 3,000 hours were common in the mid-19th century. At the same time, a qualitative transformation occurred, specifically continuous monetary evaluation of working time (see e.g., Hareven, 1982).

With an increasingly monetarized economy, better government statistical offices, more systematic tax collection, the rise of labor movements, and other forces, data were collected that now allow us to estimate hours of paid work for many countries in the latter part of the 19th century. The UK is of particular interest given its history of early industrialization, and we summarize the major trends below (drawn from Ausubel and Grübler, 1995).

Table 7.1 summarizes the decline in total lifetime hours worked in the UK between 1860 and 1980 for both women and men. The table illustrates the contributions to the overall decline from three key variables: reductions in the number of hours worked per week, longer vacations (i.e., reductions in the number of weeks worked per year), and changes in the length of a working career. The most important variable for both men and women has been the reduction of weekly working hours by shortening the working day and lengthening the weekend. Longer vacations have played less of a role, accounting for only 17% of the 1856–1981 reduction in lifetime working hours of both men and women.

The influence of changes in the number of years workers spend in the work force is more varied. For men through the 1930s there was a tendency toward longer working careers because of longer life expectancies. This trend was only partly offset by reductions in working time due to shorter working days and more generous vacations. Over the same time period, however, women experienced a slight reduction in lifetime working hours due to shorter work careers. Since the 1930s, the situation changed for men as their working careers became shorter due both to longer formal education and earlier retirement. The result was a 40% reduction (close to 14,000 hours) in lifetime working hours from 1931 to 1981. Combining the increase in lifetime working hours for men prior to 1930 with the subsequent decrease results in an overall slight decrease from 1856 to 1981.

For women similar conclusions hold. Reductions in weekly working hours predominate over longer vacation periods in contributing to a long-term decrease in lifetime working hours. Note, however, that because of the significant share of part-time work and of generally shorter work careers (owing

Table 7.1: Lifetime hours at (remunerated) work and reductions by source (in hours) for the UK, 1856–1981.

	Men	Women
Lifetime		
Hours worked 1856	149,690	62,750
Change 1856–1931		
Working less[a]	−36,760	−18,845
More vacations[b]	−1,744	−720
Shorter/longer work career[c]	+11,674	−2,675
Lifetime		
Hours worked 1931	122,860	40,510
Change 1931–1981		
Working less[a]	−12,497	−16,698
More vacations[b]	−8,534	−3,280
Shorter/longer work career[c]	−13,779	+19,268
Lifetime		
Hours worked 1981	88,050	39,800

[a]Changes in hours worked per week (lower values for women due to shorter work careers).
[b]Changes in weeks worked per year (lower values for women due to shorter work careers).
[c]Changes in years worked.
Source: Ausubel and Grübler (1995:198).

to childcare), lifetime hours at compensated or formal work are lower for women than for men. Since the 1930s the reductions in lifetime working hours for women (around 700 hours in 50 years) are also much less than those for men. The explanation is that the increase in the length of the average female work career from around 20 to 30 years has almost completely compensated for shorter working days and weeks and for longer vacations.

Still, the long-term reduction in lifetime working hours remains impressive for both genders. Since 1856, lifetime working hours have fallen by 61,640 hours for men (a 42% reduction) and by 22,950 hours for women (a 37% reduction). However, while lifetime working hours have decreased substantially, note that in the UK, over a time-scale of more than a century, the average working career has changed little: it remains at about 40 years (men and women taken together). Reductions in career length for men have been balanced by corresponding increases in female participation rates and career lengths.

The length of a working career is related to the speed at which innovations can diffuse through an economy. If the average age at which a person leaves school is assumed to be 15 years and a working career lasts 40 years, then the average lifetime of the human capital stock (its formation, integration, and use in the productive sphere of the economy) is about 55 years. The maximum speed at which innovations can diffuse is related to learning rates for individuals and groups, and whether individuals or groups have already become saturated, or locked-in to particular procedures and technical know-how and thus unable to accept new ideas or practices. From this perspective, it might be argued that 55 years are needed to replace entirely a workplace organization that is no longer satisfactory but has become fixed in the minds of the managerial and labor force.

In any event, the relatively steady length of working careers at 40 years represents a slow variable among factors affecting workplace change and performance. Jobs evolve, the work force turns over, working hours are reduced, technologies change, but the length of a working career and the length of social memory in the workplace remain roughly constant. This regularity is important for employers and governments to recognize when developing education and re-training policies, especially if there are rapid changes in technologies and corresponding skill requirements.

Several explanations have been offered for the long-term decrease in lifetime working hours. In *A Theory of Wages* (1934), Douglas views the decrease as the result of decisions by workers in response to increased pay. Workers choose to divide the benefits of productivity gains between additional income and leisure. Douglas also views the reduction in the number of years spent in the labor force as a consequence of higher family incomes and government expenditures on pensions. Owen (1978, 1979) extends pure economic arguments to explore how entrepreneurs in a competitive market try to minimize labor costs (hire the best labor at the lowest cost). The theory argues that changes in employee preferences induce employers to reduce working hours, all in the interest of minimizing the employer's labor cost, or "efficiency wage". For others (see e.g., the discussion in Dumazedier, 1989) the decrease in lifetime working hours results primarily from productivity increases driven by permanent scientific/technical revolutions. The time that is thus "generated" along with the wealth that is produced is then distributed according to the relative power of different social classes. Historians such as Thompson (1967) have consequently placed more emphasis on group and class interactions and negotiations and social movements in trying to explain the trends. Despite such differences, however, all explanations revolve around what Leontieff (1978) has described as the "inexorable forces of technological change in increasing labor productivity" (cf. Chapter 6).

Trends in other industrialized countries are similar to those discussed for the UK (cf. *Figure 6.11*, and Maddison, 1995:248). The one important exception is Japan. There, on average workers currently work 400 hours more than in other industrialized countries, laboring about as long as Americans or Europeans did in the 1950s.

Over the last decade, however, data on working hours have departed from the long-term trends just described (Marchand, 1992), and this has undoubtedly exacerbated chronic unemployment problems in most OECD countries. The formalized work contract and working time are thus becoming increasingly distributional issues. Historically, reductions in working time have been more or less uniform across the work force (with the exception noted above of different average career lengths for men and women). However, more recent reductions appear to be distributed increasingly unevenly. Thus the overall reduction in working hours for the society as a whole comes at the expense of an increasing fraction of the working population that is unemployed or is working only part-time. At the same time there is a corresponding reduction in the fraction of the work force working full-time and the traditional 40-hour week (or even more in highly qualified professions like doctors, managers, and software developers). Indeed, there are increasing sentiments of being "overworked" (Schor, 1991), as revealed by panel data, where people are asked how much they *think* they work. This data collection technique is opposed to direct observations or diary techniques that try to measure how many hours people actually *do* work (and that generally confirm the above-discussed trends of declining hours worked).

Overall, the importance of the formalized work contract, and the amount of time that it governs, are both diminishing. Indeed, there is a growing need for serious research on the portions of life that are not part of the formal work contract notwithstanding its continued importance for shaping social relations and structured time experiences (Jahoda, 1988). To measure the decline in the relative importance of formal work, we can combine data on increasing life expectancies with the data above of reduced working time. From life expectancy data we first calculate the total "disposable" hours available in a lifetime by subtracting 10 years for childhood and elementary education and 10 hours per day for sleeping, eating, and personal hygiene (so-called physiological time). How these disposable hours are then allocated to work and other activities is shown in *Figure 7.2* for the male work force in the UK.

For the overall work force in the UK, average lifetime working hours decreased between 1856 and 1981 from about 124,000 to 69,000, while disposable lifetime hours increased from 292,000 to 378,000. Thus, nonworking hours increased from 118,000 to 287,000. And while more than 40% of the disposable lifetime of adults was spent at work in 1856, we now spend less than 20% of our disposable lifetime hours working.

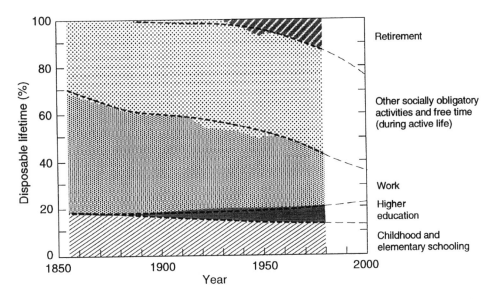

Figure 7.2: Allocation of lifetime hours to different activities for the male working population in the UK from 1856 to 1981 (excluding physiological time). Other socially obligatory activities include caring for children and household activities. Source: Ausubel and Grübler (1995:206).

For the male UK work force shown in *Figure 7.2*, the percentage of disposable lifetime hours spent at work dropped from more than 50% in 1856 to less than 25% in 1981. Even more significant are the variations among the gains in nonwork activities. Overall these increased from 31% of an average male lifetime in 1856 to 65% in 1981. The smallest relative gain in nonwork time is for other socially obligatory activities and leisure proper that take place during one's active working career. Their net increase equals 13.5% of the lifetime time budget. In contrast, nonwork activities before one's active working career (i.e., higher education) and after the working career (i.e., retirement) have increased from zero to around 20% of the lifetime time budget.

The increased time devoted to higher education reflects the growing importance of prework preparation, which can be seen both as a prerequisite for, and as a consequence of, continued technological change and productivity increases. This is the time budget equivalent of increases in "human capital" that are so important for sustaining productivity increases, technological innovation, and economic growth.

The component of the male nonwork time budget that has risen the fastest, however, is the time after one's active working career: retirement.

Until the 1930s, men worked, in effect, until they died. That has changed drastically. Currently retirement accounts for close to 13% of the average male lifetime in the UK. As the population continues to age and work occupies less and less of one's lifetime, retirement could increase to about one-quarter of the total lifespan within the next two to three decades. Indeed, if current trends continue, as much as half of the lifetime of the male work force after 2000 will be accounted for by pre- and afterwork activities and only half by the active working career. Moreover, within the half devoted to an active working career, formal working time will account for a decreasing fraction, possibly 30% or less. The result will be more time for both leisure and other socially obligatory activities such as caring for children and the home.

For women, the changes since 1856 in lifetime time budgets have been broadly similar to those for men. The exception is that female lifetime working hours have stayed roughly constant since the 1930s. Their share of the overall time budget has therefore declined only as much as life expectancy has increased. The reason for constant lifetime working hours for women since the 1930s is the pronounced discontinuity in both female participation in the work force and the length of female work careers after World War II. At that time both increased substantially. Thus reductions in working hours have been offset by more women joining the work force and working more years.

A number of explanations for the discontinuity have been offered (Harris, 1981). One category of possible contributors includes innovations in household technologies, medicine, and public health. These affect both the ease of household work and the number and health of children. When the tin can was invented in the 19th century, it was predicted that the resulting reduction in the time needed to prepare meals would lead to more time spent outside the home at formal work. Several electrical appliances were forecast to have similar effects: the iron (invented in 1882), sewing machine (1889), stove (1896), clothes washing machine (1907), and domestic refrigerator (1918) (source: Desmond, 1987). Mr. Birdseye's success with frozen foods in 1929 prompted the same statements that had been made about tin cans 50 years earlier.

Despite these predictions, women in the UK (and elsewhere) did not lengthen their working careers until many household innovations achieved widespread diffusion. One can postulate a conservation law, of sorts, according to which homemakers spend a certain amount of time on household work. Innovations lead not to less work but to quality and quantity improvements. For example, diets may be of higher quality and more varied; houses may be cleaner; and one individual may be able to care for more space. In any event, household innovations appear to have had little impact on time budgets before they became pervasive and complementary to one another (Vanek, 1974; Strasser, 1982).[3] Studies for the USA also show only minor changes between the 1930s and 1965 (Robinson and Converse, 1967). In reviewing a number of industrial societies, Minge-Kalman (1980) concluded that time spent on housework even increased while work outside the home decreased.

From the 1960s to the 1980s, however, women appear to have substantially reduced their time spent on housework (Gershuny and Robinson, 1989). Gershuny's explanation (1984) is that as long as the new domestic technologies were comparatively crude (e.g., the first generation of top-loading washing machines) and required substantial ancillary labor and constant attention, little time was saved on domestic chores. Any productivity increases went into improved quality and quantity, not time savings. That is, the first washing machines were used in much the same way and for the same amount of time as traditional washing methods. The result was more clean clothes, not less time spent washing. Only in the early 1960s did things begin to change with the introduction of modern, automatic washing machines that could be used in a "batch" mode without continual attention. This altered the organization of household work, resulting in less time spent washing clothes.

Similar arguments hold for other domestic innovations such as the microwave oven. They had little effect on the time spent cooking as long as they were used in more or less the same way as a traditional oven. As preprepared meals specifically designed for microwaves became available, reductions in the time spent preparing meals became noticeable.

The washing machine and the microwave illustrate two important points about the productivity potentials of new technologies. In order for these productivity gains to be realized, there need to be changes in how work is organized and complementary technologies and products (referred to above as "technological interrelatedness") need to become available. Without reorganization and complementary technologies, productivity gains from new technologies often fall short of projections, which have usually been estimated by (naive) engineers and technology developers rather than by potential users.

In addition to new domestic technologies, medical and public health innovations are likely to be main contributors to the post-World War II

---

[3]Some studies have reported increases since the 1930s in the time devoted to housework (Vanek, 1974). However, these focus on middle- and upper-class households and actual household members. They do not include the time of the domestic servants, which were more common in middle-class households of the 1930s than today. See in particular Gershuny (1991:7–8).

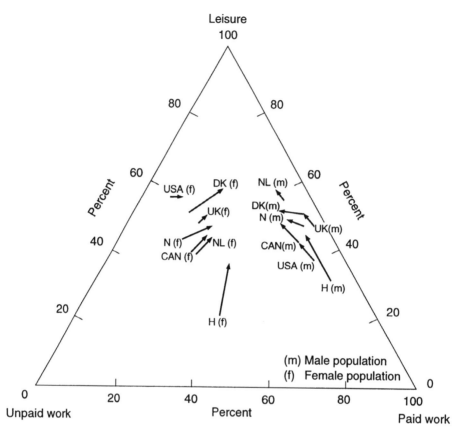

Figure 7.3: Relative allocation of time budgets to three different classes of activities for males and females in seven countries, 1960s to 1980s (in %). Country abbreviations: CAN, Canada; DK, Denmark; H, Hungary; N, Norway; NL, Netherlands. Source: adapted from Gershuny (1991, 1992).

discontinuity in both female participation in the work force and the length of female work careers. Innovations such as the oral contraceptive (1951) and measles vaccine (1953) made it possible to have fewer and healthier children. Indeed these, together with associated social innovations such as day-care centers, appear to have contributed more to the post-World War II discontinuity than new domestic technologies.

Figure 7.3 summarizes the most important changes in time budgets for both men and women in several industrialized countries. The consistent striking results are first, for men, the overall transition away from the formalized work contract toward unpaid work like household work and childcare. Second, for women the trend is in the opposite direction. Third, for both genders there is a consistent shift toward increased free time and leisure. Fourth, there is a noticeable convergence in time allocation patterns between countries and genders alike, especially as incomes rise. As we shall see below, household expenditure patterns seem to follow the same broad trends as shown here for time budgets. What is less clear is whether these persistent trends away from the formal work contract toward "nonwork" and free time are appreciated by policymakers. In the most affluent societies, the time spent on leisure activities is now approaching the time devoted to socially obligatory activities, whether formal work, informal work, or household work.

7.2.2. Spending money

The other historical impact of productivity increases, next to reductions in working time, has been rising incomes. Particularly in the 20th century, incomes (and expenditures) have risen steadily. Since 1900 disposable per capita income in the USA has increased by nearly 2% annually in real terms. Growth rates in other industrialized countries have been comparable (cf. *Figure 6.10*). For "late industrializers", such as Japan and the Scandinavian countries, which started from lower income levels, growth has been even higher.

Two percent annually may sound comparatively modest, but it translates into hefty absolute increases when sustained over a century or more. Per capita GNP, for instance, has risen in the USA by a factor of 5 from 1990 US\$4,360 to 1990 US\$23,000 between 1900 and 1990.[4] Real-term per capita personal consumption expenditures have risen by a factor of 4 over the same time period. By 1900 some 85% of the US economic output was spent on personal consumption, a share that has dropped to a figure that has stayed below 70% since World War II. The balance goes as investments into maintaining societies' ever larger industrial and infrastructural base.

To analyze how spending patterns have changed as a result of such increases in incomes we turn to several theories of consumer expenditures as a function of income differences. Perhaps the best known work is that of the German statistician Ernst Engel, who in 1857 formulated Engel's law, according to which the percentage of income that is spent on food decreases as income increases (as illustrated in *Figure 7.4*).

Again the data are best for the USA, so that is where we will focus the discussion. Our analysis draws on detailed data sets published by Stanley Lebergott (1993, 1996), including personal consumption expenditures in the USA since 1900 for 66 different categories. These we first aggregated into ten

---

[4]US\$ in this book refers to constant 1990 money and prices, unless otherwise stated.

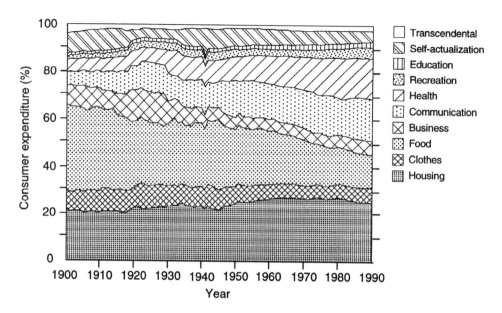

Figure 7.4: Relative shares (in %) of different consumer expenditures in the USA since 1900. Source: based on data given in Lebergott (1993:147–163, 1996:160–178).

categories inspired by Maslow's (1954) hierarchy of human needs.[5] The ten expenditure categories were grouped under three broad headings of human "needs", including:

- "Sufficiency", i.e., expenditures for food, clothes, health, and housing;
- "Identity", i.e., expenditures for personal business, education, and "transcendental" needs such as religion; and
- "Choice", i.e., expenditures furthering human choices and self-actualization (Maslow, 1954:91), including transport and communication, media, recreation, and "beautification" expenditures (including beauty parlors, jewelry, plants, flowers, etc.).

Figure 7.4 shows how the relative shares of these ten aggregated expenditure categories in the three "needs" groupings have changed since 1900.

The most important result is the apparent stability in the share of most consumer expenditure categories in total personal consumption, be it for "basic" needs such as housing, or the more ephemeral ones of self-actualization or religious experiences. *Figure 7.4* provides little evidence to suggest that spending related to "hierarchies" of human needs changes as incomes rise.

---

[5]The assistance of Andrei Gritsevskii in organizing and manipulating the very large data set is gratefully acknowledged.

Rising incomes seem to be more or less equally distributed among different needs. There is little evidence to suggest that once "basic" needs (however defined) are satisfied, incremental income is then spent overproportionally for "nonbasic" needs. Put another way, as incomes rise, "basic" needs seem to be redefined, and the money spent on housing or clothing rises roughly at the same rate as income. As an example, the roughly constant share for housing in *Figure 7.4* means that expenditures on housing, in real terms, have risen as fast as disposable incomes have.

While the relative stability shown in *Figure 7.4* is the most important overall result, there are also some noteworthy shifts among categories. The most obvious is the decline in the share of food-related expenditures from 44% of consumer expenditures in 1900 to 13% in 1990. However, despite the decline in food expenditures as a percentage of the overall total, absolute per capita expenditures on food have nearly doubled since 1900.

Compared to absolute increases in the other expenditure categories, growth in food expenditures has been relatively modest. Growth in the amount of food consumed has been even more modest considering quality improvements. Given productivity increases from new technology, fewer and fewer farmers and less and less land have been required to provide for both these modest increases in domestic consumption and increases in food exports. More generally, once the propensity of consumers to increase consumption as incomes grow, or prices fall, slows down, demand growth remains slack. This enables technology-driven productivity increases to translate into absolute reductions of resource inputs. As discussed in Chapter 5, agricultural cropland use declined in North America and Europe by 18 million ha, nearly all of it being reconverted to forests.

Figure 7.4 is thus consistent with Engel's law that food expenditures decline in importance as incomes rise. Falling prices and rising incomes result in a declining share of income being spent on food, but absolute expenditures continue to rise, although at a much slower rate than total income. The most plausible explanation for the absolute increase in food expenditures is higher quality and convenience. More variety in diets has become available through the inclusion, for example, of lettuce and tropical fruits in the winter season, or preprepared meals cooked at home, or food consumed outside the home in restaurants. Initially, some dietary changes increased food expenditures, in particular the trend toward more meat. In recent years, however, even American diets seem to be moving away from meat toward more vegetables and fruits. And within meat consumption there is a shift away from red meat (e.g., beef) toward "white" meat like fish and poultry. American beef consumption peaked at 60 kg per capita in 1976 and has fallen to 45 kg since (Waggoner, 1996b:77). Such shifts hold important environmental implications. Animals consume many more calories in feed than they supply in the form of milk, eggs, and meat. The caloric "efficiency" of beef is particularly low. A dietary shift away from meat, especially beef, toward vegetables implies less agricultural production, and hence lower land and other input requirements.

Categories in *Figure 7.4* with increasing expenditure shares are health and "communication", which includes transport, communication (mostly telephone), and media expenditures for books, newspapers and the like. The figure also shows the "bubble" in "business" expenditures (e.g., brokerage) in the 1920s, confirming that the speculative frenzy that ended with "Black Friday" in 1929 was indeed a pervasive social phenomenon, easily discernible even in aggregate consumer expenditure statistics.

Figure 7.4 suggests that new technologies have had only a limited direct impact on consumption patterns, even in the "mass production/consumption" technology cluster of the 20th century. Synthetic fibers may have revolutionized clothing, and a global fashion market may now make the latest fashion from Paris or New York available everywhere almost instantaneously, but consumers have not drastically changed their percentage expenditures on clothing. Modern houses are crammed with technological gadgets from Jacuzzis to microwave ovens. The average weight of goods that Americans carry from one home to the next has now reached three tons, not counting their automobiles (cf. *Figure 7.5*). Yet consumers still devote approximately 20% of their budgets to housing in the USA, about the same as at the turn of the century.

The endless debate on "technological determinism" is about whether technological change is so strong as to alter significantly the overall social context in which artifacts are used, or, conversely, technologies evolve entirely out of particular social contexts. The relative stability of consumer expenditure patterns suggests that the social context has remained rather stable and that whatever new consumer technologies emerged, they were integrated into that context without altering it profoundly. There have been changes, to be sure, in daily routines and how artifacts are used, but they have neither enhanced nor diminished the importance consumers attach to a nice home, fashion, and other forms of self-actualization, even the importance of religion. The long-term trends do suggest that in some instances – such as our wish to stay healthy or to communicate with others – new technologies can alter consumptive behavior. But it is difficult to argue that the telephone and the automobile, as the most visible and important communication technologies of the 20th century, have created entirely new social demands compared to a chat over the fence, a letter, or the horse carriage of the 19th century.

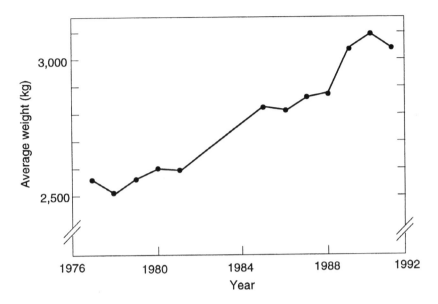

Figure 7.5: Average weight (kg) per household move in the USA. Source: US Households Goods Carriers Bureau, unpublished data, courtesy of Wernick (1997).

It is possible that the aggregation of data into the ten particular categories used in *Figure 7.4* hides interesting trends that might be seen in the disaggregated data or in alternative plausible aggregations. *Figure 7.6* therefore aggregates the data in a different way. The aggregation in *Figure 7.6* is based on the dynamics of each of the original 66 expenditure categories over the 1900–1990 period. Each was analyzed separately and then classified into one of 13 dynamic patterns (e.g., continuously strong/weak declining/increasing, inverted U-shaped, etc.). The 13 were then aggregated into four final higher-level categories, and these are shown in *Figure 7.6*. Expenditures that remained stable over the time period, which were mostly related to housing and accounted for 30% of all expenditures, were treated as a residual and excluded in the analysis and in *Figure 7.6*.

This second aggregation, which looks simply at the *dynamics* of consumer expenditures, is similar to marketing research methods for identifying particular niche markets that grow faster or slower than aggregate consumption. Our interest is principally to illustrate that *Figure 7.4* somewhat masks important changes at the level of individual consumer items. Important product or service substitutions occur within each broader expenditure category. On a more formal level, the changing shares of the four expenditure categories in this second aggregation can be quite accurately described by the

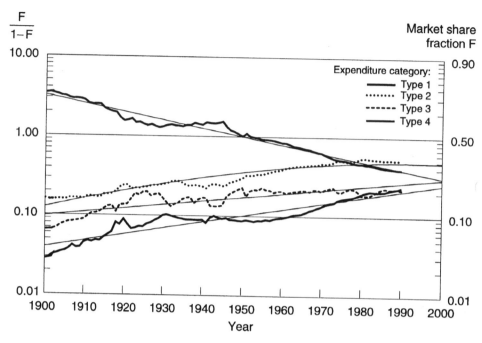

Figure 7.6: Relative shares of four different classes of consumer expenditures aggregated from 66 consumption expenditures (USA, 1900–1990) based on their dynamic behavior. These four categories account for a stable share of 70% of total consumer expenditures. Jagged lines are historical data, and smooth lines model estimates. Data are shown in fractional shares of total consumption expenditures (excluding residual), logit [i.e., log(F/1-F)] transformation. For an explanation of the categories see *Table 7.2*.

same simple multiple logistic substitution model first suggested by Marchetti and Nakićenović (1979) for the description of competing technologies. The dynamics of the four "competing" categories (and excluding the stable residual of 30%) are shown in *Figure 7.6*. The four categories are described in *Table 7.2*.

The regularity of the structural change process, particularly since World War II, is striking and suggests possibilities for forecasting. The most interesting result is that the shares of consumption expenditures for Type 2 (domestic appliances) and Type 3 (motorized transport) items, after steadily increasing after 1900 now appear to be saturating, and, based on the model's extrapolations, headed for long-term *relative* decline. This suggests that the diffusion of the technologies and consumer products most characteristic of the "mass production/consumption" cluster is indeed approaching saturation in relative terms, at least in the USA. After more than seven decades

Table 7.2: Four categories of consumer expenditures and corresponding items classified according to their dynamics underlying *Figure 7.6.*[a]

Type	Dynamics of shares	Summary characteristics	Items included in category type
1	Continuously falling	"Food and shoes"	Purchased food, furnished food, food consumed on farms, alcohol, tobacco, shoes, (dry)cleaning, toiletries, domestic services
2	Growing, currently saturating	"Domestic appliances"	Kitchen appliances, durable equipment, nondurable toys, electricity, water, cleaning supply, jewelry
3	Growing, forthcoming saturation	"Motorized transport"	Purchase of (new or used) motor vehicles; operating costs of vehicles including: tires, repair, gasoline and oil, insurance; purchased land transport services, purchased intercity transport services
4	Continuously growing	"Health and virtual reality"	Hospitals, recreational services, telephone, radio, TV, audio and computer

[a]Note: all expenditure items not listed separately here (most prominently housing-related) have shown stable shares in US consumer expenditures, accounting persistently for 30% of total expenditures over the 1900–1990 period. They are treated as a residual and are omitted from *Figure 7.6.*

where expenditures for these items grew faster than aggregate consumption, they no longer do, and in all likelihood will grow slower than total consumer expenditures in the future.

This result brings into question several long-term scenarios (e.g., those of the IPCC, cf. Pepper *et al.*, 1992) that postulate that motorized transport and other energy-intensive items like domestic appliances will grow faster than aggregate consumption. Our analysis indicates instead that the growing categories will be health, "virtual reality" media (telephone, audio, video, and computers) and recreational services. These expenditures could reach 20% of total US consumer expenditures shortly after 2000, and approach 40% after 2050. At that level they would be as important as food expenditures were at the beginning of this century.

From an environmental perspective, such changes toward "dematerialization" in the structure of consumer expenditures in the most affluent consumer societies would be relatively welcome news. It is only relatively welcome news because absolute expenditures are likely to continue growing as rising incomes allow for a second home, a third family car, or more travel by airplane. Absent radical changes in consumer preferences or policies, or both, even a "virtual reality" consumer society, is unlikely to be frugal in its resource and energy use, and that means possible substantial future environmental impacts.

On a more speculative note, let us extrapolate the patterns in *Figure 7.6* for absolute consumer expenditures. If long-term historical trends (i.e., since 1900) continue to unfold as in the past, per capita GDP in the USA would reach some US$100,000[6] by circa 2075. Assuming that as the case today, about 70% of this total income is spent on personal expenditures, these could reach US$70,000, nearly five times higher than the US$15,000 in 1990. The remaining 30% of GDP would primarily go to investments to maintain production (capital stock replacement and expansion) and for public services (infrastructures).

How then would personal expenditures of our speculative US$100,000 per capita US economy look like based on extrapolating the trends from *Figure 7.6*? Type 1 expenditures (mostly food) would stay with US$2,500 roughly constant in absolute amounts. Type 2 expenditures (mostly household appliances including energy) would grow from current US$1,900 to US$5,400, and motorized transportation, despite a saturating share of total expenditures, would reach a phenomenal US$16,000 per capita, compared to US$3,500 in 1990. The most surprising result from this highly speculative exercise, however, concerns Type 4 expenditures on health, recreation, and "virtual reality". They would increase by a factor of more than ten to about US$25,000 per capita per year.[7] From such a perspective, the next technology cluster might be termed most appropriately "health and information".

7.3.   Lifestyles, Services, and the Environment

There are three principal reasons for looking closely at lifestyles and services in a book dealing with technology and global change. First, at least in highly industrialized societies, "consumption" and services have become a more powerful agent of global change than traditional economic activities

---

[6] All US$ are in constant 1990 prices. The original Lebergott personal expenditure data were expressed in 1987 US$ and have been converted to 1990 US$ using the GDP deflator (of 1.132).

[7] The constant residual expenditure category (mostly housing) would maintain its 30% share while growing in absolute terms to US$21,000 per capita per year, more than four times higher than in 1990. Considering an average family size of three persons, this would imply annual housing costs per family of some US$60,000, equivalent to the present costs of a three-room apartment in downtown Tokyo.

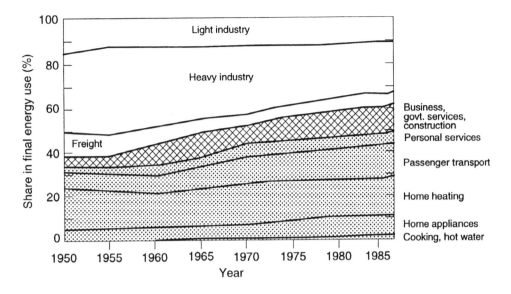

Figure 7.7: Changes in the structure of energy end use (in %) by sector/human activity for Germany (FRG territory only). Source: adapted from Schipper *et al.* (1989:298).

like agriculture and industry. This transformation can be seen most notably in statistics on energy consumption and CO_2 emissions as shown in *Figure 7.7* for Germany. Whereas two-thirds of Germany's final energy consumption in 1950 was for industrial activities (including freight transport), industry's share today is only one-third. Private consumption (most notably transportation) and services account for two-thirds of energy use.

Second, if we are indeed at the brink of a new information technology cluster, we need to look at how these technologies affect service industries such as banking, travel, retail, etc., instead of maintaining the traditional focus on manufacturing industries as "new technology carriers". With the rise of services and the transition from vertical to horizontal integration of economic activities, the borders between economic sectors and between suppliers and customers of new technology will become more indistinguishable. A small service company working in genetic engineering might be the source of important future agricultural innovations, and a software firm might revolutionize the entire airlines reservation business (and that of travel agents) with new internet-based, and customer operated, software.

Therefore when scouting for potentially important technology development, whether hardware or software, one should not only look "where the

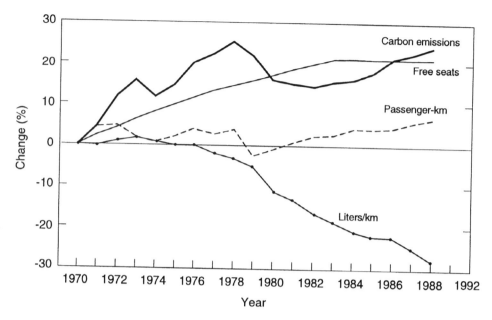

Figure 7.8: Relative change compared to 1970 (in %) of total carbon emissions from US passenger cars and its main driving forces. Source: Grübler (1993a:120).

light is",[8] i.e., in large industry and government R&D labs. We must also look elsewhere to places still "in the dark". Even the military–industrial complex is increasingly speaking about "spin-on" effects, i.e., military applications of civilian technologies as opposed to the usual justification of military-related R&D in terms of their potential "spin-offs" into civilian applications.

Third, and most important from an environmental perspective, is how consumers use products and how services are organized. These factors matter more and more in determining environmental impacts. For technology, this means that attention should be paid both to potential new fields of application, and to consumer behavior that either promotes environmental improvements or exacerbates problems. An illuminating example of the latter is the way that automobiles have been used in the USA (see *Figure 7.8*).

Since 1970 specific fuel consumption for US passenger cars has declined by 30% and total passenger-kilometers driven have increased only slightly. Nevertheless total carbon emissions have increased by 20% compared to

---

[8]The allusion is to the story of the drunkard who lost his keys one night. When asked why he was looking for them under a street light, even though he had not lost them there, he answered, "because that is where the light is".

1970 levels. Although the technological artifact – the car – has become more energy efficient, that improvement has been more than offset by the way consumers use the technological artifact. The load factor has decreased, meaning fewer and fewer people per car. Correspondingly the number of "free seats" transported has increased by 20%. This change has been sufficient to more than negate all the energy savings and carbon emission reductions that would have otherwise come from the improved fuel efficiency of US cars. Despite fuel efficiency improvements of 30%, total emissions are up 20%.

This conclusion is confirmed by a similar decompositional analysis reported in Davis and McFarlin (1996:2–24). From 1972 to 1994 increased fuel efficiency in US passenger cars should have resulted *ceteris paribus* in an energy demand reduction of 180 Mtoe, i.e., over 60% of 1972 fuel use. However, 58% of these potential savings were offset by declining car occupancy rates, and another 5% were offset by a shift to more energy consuming vehicles (i.e., the increasing popularity of light trucks). Together, these two factors have negated 120 Mtoe of the potential gains from improved fuel efficiency, leaving a net potential savings of 60 Mtoe. Even that was never achieved because of an increase in the number of passenger-kilometers traveled. The overall result was a net increase of 66 Mtoe in car fuel use. Regulatory policies such as the CAFE (corporate average fuel efficiency) standards have thus indeed been effective in stimulating technology changes in the form of improved car fuel efficiencies. However, the resulting gains have been more than eliminated by changed consumer behavior and demand growth. As we will see when we examine motorized personal transportation in the next section, the enforced diffusion of catalytic converters "only" stabilized total automobile emissions, and did not reduce them significantly.

The growing importance of services and consumer behavior poses numerous challenges for environmental policy. First, the incentive structure is more complex. Consumers are more difficult to influence through policy decisions than the traditional targets of environmental policy: large industrial enterprises or public bodies. As the energy price increases of the 1970s and early 1980s demonstrated, industry can react rather swiftly to changing price signals by restructuring, conserving energy, and diversifying supply. Service industries and especially consumers reacted quite differently. Despite hefty price increases, demand continued to rise, particularly in transportation, but also in banks, supermarkets, tourism, and other high value-added services where, unlike industry, energy costs and environmental taxes are an insignificant part of overall costs. The second challenge for environmental policy is that emissions due to services and consumer behavior are much more diffuse and decisions are much more decentralized than in the case of industrial activities. It is one thing to regulate pollution from a few large industrial enterprises. It is another to try to influence millions of individual consumers. Initiatives to change public awareness may play an increasingly important role in environmental policy. Some success has already been demonstrated by the elaborate consumer waste separation programs in a number of European countries and their high, more or less voluntary, participation rates.

There may also be large opportunities for industry to extend its traditional focus on producing consumer goods to providing integrated consumer services packages, e.g., for information services rather than a telephone, or for mobility rather than a car. Such integrated packages may be able to better address environmental objectives of regular maintenance, material recycling, and waste minimization (cf. Stahel, 1994).

7.4.   Mobility: Growing Demands and Emissions

7.4.1.   Overview

Transportation is perhaps the human activity that has been the most pervasively affected by technological change since the start of the Industrial Revolution. Transportation is the key to virtually all economic activity and affects nearly every aspect of our daily lives. Technological change in transport systems – both in infrastructures and vehicles – has had far-reaching impacts on the spatial division of economic activities, international trade, settlement patterns, urbanization, and individual access to jobs, services, and entertainment. Together, these impacts have been important indirect and direct sources of global environmental change.

Like all examples of pervasive technological change there has not been a simple unidirectional chain of cause and effect. Take the examples of railways and automobiles. The diffusion of steam railways and steam ships enabled unprecedented spatial concentrations of economic activities and of urban populations. These provided an additional stimulus for further developing transport systems. Steam locomotives brought energy (coal), and steam ships brought food from distant climatic zones to the rapidly growing population of cities such as London. At the same time, increased traffic congestion in growing cities created powerful incentives to improve local urban transport, traditionally comprising walking and horse carriages. The steam railroad went underground and became the indispensable Metro. Regions that did not participate in these mutually reinforcing developments of the 19th century's "railway boom" never built extensive railway networks comparable to those of the "railway core" countries (cf. Grübler, 1990a). By 1930, railways were approaching saturation in all industrial countries.

When the first automobiles[9] appeared at the end of the 19th century, they were much less revolutionary than one might think. (They were also unreliable, expensive, and cumbersome to use.) The market for individual transport vehicles already existed around horse carriages – including private vehicles, buses, and vehicles for transporting goods – and of course in the form of the bicycle. (Paved) road infrastructures were already put in place for these automobile precursor technologies, further enhancing the initial diffusion potentials of cars. Despite there continuing to be a close relationship between infrastructure developments, e.g., in the form of *Autobahnen* and interstate highways, and increasing car densities, it remains an undeniable fact that road infrastructure development significantly preceded the spread of the automobile (see especially the data given in Grübler, 1990a).

The automobile diffused initially by displacing traditional individual road transport modes, in particular horse carriages (cf. *Figure 2.11* in Chapter 2), and only much later expanded into new market niches and applications. These included long-distance transport in competition with the railways, and the mass-motorization associated with suburbanization. In turn, changing spatial settlement patterns and their associated long commutes, "hypermarkets" and shopping malls further increased dependence on automobiles and eroded the market niche of alternative public transport systems like buses or tramways. Large car manufacturers also took a more active role in the demise of public transport in North America by purchasing and closing bus or tramway lines.

In Europe and Japan the patterns were different. The automobile diffused much later, and large-scale suburbanization as in the USA did not occur. This was partly due to higher population densities, higher land prices, less time for large-scale urban spatial restructuring, and explicit policies and subsidies to maintain or even improve public transport systems. These differences relative to the USA emphasize the role of positive feedback mechanisms and path dependency in alternative development trajectories. These caution against quickly extrapolating experiences between countries.

---

[9]The history of the automobile includes many innovations and early technology pioneers. The date we will use for the *invention* of the automobile is 1875 when Siegfried Marcus built the first combustion engine powered automobile (0.7 horsepower) in Vienna. This was slightly earlier than Benz's three-wheeler. Marcus' vehicle can still be seen in Vienna today. At the turn of the century hundreds of manufacturers of "horseless carriages" existed in Europe and the USA. As the term implies, they produced extremely small series of handmade, largely wooden carriages. The automobile industry as we know it today was conceived by Henry Ford with his Model T design and mass production manufacturing plant in 1908. Therefore 1908 is perhaps the best point estimate for the *innovation* date of the automobile.

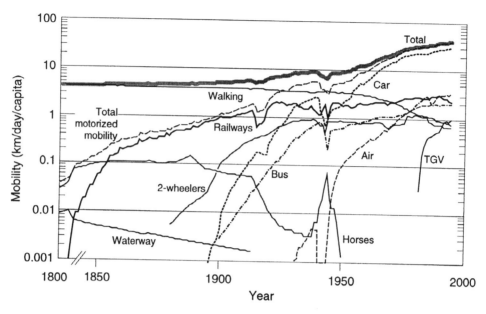

Figure 7.9: Daily mobility (passenger-kilometers traveled per person) in France since 1800, mobility by mode, and total motorized and aggregate mobility. Source: adapted from Grübler (1990a:232). For the data of this graphic see the Appendix.

7.4.2. Mobility: Motorized and total

Figure 7.9 shows the average daily mobility of the French population since 1800. It is the longest continuous historical time series available. It shows the sequential diffusion of successive individual transport modes, as well as stable growth in total motorized mobility, which has followed an essentially continuous, exponential growth path since the beginning of the 19th century.

Motorized mobility has increased by a factor close to 1,000 since the beginning of the 19th century, from 40 meters per day per capita in the 1820s to over 40 kilometers per day per capita in the early 1990s. This growth was sustained primarily by the expansion of railways in the 19th century, and by motorized road transport in the 20th century, initially two-wheelers and buses and later cars. The growth in *total* mobility (including walking) has been less dramatic. There the increase is by about a factor of 10, from 4 to 40 kilometers per person per day.

Growth in car mobility has slowed down significantly since the early 1970s, indicating possible saturation. Growth in fast, long-distance mobility by aircraft but also by the successful new fast train system TGV (*Train à Grande Vitesse*) has been spectacular. As of the early 1990s, the French

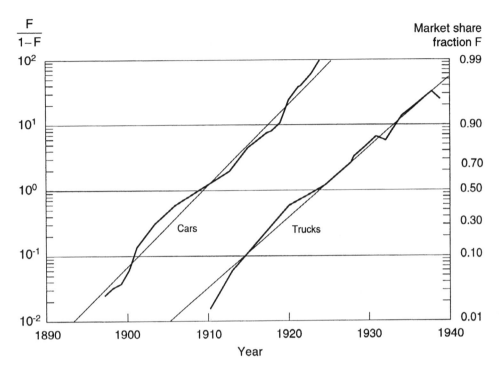

Figure 7.10: Share of motor vehicles for passenger and goods transport in France, measuring relative shares in passenger- and ton-kilometers, empirical data and model estimates with a simple three-parameter logistic equation, logit [i.e., log(F/1-F)] transformation. (Note that the symmetrically declining shares of horse-drawn vehicles are omitted in the graphic). Source: Grübler (1990a:140).

travel more passenger-kilometers with the TGV than they collectively cover by walking.

There is both complementarity and substitution between successive transport modes. An example of substitution is the precipitous decline in passenger transport on the elaborate inland system of artificial canals caused by the diffusion of railways in the 19th century. Conversely, horse vehicle traffic, which provided complementary "feeder" functions to the railway stations, was unaffected. However, horse vehicles rapidly disappeared with the advent of automobiles and buses, and later trucks (cf. *Figure 7.10*). The substitution of horse vehicles by automobiles also significantly improved urban environments, at least as long as there were not more automobiles on the road than the horses they were replacing (see *Box 7.1*).

After replacing its most direct competitor, the horse, the automobile found new markets and many new customers. Private motorization in the

Box 7.1: Horses and the Automobile:
An Early Environmental Story

The automobile today is rightly associated with numerous environmental impacts. However, it is frequently forgotten that the first automobiles, as energy-inefficient and polluting as they were when compared to their modern counterparts, nevertheless *improved* urban environments considerably. Contemporary complaints about horse manure were widespread (cf. Thompson, 1976), even if giving rise to employment. In London alone thousands of "crossing sweepers" cleaned the street, for a halfpenny, before pedestrians would pass (Montroll and Badger, 1974:225). After all, urban horses at their peak in England are estimated to have produced 6 million tons of manure per year (Thompson, 1976:77).

The generic environmental advantage of the automobile over the horse was due to two factors: higher energy conversion efficiency and lower emissions.

Energy efficiency, and emissions (grams per km) for horses, and early and contemporary automobiles.

	Horses[a]	Cars (ca. 1920)[b]	Cars (1995)[c]
Engine efficiency	4	10	20
Wastes			
Solid	400	–	–
Liquid	200	–	–
Gaseous, including			
Carbon (CO_2)[d]	170	120	70
Carbon (CO)	–	90	2
Nitrogen (NO_x)	–	4	0.2
Hydrocarbons	2[e]	15	0.2

Data sources:
[a] Montroll and Badger (1974:225), Thompson (1976:77), and Crutzen *et al.* (1986:275).
[b] MVMA (1985:88) and EPA (1995:ES-8 and 3.1-3.29).
[c] Based on EPA 1994–1996 "Tier 1" standards (AAMA, 1996:88), and 1994 average fuel consumption (Davis and McFarlin, 1996:3–43).
[d] Total carbon content of fuel.
[e] Methane.

Despite important qualitative differences, especially the different toxicity of emissions, quantitative differences are nevertheless significant. The first cars used energy more than twice as efficiently, and a modern car about five times as efficiently as a horse. The significance of this efficiency improvement is easily grasped considering that feed (i.e., energy inputs) for urban horses in England cost the equivalent of the entire agricultural production of England around 1900 (Thompson, 1976:78). Feeding US farm and urban horses in 1910 required nearly half as much agricultural land as was required for feeding the entire US population (Waggoner, 1996b:76). Early cars reduced emissions per km driven by about a factor of three compared to horses, and modern cars with catalytic converters, reduce them by a factor of 10.

Thus, as long as the number of cars did not grow beyond the number of horses they replaced, the automobile did improve urban environments. But the catch of the automobile story is, of course, that the car population continued to grow. By the late 1920s, some 25 million cars had replaced practically all of the urban and farm riding horses in the USA (Nakićenović, 1986:319–320). In 1995 US cars numbered more than 130 million, and total motor vehicles, including trucks and buses, numbered 200 million (AAMA, 1996:32).

second half of the 20th century not only started to rival railway transport, which had been largely saturated since the 1920s, but increasingly also substituted for the oldest transport mode known, walking. The French, as is true in all industrialized countries and among the affluent elite in developing countries as well, increasingly drive more and walk less. The TGV appears not to have eroded car travel significantly, challenging instead the position of traditional railways and, to a lesser extent, domestic air travel. Both trends can be understood in light of the dominance of short-distance trips in total mobility. In 1982, for example, the average French person took about nine long distance (above 100 kilometers) trips per year for business and holiday travel, but these trips accounted for about one-third of total passenger-kilometers traveled (Orfeuil, 1993:254).

Altogether, the daily French mobility data in *Figure 7.9* and their spectacular historic growth are representative of the range observed among other industrialized countries, although the timing of growth, absolute mobility levels and the modal split among different transport modes and technologies may vary.

In view of the possible advent of an "information society", there has been much speculation on the possibilities of "virtual communication", e.g., via the internet. This could replace a great deal of physical travel. However, even the telegraph and the telephone did not affect the steady exponential growth in motorized mobility of railways and cars (see *Box 7.2*). Thus transport and communication technologies appear more complementary than substitutive, at least in the societal aggregate. In specific applications, such as video conference calls, however, there are limited substitution possibilities of electronic communication for physical travel. Overall, to the extent that an "information society" would reduce travel needs, it could make a significant contribution to reducing adverse environmental impacts.

Although motorized mobility has grown spectacularly across a wide range of countries, there remain important differences as noted previously. Typically, public transport modes like buses and railways have retained a stronger role in Central and Eastern Europe, and to a smaller degree in Western Europe and Japan, than in the USA. In the USA personal mobility is dominated by the automobile for short distance trips, and by cars and aircraft for longer distances. In addition, US mobility (over 70 kilometers per day per capita) is nearly twice as high as in Western Europe due both to higher car ownership and longer distances traveled. *Figure 7.11* shows the long-term structural change in US long-distance (intercity) passenger transport since 1950.

In the USA, by the 1960s cars had largely displaced railways as the dominant long-distance transport mode. Then at the height of the car's

Box 7.2: Transport and Communication:
Complementarity or Possible Substitution?

The advent of new communication technologies that combine features and communication styles of existing artifacts opens new frontiers for both quality and quantity of interpersonal communication. The marriage of the telephone with the TV and of both to video, computer animation, etc., in its first (and still crude and rudimentary) form of multimedia PCs is often heralded as a possibility for substituting face-to-face communication and thus travel. Let's contrast this with a historical perspective. The figure below shows the growth in total motorized mobility (from *Figure 7.9*) and the growth of total communication with technology aids (i.e., the total number of messages exchanged in the form of letters, faxes, telephone calls, etc.) for France since 1800. Both indicators are renormalized to an index, where 1985 output levels equal 100. (Source: Grübler, 1990a:256.)

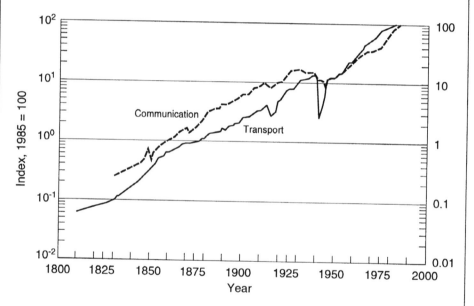

The surprising historical fact is that both motorized transport and communication via technologies (i.e., excluding verbal exchanges, for which there are no available estimates) grow at roughly the same, and exponential, rate. Both have increased by about four orders of magnitude since the beginning of the 19th century. The invention of the telephone does not seem to have resulted in less travel by railroads, bicycles and later cars. We feel that the same may also apply for more recent communication innovations. "Travel and communication are better seen as interrelated elements in a social context which they help to create" (Albertson, 1977:43).

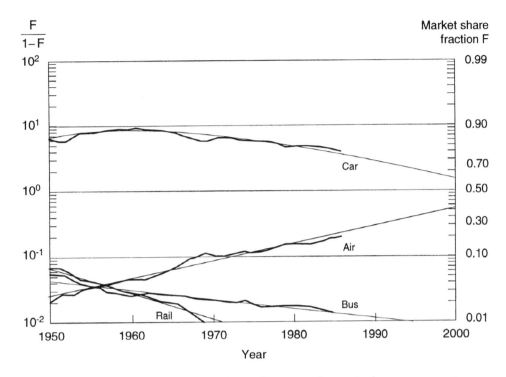

Figure 7.11: Changes in US long-distance (intercity) passenger transport, in fractional shares (F) of total intercity passenger-kilometers, logit [i.e., log(F/1-F)] transformation. Source: Grübler (1990a:201) based on Nakićenović (1986).

market dominance, when some 90% of all intercity passenger-kilometers were driven by cars, a new competitor, the aeroplane, began to challenge the automobile. Apart from some formalistic mimicry (see *Figure 7.12*), there was no real technological response by the automobile industry to the speed and convenience of aircraft for long-distance travel.

Based on the simple model shown in *Figure 7.11*, cars and aircraft will have equal shares of long-distance mobility in the USA within the next few decades. Similar structural changes (albeit lagged by several decades) can be observed even in countries such as the former USSR and China where the car has never held a dominant position in long-distance travel. There buses account for close to 50% of all intercity passenger-kilometers (Grübler *et al.*, 1993a) and railways for much of the remainder. Nonetheless from a purely technological perspective, long-distance mobility in all three countries may be considered similar – roads are the dominant infrastructure, and vehicles powered by internal combustion engines are the dominant transport

Figure 7.12: One response of the automobile industry, when challenged by a new competitor for long-distance travel: the 1959 Cadillac Cyclone, a concept car with an automotive heart but designed to look like an aircraft. Source: General Motors (1983:145).

technology. The three differ in how they organize the use of technologies: privately owned, individual cars in the USA versus publicly owned collective buses in the case of the former USSR and China. But in all three countries, the share of aircraft in long-distance passenger transport grows at comparable rates. (In the 1980s, for example, aircraft accounted for about 20% of all intercity passenger-kilometers both in the former USSR and the USA.)

Table 7.3 contrasts the results of a 1990 US passenger transport survey (NPTS, 1992) with mobility data for selected countries. The most significant result is the dominance of nonwork trips and mobility. Household and family chores, as well as leisure trips (vacation, seeing friends, and other social and recreational activities) account each for about 40% of US passenger-kilometers. Both nonwork categories exceed in absolute terms the total mobility of the Russian and Chinese population taken together. Even trips for which respondents could give no other reason than "to go for a ride" exceed the total mobility of Norway. These statistics reinforce the transition "from work to pleasure" cited in the introduction to this chapter.

7.4.3. North versus South, linear versus nonlinear futures

Comparisons between industrialized and developing countries, or the "North" and "South", are difficult. First there is a formidable data availability problem, particularly for nonmotorized transportation. There

Table 7.3: Individual (motorized)[a] mobility in the USA by trip purpose in 1990 and comparison with mobility in selected countries (in billion passenger-kilometers and percent).

Trip purpose	10^9 pass-km	Percent
Work	1,190	17.7
Household and family	2,981	44.4
Leisure[b]	2,484	37.0
"Go for a ride"	55	0.8
Total	6,710	100.0
Total mobility		
China	607	
Former USSR	1,770	
France	704	
Norway	47	

[a]Excluding walking and cycling (no data reported).
[b]Vacation, seeing friends, other social and recreational activities.
Source: Grübler (1993b:184) based on NTPS (1992).

are also substantial variations among different countries, cities, and social strata.

As a generalization, however, walking and bicycle trips dominate urban and rural mobility in developing countries to the same extent that they dominated mobility in industrialized countries 50–100 years ago. Mobility surveys for Indian cities suggest that between 40% and 50% of all trips are done by walking and another 10–20% by bicycle and cycle rickshaw (TERI, 1993:49). Thus, about two-thirds of all trips are by nonmotorized transport. The share of nonmotorized transport modes in terms of absolute distance traveled (passenger-kilometers) is, of course, much smaller. An interesting comparison is between the average occupancy rates for two-wheelers in India and automobiles in the USA. India's rate of 1.6 passengers per bicycle (TERI, 1993:53) roughly equals the US rate for cars (Davis and McFarlin, 1996:4–11).

Overall mobility patterns and their resulting energy use and environmental impacts are influenced principally by accessibility to transport technologies and how these technologies are used (e.g., load factors). Accessibility is largely a function of disposable income, but depends also on the infrastructure (are bus routes available, or even a metro?) and settlement patterns. Different patterns of accessibility mean that even at similar motorized mobility levels, the ratio of individual transport technologies (cars and motorcycles) to collective technologies (buses, trams, and metros) may be very variable in different countries. Compared to the importance of accessibility

Table 7.4: Urban transport in selected cities, mid-1980s. Passenger cars registered (million), cars per 1,000 population, and public transport trip rate (per 1,000 people). The two motorized mobility indicators become comparable assuming on average that each car is used for a trip per day.[a]

	Passenger cars registered		Public transport[b]
	10^6	Per 1,000 people	trips/day/1,000 people
Beijing	0.06	2	1,641
Dallas	1.53	1,624	160
Hong Kong	0.16	29	1,274
Los Angeles (MA)[c]	4.18	1,409	402
Mexico City (MA)	1.85	96	1,570
Paris	1.00	460	1,779
São Paulo	2.29	619	1,878
Tokyo (MA)[c]	2.15	182	2,139
Warsaw	0.32	193	2,609

[a]Traffic surveys in European cities suggest that in fact every second car is used for two trips per day.
[b]Bus, taxi, metro, and railways.
[c]1980 data.
Abbreviation: MA, metropolitan area.
Source: Grübler (1993b:185) based on IEM (1986).

and how technologies are used, particular technological characteristics – such as whether a car or bus is well maintained or has a catalytic converter – have only second-order influence on environmental impacts.

Table 7.4 compares motorized mobility indicators for selected metropolitan areas. At one extreme is Los Angeles with more than 4 million cars registered in the mid-1980s, over 1.4 cars per inhabitant. The car density in Dallas is similar. In such "automobile cities", the market niche for public transportation is obviously small. But low car densities in cities such as Beijing or Hong Kong do not necessarily mean less mobility. Public transport is simply used instead of cars. Such differences do not result from income differences alone, as surveys of car ownership as a function of income indicate decisive differences between cities (Grübler, 1993b). Thus transport policies, infrastructure availability, and urban form make a large difference in urban mobility choices.

We focus on urban mobility because cities amplify both the functional and environmental problems inherent in transportation. Space limitations and high densities of land-intensive individual transport modes (cars) result in congestion, a breakdown in the transport system's ability to fulfill its fundamental role of moving people and goods. High population densities and concentrated trips emphasize another constraint: the limited capacity of the environment to assimilate emissions. Even where the entire car fleet has catalytic converters, as in Los Angeles, smog episodes continue. Cities in developing countries share the congestion and many environmental problems of OECD cities, but have additional problems resulting from capital shortages that limit infrastructure improvements. Given these constraints, the past development pattern of motorized car mobility in developed countries is not necessarily even a feasible, much less a preferred, scenario for developing countries as their incomes rise and cities expand.

Despite their differences, industrialized and developing countries share one significant experience over the last 25 years: demand growth, even in periods of oil price "shocks". Transport energy requirements in the OECD have increased over 50% since the early 1970s, despite significant improvements in fuel economy. And in the developing countries transport energy demand has increased three-fold over the same period (Grübler, 1993b:180).

In light of consistent historical demand growth, it is not surprising that most long-term scenarios are particularly "bullish", assuming continuing transportation demand growth and ever rising environmental impacts. Most scenarios assume a future that is "like today, just much more" (for a review cf. e.g., Michaelis *et al.*, 1996:684–686). One example is the EPA's "Rapidly Changing World" scenario (1990) shown as the top line in *Figure 7.14*. As a contrast, we offer an alternative scenario of our own.

The scenario is based on the innovation diffusion patterns discussed in Chapter 2. The diffusion rates for a new technology are slowest in the regions that lead in introducing the new technology. The diffusion process therefore lasts longest in these regions, and they end up with the highest adoption rates. Conversely, regions joining a particular "diffusion bandwagon" later on show much faster growth rates: they are "catching up". Their diffusion process is, however, shorter, and ultimate adoption levels are lower. *Figure 7.13* illustrates these observations for the automobile.

Our alternative scenario uses the estimated relationships between diffusion rates and ultimate diffusion levels from *Figure 7.13*.[10] In the scenario the number of cars in OECD countries reaches saturation shortly after the year 2000. Non-OECD countries, while still experiencing substantial growth, do not reach similar high levels of car ownership. This resulting "saturation scenario" can then be combined with corresponding scenarios of fuel efficiency improvements and new propulsion technologies for automobiles. Three illustrative scenarios have been developed. The first keeps fuel efficiency constant at 1990 levels, a rather improbable conservative assumption.

---

[10]The methodology, data, and resulting scenario is described in more detail in Grübler and Nakićenović (1991).

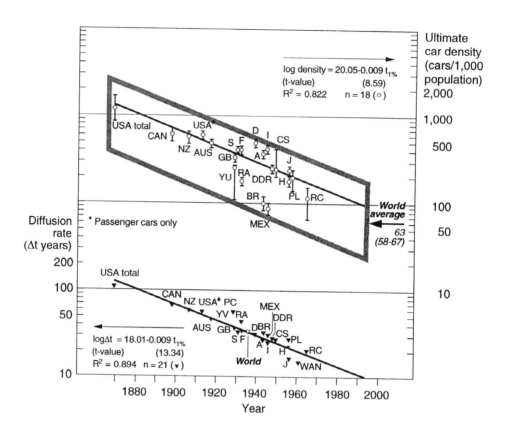

Figure 7.13: Car diffusion. The upper curve shows estimated saturation levels of car ownership rates (per 1,000 population) and associated uncertainty bands. The lower curve shows the corresponding diffusion time (Δt, the time required to grow from 10% to 90% of the estimated saturation level) in years. Diffusion rates and levels have been estimated from historical data based on a logistic model. Both are plotted versus the introduction date of the car (time when diffusion had reached 1% of the ultimate saturation density). Diffusion times in "innovation centers", e.g., the USA, are the longest and lead to the highest saturation levels. Diffusion times for late-starters, e.g., Japan, are much faster (smaller Δt), but saturation levels are also lower. The alphabetical codes for countries follows the United Nations 1949 Convention on International Vehicle Registration Plates. The two values for the USA refer to the diffusion of all road vehicles and to passenger cars only (denoted by an asterisk). Source: Grübler (1990a:152).

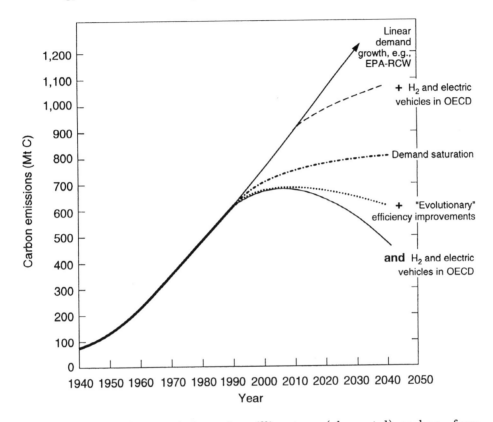

Figure 7.14: Carbon emissions, in million tons (elemental) carbon, from passenger transport, historical data to 1990 and alternative scenarios for the future. For an explanation of the scenarios see the text. Source: adapted from Grübler *et al.* (1993a:564).

The second assumes "evolutionary" efficiency improvements at the historically observed rate of 1% per year. The third adds the possible introduction of new fuels and new vehicles in the 21st century, such as hydrogen fuel cell or electric vehicles, starting first in the affluent OECD countries.

The resulting carbon emissions are shown in *Figure 7.14* and contrasted with the EPA (1990) "Rapidly Changing World (RCW)" scenario. Also shown is a scenario combining the linear demand growth of the RCW scenario with assumed penetration of hydrogen and electric fueled vehicles in the OECD countries.

The results in *Figure 7.14* are not intended as forecasts. Rather, their purpose is two-fold. First, they illustrate that consistency with historical diffusion patterns of technologies leads to very different results from linear extrapolations of historical trends. The technology cluster of which the car was a central feature may be approaching limits similar to those reached in the 1920s and 1930s by the technology cluster epitomized by the railways.

Second, the scenarios show the possibilities for, and limits on, a drastic change in the persistent growth of environmental impacts from our present transport technologies. New technologies can do a lot of tricks, *inter alia* in reducing emissions, but as long as demand growth remains unabated, the environmental benefits are likely to be limited. The best we might realistically achieve may be stabilizing impacts at high levels (cf. the illustrative scenario with H_2 and electric vehicles in the OECD on top of the RCW scenario in *Figure 7.14*). Conversely, the same holds true if demands saturate but transport technologies remain unchanged. Drastic reductions in emissions require responses from both the demand *and* the supply side. Slower demand growth needs to be combined with both incremental changes (efficiency improvements) and radical technological advances.

7.5. Transport and the Environment

Environmental impacts from transport systems have reached global dimensions on both accounts of our definition of "global change". First, the energy use and CO_2 emissions related to the global diffusion of passenger cars, trucks, buses, and aircraft have reached truly global dimensions. Traffic-related CO_2 emissions are estimated at 1.3 billion tons of carbon (Gt C), (Michaelis *et al.*, 1996:683), rivaling the 1.6 (±1) Gt C estimated for net carbon releases from land-use changes (IPCC, 1994:11). At the local and regional levels, traffic pollutants, in particular nitrogen oxides, are (together with sulfur emissions) the principal precursors of acid rain. Once only a phenomenon in Europe and North America, acid rain has emerged as a serious environmental issue in Asia. Thus it is beginning to spread globally.

Second, local environmental impacts from transport are ubiquitous. Traffic jams occur all over the globe (with the exception of Antarctica), as do smog episodes in large urban agglomerations in both developed and developing countries. Such adverse impacts damage the environment and human health. Human health also suffers directly from transportation accidents. Globally, over 500,000 people are killed annually in road accidents.[11] Adding fatalities associated with trains and aircraft, which are much safer than road vehicles, would increase this death toll only slightly.

Accidents, congestion, and urban environmental pollution are not new phenomena of our technological age. Traffic congestion has plagued large

---

[11] This estimate is based on incomplete IRF statistics (1993:108–118), that do not cover all the countries of the world.

cities since antiquity, and as discussed previously, the environmental impacts of hundreds of thousands of urban horses in 19th century London and New York were substantial.

Traffic casualties during the horse era were in fact very high. Estimates (Lay, 1992:176) indicate that in the USA about 6 million km were traveled on horseback per traffic fatality. The toll from early automobiles was similar. Today on average more than 60 million km are driven in cars per traffic fatality in the USA.[12]

However, with the exception of large metropolises, transport technologies did not pose substantial environmental problems prior to the pervasive diffusion of the car. Before the advent of the car, environmental impacts were primarily indirect. Steam locomotives and ships in the 19th century expanded the spatial divisions of economic activities which, in today's language, means they made it possible to "export" environmental impacts. Food and raw materials production could move to distant locations, e.g., to more favorable climatic zones, thereby dispersing the environmental impacts of vastly raised demands for food and raw materials. As a representative example we mentioned in Chapter 5 the land-use changes outside Europe resulting from increased European food imports and the establishment of large plantations in many developing countries to export "cash crops" such as sugar, tobacco, cotton, and coffee.

Some of these indirect environmental impacts still persist, but the focus has moved to more direct transport impacts affecting the environment at local, regional, and global levels. These include land requirements that are particularly critical in densely populated areas. They include the direct human health impacts of accidents. And they include impacts on human health and the environment through airborne pollutants.

Table 7.5 lists major traffic-related pollutants and their air concentration standards as suggested by the World Health Organization. It contrasts these with representative air quality values in four megacities, two in the industrialized world and two in developing countries. The pollutants include carbon monoxide (CO) and nitrogen oxides (NO_x), notably nitrogen dioxide (NO_2), that originate from all combustion processes but are dominated by the transport sector. Also included is ozone (O_3), a secondary air pollutant resulting from complex atmospheric reactions of NO_2 and volatile organic compounds (VOC) under the influence of sunlight. Together these form the characteristic urban, traffic-related "photooxidant smog". Also included are lead and so-called SPM-10s, solid particulate matter with a diameter of less

---

[12]At the height of the horse era, riding horses in the USA may have caused as many as 25,000 traffic fatalities per year. Today's death toll from motor vehicle accidents in the USA is about 50,000 per year for at least ten times as many road vehicles.

Table 7.5: Traffic-related pollutants: WHO standards versus metropolitan experience, ca. 1990.

	WHO standards (μg/m^3)		Typical concentrations[c]			
	Short-term[a]	Long-term[b]	Cairo	Los Angeles	Mexico City	Tokyo
Carbon monoxide (in mg = $10^3 \mu$g)	30	10	10–25	11–15	15–30	1–?
Nitrogen dioxide	400	150	400–1,400	88–526	400–800	60–?
Ozone	<200	<120	100–1,000	?–400	100–400	40–250
Lead	–	<1	1–3	~0.1	1–3	n.a.
Particles < 10μm (SPM-10)	–	70	n.a.	50–150	?–300?	n.a.

[a] Average time one hour.
[b] Average time eight hours for CO and O$_3$; 24 hours for NO$_x$ and SPM-10; one year for lead.
[c] The range corresponds to typical long-term (lower values) and short-term (higher values) concentrations, respectively. Concentrations can vary enormously between different city districts, therefore values are only indicative.
Source: WHO and UNEP (1993).

than 10 micrometers. This is the respirable size range, making SPM-10s potentially hazardous, particularly for human health.

The first observation from *Table 7.5* is that concentrations generally exceed WHO standards in cities of both industrialized and developing countries alike. However, there are important differences. First, Cairo and Mexico City exceed the WHO standards more dramatically than Los Angeles and Tokyo. This is particularly true for nitrogen oxides and ozone. A second important difference is that the phase-out of lead as a gasoline antiknock additive in most OECD countries has drastically reduced lead pollution. The difference is most pronounced when comparing Los Angeles and Mexico City. The impact of catalytic converters is most notable in Tokyo. Together with de-NO$_x$ units in power plants, they have greatly improved air quality. Compared to the 1960s, mean CO concentrations in Tokyo have been reduced by a factor of 4, and photooxidant concentrations by a factor of 2 (WHO and UNEP, 1993:215–216). This is in striking contrast to Los Angeles, where CO, NO$_x$ and O$_3$ concentrations remain high, exceeding WHO standards, despite the complete replacement of the car fleet by catalytic converter cars.

This is a somewhat surprising result, and in the next few paragraphs we therefore look more closely at transport emissions for the USA as a whole (see *Table 7.6*). Since the 1960s tightened regulations have reduced the exhaust emissions standards of cars in the USA. Hydrocarbon and carbon monoxide emission standards for passenger cars were reduced by a factor of 10 between

Table 7.6: Transport-related emissions of five pollutants (CO, NO$_x$, VOC, SPM-10, and lead) in the USA (in million tons), 1970–1994.

Pollutant	1970	1980	1990	1994	Change (%) 1970–1994
CO					
Road	79.8	70.8	57.0	55.4	−31
Other transport	9.6	11.5	13.3	14.2	+48
Total	89.4	82.3	70.3	69.6	−22
NO$_x$					
Road	6.7	7.8	6.8	6.8	+1
Other transport	1.5	2.2	2.6	2.8	+87
Total	8.2	10.0	9.4	9.6	+17
VOC					
Road	11.8	8.1	6.2	5.7	−52
Other transport	1.4	1.7	1.9	2.0	+43
Total	13.2	9.8	8.1	7.7	−58
SPM-10					
Road	0.4	0.4	0.3	0.3	−25
Other transport	0.2	0.3	0.3	0.4	+100
Total	0.6	0.7	0.6	0.7	+17
Lead					
Road	0.16	0.06	<0.01	0.001	−99
Other transport	0.01	<0.01	~0.00	~0.000	−98
Total	0.17	0.06	<0.01	0.001	−99

Source: EPA (1995:3-11 to 3-16).

1971 and 1993, from 4.1 to 0.41 grams/vehicle-mile for hydrocarbons and from 34 to 3.4 g/mile for CO. Emissions standards for NO$_x$ were also reduced substantially, albeit at a slower pace, from 4.1 g/mile (precontrol levels) to 2 g/mile in 1980, to 1 g/mile in 1983, and to 0.4 g/mile in 1993.[13]

Taking into account the turnover of the automobile fleet (cf. *Figure 2.12*), by the mid-1990s 90% of the cars on the street were post-1980 vintage. *Ceteris paribus* automotive emissions should have fallen in proportion to the mandatory emissions standards for cars. As *Table 7.6* indicates, this has not been the case. The tightening of emission standards for hydrocarbons and CO by a factor of 10 for cars, or by 90%, compares with emission reductions of 30–50% for CO and VOCs, respectively. For NO$_x$ there has been no emission reduction at all compared to 1970, despite tightening emission standards to 25% of 1970 levels.

---

[13]The original nonmetric units of the US standard are retained.

Although road transport emissions include those from vehicles (e.g., heavy trucks) that are not subject to the same standards as cars, standards for these vehicles have also been significantly tightened. Standards for NO_x emissions from light duty trucks were reduced by 25% from 3.1 g/vehicle-mile for 1975–1978 models to 2.3 g/mile for 1984–1987 models. Currently the standard is between 0.4 and 0.7 g/mile depending on truck weight.[14] Similar decreases have taken place for heavy trucks, although due to their longer life on the road it takes more time for new standards to work their way through to average fleet and aggregate total emissions. Thus one expects slower progress for NO_x emissions reductions, but it is still surprising to see no impacts at all on total road vehicle emissions.

The EPA (1995:3–12) NO_x emissions inventory indicates that since 1970, total car emissions declined by 10%, those of trucks increased, and total road vehicle emissions basically remained at 1970 levels. Adding nonroad transport emissions (e.g., from farm tractors, railways, and aircraft) results in a 20% increase in transport NO_x emissions since 1970, and a similar increase for all NO_x emissions from all sources taken together.[15]

Thus, although environmental regulations have achieved some progress for CO and VOCs, and spectacular progress for lead emissions (cf. discussion below), the case of NO_x is a reminder that it is a long and winding road between the introduction of environmental standards and effective reductions in emissions. Counterbalancing factors include continued demand growth (more cars and more driving), changes in vehicle fleet composition (more light trucks), and above all differences between actual emissions in everyday operations and emissions under test conditions as defined in regulatory standards. These differences relate to the fact that actual driving cycles are different from those defined for setting environmental standards and due to malfunctions in the emission control equipment.

For example, catalytic converters are effective only after they have reached high operating temperatures. On the short-distance trips of less than a few kilometers that make up the majority of urban trips, the converter basically does not work. Maintenance is also an issue. A single fueling with leaded gasoline ruins the catalytic converter. Physical damage to the converter and a breakdown of controls can also stop the functioning of catalytic converters. If they are not replaced, or the car is scrapped, there is obviously no emission reduction. Taken together, it is estimated that there is a factor difference of 2 (NO_x), 4 (HC), and 5 (CO) between the actual

---

[14]All data based on AAMA (1996:88).

[15]In 1994, transport accounted for 45% of all NO_x emissions, 38% of all VOC emissions, and 78% of all CO emissions in the USA.

average emissions over a vehicle's lifetime and those postulated by the 1993 tailpipe automobile emission standard in the USA (Ross *et al.*, 1995).

In combination with continued demand growth (albeit at considerably slower rates than in the 1950s to 1960s), these factors explain the apparent difference between actual emission reductions and those expected based on the diffusion of vehicles with new and improved environmental standards. The result cautions against overoptimism regarding the pace and effectiveness of pollution-reduction measures based on "add-on" devices where their environmental performance is largely determined by consumer usage and maintenance, and energy demand continues to grow. It also suggests that mandatory emission standards have to leave large margins for these counterbalancing effects. A standard might have to fall one order of magnitude below those levels deemed *a priori* required for a particular emission reduction. For instance, if a 50% reduction in absolute emissions is the theoretical requirement, emission standards should perhaps be set as low as 5% of pre-regulation levels. If a 90% reduction is required, the standard would have to approach the proverbial "zero-emission" car, such as that proposed for post-2000 Californian legislation.

Ross *et al.* (1995:36) provide yet another interesting interpretation. Instead of confirming the old cliché that the problem in emission control is not technical, but rather institutional or behavioral, they see the divergence between regulatory standards and actual emissions mainly as a result of a lack of accurate emissions-measurement instrumentation. This consequently leaves both regulators and car manufacturers without adequate information on which to base standards and R&D in emissions control technologies.

This somewhat sobering conclusion has to be contrasted with the experience in lead emissions, which have been reduced tremendously. In the USA in 1970, cars and trucks released some 155,000 tons of lead. In 1994 the corresponding figure was 1,550 tons (EPA, 1995:3–16), 100 times lower. The effectiveness of emission reductions in this case results from a "zero standard" (lead-*free* gasoline)[16] and the centralized implementation of emission-reduction measures "upstream", i.e., in refineries, which reduces the potential for offsetting consumer behavior.

The bottom line appears to be that dramatic reversals of prevailing emission trends require several things to happen simultaneously. Demand growth

---

[16]For a detailed history of the diffusion of tetraethyl-lead as an antiknock additive, see Ayres and Ezekoye, 1991:433–450. Tetraethyl-lead was invented by Charles Kettering in cooperation with Thomas Midgley. Kettering was also the inventor of the electric starter, first introduced into a Cadillac car in 1912, an innovation considered by many as the final blow against the comparative advantage of the electric car at the beginning of this century. Midgley, in turn, subsequently invented the first CFC, Freon.

must be slowed. Progress must be made on zero- or near-zero emission vehicles, and clean energy must be provided upstream. Incremental "add-on" innovations are not enough to reverse emission trends and reduce the environmental impacts of transport systems. We are tempted to paraphrase Joseph A. Schumpeter at this point: "Add as many catalytic converters as you please; you will never get zero emissions by so doing". Even if the present challenge of dislodging the internal combustion engine as the dominant transport technology of the "mass production/consumption" technology cluster seems daunting, there appears to be no other alternative.

Some Suggestions for Further Reading on Part II

The literature on various aspects of global change has grown tremendously. Few publications provide a comprehensive picture, and all either leave out technology altogether, or treat it as an awkward stepsister.

The most comprehensive treatment continues to be the monumental:

Turner II, B.L. *et al.* (eds.). 1990. *The Earth as Transformed by Human Action*. Cambridge University Press (713 pp.). Paperback: ISBN 0521446309, price: US$52.00.

"ET", as it has come to be known, was inspired by and, to a degree, modeled after the classic *Man's Role in Changing the Face of the Earth*, edited by W.L. Thomas (University of Chicago Press, 1956), which still provides extremely interesting reading. "ET" combines a breadth of coverage and excellence in individual chapters that is only possible in a carefully designed and edited multiauthor volume. Its sheer size does not suggest cover-to-cover reading, but the volume continues to be an excellent reference book to consult on specific topics, or to start reading on a new subject. "ET" builds and extends on an earlier volume, *Sustainable Development of the Biosphere*, edited by W.C. Clark and T. Munn, published in 1986 by the same publisher, which can still be recommended as good reading. A very readable summary of the "ET" volume is written by one of its coeditors:

Meyer, W.B. 1996. *Human Impact on the Earth*. Cambridge University Press (253 pp.). Paperback: ISBN 0521558476, price: US$23.90.

An excellent introductory text on global nutrient cycles (in particular of carbon, nitrogen and sulfur) in the form of a straightforward beautiful book has recently appeared in the *Scientific American Library* series. It is competently and well written by Vaclav Smil, and the illustrations are exceptionally clear and the numerous photographs highly aesthetic.

Smil, V. 1997. *Cycles of Life: Civilization and the Biosphere*. Scientific American Library, W.H. Freeman and Co. (221 pp.). ISBN 0-7167-5079-1 (hard copy), price: US$32.95.

For recent overviews with a more sympathetic and detailed treatment of technology we suggest two special journal issues that should be found in any reasonable library. First is a candid collection of papers under the thought-provoking title *The Liberation of the Environment*. This was published as a special issue of the journal of the American Academy of Arts and Sciences *Dædalus* (summer 1996 issue). Eminently well written and edited, the issue assembles a wide range of topics, ranging from global change drivers such as demographics, technology, and culture, to cautiously optimistic visions of what could be done to liberate the environment from undue human intervention. The issue contains contributions from such distinguished scholars as Robert A. Frosch, Bob Kates, and Paul Waggoner, among others. Areas covered include *inter alia* agriculture, energy, materials, and wastes. Carefully chosen and beautiful quotations from classical works are interspersed among the individual papers, giving the issue a much welcomed humanistic dimension in contrast to the usually dry, ecological global change literature.

More in standard academic prose is a special issue on *Technology and the Environment* that appeared in the journal *Technological Forecasting and Social Change* (Volume 53, Number 1, September 1996). Each contribution in the issue addresses theoretical or empirical aspects of the relationships between technological and environmental change. The issue extends its conceptual discussion to technology policy and of policy making in general. Two case studies in the areas of energy and materials overlap in authorship and subject treatment with the *Dædalus* issue cited previously.

Finally, we would like to recommend by now a classic volume, the first really to respark the interest in technology as an important driver of environmental change, and as an important starting point for finding solutions.

Ausubel, J.H., and Sladovich, H.E. (eds.). 1989. *Technology and Environment*. National Academy Press (219 pp.). Paperback: ISBN 0309040752, price: US$35.10.

Although it was published almost a decade ago, nowhere does the volume read as outdated. It also includes a particular favorite of the author, the contribution of Paul E. Gray on *The Paradox of Technological Development*. This contribution provides a differentiated view of the competing dynamics of technological change as a source of global environmental change and also as its possible remedy.

Price quotations are from *British Books in Print*, February 1997. Conversion rate used: £1 = US$1.60.

Part III

The Balance of Evidence

Chapter 8

Conclusion

Synopsis

The final chapter summarizes the technology–environment paradox – technology as both source and remedy of environmental change – and mentions technology's additional critical role as an instrument for observing and monitoring environmental change. Examples are presented of how the technology–environment paradox has been resolved, and has reemerged, throughout history. The critical questions are, first, which aspect of technology – as source or remedy of environmental change – currently has the upper hand and, second, how to tilt the scales toward the latter? To answer the first question, the chapter summarizes the balance of evidence from agriculture, industry, and the service sector. Answering the second question requires better models of technological change than we have today. The chapter reviews the major insights from the previous chapters that should be incorporated in improved models and lays out the major challenges that remain. The chapter concludes with a discussion of open issues that remain for a deeper understanding of the interactions between technology and global (environmental) change. Technology's most important historical role has been to liberate humanity from environmental constraints. That job is not complete, and the immediate challenge is to include the billions of people who have so far been excluded from the benefits of technology. The next challenge is to wisely use the power of technology to "liberate" the environment from human interference.

8.1. The "Paradox" of Technology and the Environment

Paul Gray (1989) describes as a "paradox" of technological development the fact that technology is both a source *and* remedy of environmental change. We should also add that it is technology that allows us to assess the numerous unintended environmental consequences of technological choices and develop strategies for their remediation. Ultimately it is only technology (in a large sense) that will empower us to liberate the natural environment from adverse

341

human interference that, in itself, has resulted in "global change" because of the power of technology.

8.1.1. Technology as a source of global change

Technology's impacts on global change have been both direct and indirect. First, new technologies have made possible the creation of entirely new substances (e.g., CFCs and DDT) with novel, direct environmental impacts. But more often technology's impacts have been indirect, through its ability to mobilize vastly more material and greatly expand economic output due to productivity and efficiency gains from continuous technological change.

Environmental impacts depend not only on *what* technologies are used, but also on *how* they are used. Technology by itself is not the main cause of the tremendous expansion in human numbers and activity. It can amplify or moderate the environmental impacts of human activity, but technologies are not conceived, selected, and applied autonomously. Thus, as an agent of global change, technology is an *intermediary* rather than a prime cause. The design, selection, and application of technology are matters of *social choice*.

It is true that symbiotic relationships among technologies, and the longevity of technological infrastructures (e.g., buildings and transportation networks), create great inertia. At any moment in time, social choices are constrained. Change, however, is also continuous. The lesson of history is that, given the right incentives, pervasive technological transformations can be implemented in one or two human generations. In 50 years we have largely left behind, with all its pollution, an economy based on steam and coal. In roughly the same period we have spread modern agricultural varieties, technology, and farming practices around the globe and thereby kept abreast of population growth.

8.1.2. Technology as a remedy for global change

Only a long-term perspective is capable of revealing the tremendous scale of historical improvements in land productivity, labor productivity, energy productivity, and even the productivity of using the carbon atom. Increasing productivity – "doing more with less" – is the most generic and important of technological strategies to lessen environmental impacts. Other strategies center on specific technologies to reduce particular environmental impacts, by "fixing" them with clean-up technologies. Catalytic converters in cars, sulfur removal from stack gases, recycling, and clean waste incineration are all recent examples of such *"end of pipe"* environmental technologies. More generic strategies focus on pollution prevention in the first place, i.e., by radically redesigning production processes and by "cradle to grave" management of entire product cycles. To date, however, these strategies have ranked second order in their importance for global change, compared to the more generic option of improved efficiency and productivity.

Only history provides an appropriate perspective on the order of magnitude of productivity increases from continuous technological change. Since the 18th century, labor productivity has risen by a factor of 200 in industry and at least a factor of 20 in agriculture, and it continues to improve. Productivity in the use of natural resources and in energy use per unit of output has risen by about a factor of 10.

Productivity increases, however, can also lead to increased output (and consumption), and heavier tread on the environment. In some cases, productivity has increased faster than output, thereby reducing net environmental burdens. In industrialized countries, for example, agricultural productivity increases have enabled large-scale reversions of farmland back to forests and natural ecosystems, despite output increases. But in most of the examples examined in this book, productivity gains have been outpaced by output increases. Energy and carbon productivities, for example, have risen at a combined rate of 1.3% per year, an improvement by more than a factor of 10 since 1800. But this has been more than offset by economic growth of 3% per year, an increase by a factor of 200. Further output growth appears both inevitable and socially desirable given the low living standards of most of the world's population. Such growth may differ from the past experience of today's industrialized countries, but it will in any event be substantial. Making such growth possible without increasing environmental stress requires solving the technological paradox: it is the ultimate aim of using technology to liberate the environment.

How Far Can We Go?

The substantial environmental productivity gains that have been achieved in the past give reason for cautious optimism. Strong evidence suggests that the material intensity of human society could be further reduced by a factor of 10 or more. Indeed it may have to, considering that the global population is expected to at least double by 2100 and material consumption might rise by a factor of 10.

Reducing material intensity by a factor of 10 remains feasible because, despite all our technological progress, we continue to use resources inefficiently. In the case of agriculture, Waggoner (1996) cites the example of corn yields. Global average yields are around 4 tons per hectare. The US average is around 7 tons. The best Iowa farmer grows 17 tons per hectare, and the best irrigated US field yields 21 tons per hectare, five times the world average and 13 times the average yield in Africa. For energy, calculations based on the second law of thermodynamics show that as little as 10% of primary energy actually ends up in the final services demanded by consumers (Nakićenović *et al.*, 1990). Thus there is no thermodynamic principle preventing us from using energy ten times more efficiently than we do today. Indeed, we would immediately halve primary energy needs if we were to instantaneously implement available "best practice" technology everywhere (Nakićenović *et al.*, 1990). There is also no obstacle, in principle, to long-term energy systems that produce no harmful emissions and make no use of the carbon atom (Häfele *et al.*, 1986; Ausubel *et al.*, 1988).

Gaps between average and best practice, or between current practice and calculated physical limits, indicate vast scope for improvements. But exploiting such opportunities is not a trivial matter. Developing the necessary technologies and assuring their wide diffusion are uncertain propositions at best, especially in view of current declines in research and development, inadequate institutions supporting technology diffusion, and limited incentives for adopting new technologies and *techniques*. In the words of Leach (1995:86) "... while the potential for reaching a benign state of affairs is indeed enormous, the path toward it will certainly not be smooth. In a future of very rapid change and large uncertainties, we can expect both the Malthusian pessimists and the technological optimists to be right in different places and at different times, while we have to remain ignorant about the total outcome".

8.1.3. Technology as instrument

Technology is also an important instrument for observation, especially when combined with its nonmaterial sister, science. In this combination, technology has tremendous power to improve our understanding of the environment and the impacts of our activities. Huygen's invention of the telescope made it possible for Galileo to observe Jupiter's moons and challenge the Aristotelian heliocentric view of the world. Foucault's pendulum made the rotation of the earth visible to every Parisian. Basic technological advances in measurement have always been central to industrialization and to an improved understanding of environmental issues.

Monitoring and measurement are indispensable for analyzing pollution. Technology now makes it possible to measure pollutants at levels of one part per trillion (10^{12}) for CFCs and even less for some pollutants, such as dioxin. Measurements, observations, and advances in scientific knowledge and simulation techniques (models) are the key to improving our understanding of complex environmental phenomena, and technological sophistication has increased steadily. Compare, for example, the first attempts to measure London's smog with carboscopes (optical devices to measure the darkness of smoke)[1] to today's chemical analyses through gas-chromatography and satellite measurements of land-use patterns, stratospheric ozone concentrations, and atmospheric temperatures.

New technology, in the form of modeling and simulation, is also the only way to analyze possible undesirable outcomes *before* large scale and irreversible changes occur. Simulation techniques have a long history in the military and in the aircraft industry, where the large consequences of failure call for particularly cautious design and implementation and where actual testing can be either too expensive or infeasible. Live nuclear test explosions can now be reliably replaced by computer simulations, which have made possible the recent adoption of a comprehensive nuclear test ban treaty. Simulation techniques in the aircraft industry are now at the point where the first prototype is already the final test flight model, as was the case for the Boeing 777.[2] In environmental research the techniques of simulation, scenario analysis, and anticipatory impact assessment are becoming increasingly important.

8.2. Technology, Productivity, and the Environment

8.2.1. The paradox in a historical perspective

To evaluate current concerns that population growth and output increases will outstrip technological productivity growth, we need to review historical concerns from the perspective of their day, and then with the benefit of hindsight. In 1798 Thomas Malthus feared population growth in England would outpace agricultural productivity and cause mass starvation. At the time,[3] England had a population of about nine million and average wheat yields of less than 1 ton per hectare. Throughout 1950 productivity increases did indeed lag behind population growth (population rose by nearly a factor of 5, but wheat yields, for example, increased only by a factor of 2.5). Pressure on the system was relieved by expanding its boundaries – both through food imports (financed by industrial exports and brought in by improved

---

[1]Cf. Brimblecombe (1987:164–170).

[2]Test pilots and safety testing are still required before passengers can board a new aircraft. But simulation techniques help identify problems early in the design stage and can speed development cycles and lower development costs.

[3]Population data (England and Wales) from UK Central Statistical Office (1996:16); wheat yield data are from Slicher von Bath (1963b:173), Mitchell (1980:219–290), and FAO (1991:68).

transport technologies) and through emigration (particularly important for Ireland). Only in the second half of the 20th century, after England's demographic transition was complete, did the race between agricultural productivity growth and population growth tilt in favor of productivity. By 1990 the population had risen almost to 51 million (i.e., by a factor close to 6), but wheat yields had reached 7 tons/ha (i.e., a seven-fold improvement compared to 1800). What made this possible was the widespread application of industrial technology to agriculture in the form of fertilizers, tractors, and new high-yield varieties developed through systematic agricultural R&D.

History provides other examples of fears about resource and environmental limits that fortunately were never realized. Jevons (1865), who feared England would run "out of steam" as coal resources were depleted, could not have foreseen the shift from coal to oil and its discovery in the North Sea. Indeed, England has a continued abundance of coal for which the economy has little direct use. Industrial, transport, and direct residential demands have receded with the disappearance of steam power and residential coal fires. In the 19th century concerns voiced by US presidents about a "timber famine" confronting steam railways disappeared with the chemical preservation of railway ties and the replacement of wood-fired steam engines by diesel and electric locomotives. One hundred years ago urban pollution from horse manure was a major environmental problem. Today it is a largely forgotten historical anecdote.

But the resolution of one technology–environment paradox has often created another. Replacing 25 million urban and farm horses in the USA with cars and tractors resulted in cleaner streets and freed 40 million hectares of agricultural land (five times the area of Austria) that had been required to produce animal feed. But as the number of cars has grown to nearly 150 million, they have exacerbated old problems (e.g., congestion) and created entirely new ones (e.g., urban smog). In general, success in reducing the environmental impacts of human activity often enables a further expansion of material wants, which frequently create new environmental challenges.

Thus while Malthus' original forecast did not materialize because both population and technology evolved differently than predicted, there is still concern that population and economic growth will outstrip productivity growth or encounter absolute limits. Since the early 1970s the "limits to growth" debate has raised anew questions of whether we are nearing limits on the productivity of agricultural soils, the availability of energy and material resources, and the capacity of the environment to absorb the wastes generated by vastly expanding human activities.

How great a concern is justified by current evidence? We have argued that technology must be explicitly considered at the center of any discussion of limits to growth, or how to resolve the technology–environment paradox. More particularly we must abandon the idea that technological change is exogenous to social and economic systems. We must recognize it as inherently dynamic and more fully appreciate that human ingenuity, with all its technological and institutional innovations, is the key to resolving the technology–environment paradox. With this in mind then, what is the evidence? We look in turn at agriculture, industry, and services.

8.2.2. The balance of evidence

Agriculture

Next to fire, agriculture is the oldest of human technologies to impact the natural environment. It is the largest user of land and water resources. Intensive soil cultivation, reservoirs, and irrigation have been part of "hydraulic civilizations" since antiquity, and land use and land-use change have been dominated by agriculture. Anthropogenic mobilization of nitrogen and phosphorus, principal agricultural nutrients, rival natural flows.

Even before industrialization (ca. 1700), agricultural productivity was sufficiently advanced in China that five people could be fed from each hectare of agricultural land. (This substantially exceeds current land productivity in Europe where "high-tech" agriculture feeds about three people per ha.[4] Since 1700 the world population has risen by about 4 billion people and some 1.2 billion hectares have been converted from forests to cropland, i.e., about 3,000 m^2 (0.3 ha) per additional head. Currently agricultural fields cover 1.5 billion hectares; 5 billion hectares of land remains as forest.

Although much land has been converted to feed a growing population, technology has helped substantially to decouple agricultural land requirements from population growth. Absent productivity increases, cropland areas in developing countries would have had to expand by some 700 million ha between 1950 and 1980 alone. Actual land conversions were, however, only 350 million ha. Technological change in the form of increased agricultural land productivity thus spared some 350 million ha of forests from being plowed. In the industrialized countries, such as in Europe and North

---

[4]Part of the difference is diet-related. The 18th century Chinese diet was based on rice, whereas that of Europe was, and still is, based on meat.

America, productivity increases have made possible the reconversion of agricultural areas to forests, while simultaneously increasing agricultural production (and surpluses). About 18 million ha of agricultural land have been
reconverted to forests since 1950 in Europe and North America.

Part of the change has been due to transportation. Revolutions in transport technologies have made it possible to cover larger and larger distances
at lower costs, thereby facilitating world trade in food. This has expanded
the agricultural resource base of densely populated areas, diversified diets,
and contributed to further global diffusion of agricultural crops. Part of the
change has been in labor productivity, which has seen tremendous increases
since the onset of the Industrial Revolution. Today in industrialized countries, only a minor percentage of the population supply food for all, compared
to 70% or 80% some 300 years ago. Employment has shifted from agriculture to manufacturing and services, and settlement from rural to urban.
These changes contributed in turn to the Industrial Revolution, mechanization, trade, and other developments that fostered increases in agricultural
productivity.

At the beginning of the Industrial Revolution, Malthus was concerned
that advances in agricultural productivity would not keep pace with population growth, whereas Boserup (1981) saw increasing population density as
fostering the development and diffusion of more productive technology and
social organization. Such developments would, in turn, facilitate increases in
population and living standards. The expansion of agricultural areas since
1700 tends to partially confirm Malthus' views. However, the expansion
was much smaller than it would have been in the absence of technological
change (and has even been reversed in industrialized countries), thus supporting a Boserupian view. Overall, in light of the tremendous expansion of
agricultural labor productivity and the transformation of basically agrarian
societies into now postindustrial societies, we believe the balance of evidence
supports more a Boserupian than a Malthusian perspective.

This is reassuring as agriculture lies at the center of the debate over the
ultimate carrying capacity[5] of planet Earth. Is it 10, 30, or even 1,000 billion
people as provocatively argued by Marchetti (1978)? As this book has sought
to reinforce, the answer depends critically on the direction and pace of future
technological change. Therefore, the real question is whether we develop and
implement technologies so that we feed, house, and employ *whatever* global
population level, in an environmentally compatible, adequate, and equitable
way.

---

[5]For a review see Cohen (1995).

Industry

Industrialization is at the core of global change. Industry is the largest transformer of matter and energy. Because of the success of industrialization, these transformations have assumed dimensions of truly global change. Industry mobilizes[6] about 20 Gt (billion tons) of materials annually in the form of fossil fuels, minerals, and renewable raw materials. The extraction, conversion, and disposal of these vast quantities produce more than 40 Gt of solid wastes per year. In comparison, total materials transport by natural river runoff is only 10–25 Gt per year (Douglas, 1990:231).

In addition to quantity, quality also matters. The amounts of new substances introduced by industry may be small compared to natural materials, but their environmental impact may be great. A mere million tons of CFCs has been sufficient to threaten the planet's ozone layer. The total US release of dioxins and furans is less than 1 ton per year; however, that ton is responsible for major human health and environmental concerns.

The vast scale of industrial metabolism is due to a series of technological revolutions. These have drastically changed the materials and energy that are available to be transformed into ever more and increasingly varied products. The replacement of natural by manufactured materials, of scarce by more abundant materials, of human and animate energy by fossil fuels, are all generic characteristics of technological change since the onset of the Industrial Revolution. Technology clusters have drawn on different principal raw materials and different energy sources, ranging from iron and coal in the 19th century to plastics, petrochemicals, oil, and natural gas in the 20th. However, in all cases there is a persistent pattern: the perennial race between productivity growth, which lowers resource requirements per unit of production, and output growth, which raises resource requirements. In its fundamental properties, the race, which has a long history, is the same as the technology–environment paradox of today.

The cumulative effects of technological change on industrial productivity and output growth are simply astounding. In the last 200 years industrial labor productivity has risen by about a factor of 200, and industrial output (in monetary terms) by at least a factor of 100. Material productivity and energy productivity have also risen tremendously. Producing a ton of steel today requires only one-tenth of the energy inputs of 100 years ago. These improvements all concern factor inputs that are valued (i.e., are costly) in industrial production. Therefore, there is an incentive to minimize inputs

---

[6]We use the term "mobilization" rather than "consumption" because neither matter nor energy can be consumed, but only transformed and "downgraded" by human use.

in order to minimize costs. Evidence suggests that if input costs were to rise further – in particular, if they were to be increased to reflect the costs of environmental externalities – then productivity and the efficiency of resource use would rise further as well. However, there are also examples of productivity improvements in connection with environmental factor inputs that are not subject to market transactions. The persistent historical trend toward decarbonization is one of the most striking examples. Current use of one unit of primary energy results in carbon emissions that are one-third less than in the mid-19th century. While decarbonization has not "solved" the carbon problem (it proceeds at 0.3% per year, which is significantly short of the 2% per year growth in energy use since the onset of the Industrial Revolution), it might be substantially accelerated were the carbon externality to be factored into the cost calculus of industry.

Higher productivity and more output have enabled higher wages and shorter working hours. Both are core elements of a "consumer" society. High consumption is the necessary counterpart to high production of the industrial sector. Over the last 100 years, real wages in industry have risen by more than a factor of 10, and working time has dropped by more than a factor of 2. In short, industrialization has brought affluence and leisure. And these have driven consumption and output growth, the other side of the technology paradox. For industry, the balance of historical evidence is more grim than it is for agriculture. Industrial output (and consumption) growth have outpaced productivity growth. Indeed, industry uses resources more sparingly per unit of output; it is "dematerializing" and "decarbonizing". However, these overall improvements have been more than offset by the growth in output and consumption that productivity growth has made possible.

Services

Unfortunately no comprehensive history of technological change in services or of the global-change implications of the service sector has been written. Data, analytical studies, models, and scenarios are only available for a few service activities such as transportation. Little research has been done on technology and global change in connection with health services, entertainment, or tourism. Perhaps the main reason for this is the extreme variety across different services and the widespread conception that services are a technology "taker" or are generally "low tech". As a result research on technological change has focused on industry and agriculture.

Available statistics contradict such conceptions. In the advanced industrialized countries services are the most important economic activity. They typically account for about two-thirds of economic output and employment. This partly reflects changes in industrial organization that have led to many traditional agricultural and industrial activities now being performed within the service sector. These range from design and advertising to distribution and retail. Services have also emerged as the largest user of new technology, particularly information and communication technologies. In the USA the service sector now accounts for approximately 80% of all investments in information technology hardware and most likely a similar or higher percentage of investments in software.

Growth in services is partially fueled by the substantial transition from work to pleasure made possible by productivity increases. We have higher incomes and more free time. Compared with the beginning of the Industrial Revolution, we work less and live longer by about a factor of 2 in each case. The fraction of our lifetime that we spend involved in producing goods and services has dropped to less than one-quarter. Growth in disposable income has been even more dramatic – by more than a factor of 10 since the beginning of the Industrial Revolution.

The global change implications of the shift from work to pleasure are ominous. Historically, increasing material wants have more than offset environmental and resource productivity increases. We accumulate ever more items to decorate our homes, we need more energy to heat and cool larger living areas, and we travel greater and greater distances for work and, increasingly, for pleasure. Personal mobility has continuously increased through faster transport modes that enlarge the range of human activity. Motorized mobility, i.e., travel by technological "prostheses", has risen by approximately a factor of 1,000 since 1800. It has begun to substitute for even the oldest and environmentally most benign transport technology – walking.

Transportation is particularly important for global change for two reasons. First, it changes the spatial division of labor and facilitates the exchange of goods, people, and ideas. Without the progression of transport technology clusters from steam ships and railways in the 19th century, to buses, cars, and aircraft in the 20th, many global change phenomena would not exist. These include urbanization, land-use changes due to export crop production, and the extraction, processing, and delivery of mineral and energy resources. Second, transportation is an important cause of human deaths and an important source of pollution at all levels: local, regional, and global.

The overall balance of evidence in the service sector suggests that consumer choices can eat up a large part, if not all, of the productivity gains and environmental improvements arising from new technologies. An encouraging counter-example is that of saturating consumer demands for food at high income levels. In the event of such saturation, additional environmental productivity increases from technological change really spare the environment. Unfortunately food demand saturation remains the only example of its type that can be clearly identified in the empirical data assembled for this book. Similar saturation phenomena may arise for material goods or mobility, but solid evidence has so far been elusive. This does not imply that current patterns of consumption and mobility will rise indefinitely, or that existing OECD patterns will be adopted globally. As was true for agriculture and industry, consumption and mobility patterns provide ample empirical evidence of path dependency. History matters, as does the cumulative nature of technological change. Changes can therefore go in alternative directions, as exemplified by development paths spanning the extremes between "high intensity" and "high efficiency" in resource use and environmental impacts, that are clearly discernible from empirical data.

8.3. Patterns and Rates of Change

8.3.1. Patterns of change

Our historical analysis has identified characteristic patterns that are remarkably stable across time, different technologies, and even different social and economic contexts. The most pervasive is the S-shaped pattern that, as a rule, describes the diffusion of new technologies and the replacement of the old by the new. Its specific parameters may vary substantially across technologies, but not the basic pattern.

In their embryonic phase, new technologies are surrounded by a maze of creativity and uncertainty. Many alternatives need to be explored, performance needs to be improved, and niche markets need to be developed. From an evolutionary perspective, this initial uncertainty and experimentation fulfill the important role of qualifying and prefiltering proposed new technological alternatives. The process is painful, but shortcuts are not a good idea, especially for radical technologies that can have the largest impact on society and the environment.

Once past this initial qualification, technologies may eventually spread. Like all historical analyses, this book has focused – with the benefit of hindsight – largely on technologies that have been successful. This should not give the impression that the direction of technological development and the pace of change are predictable in the early phases of a technology's life cycle. In fact, the history of technological change is mostly a history of nonstarters that have never diffused, e.g., wind-powered locomotives, zeppelins, and nuclear powered aircraft (cf. Rennie, 1997). The important message is that uncertainty and experimentation are inherent features of technological evolution rather than a hindrance, and that decentralized decision-making structures (markets) are, as a rule, more efficient in sorting out successes and failures than centralized decision making.

Once qualified, new technologies may spread. And this is where important stable patterns can be observed. Stability stems largely from increasing returns and path dependency. Because technological change is cumulative, characteristics (e.g. performance or costs) build on previous experience. The more a technology is tried, the more R&D that goes into improving it, and the more a technology proves useful in various applications and economic and social contexts, the more attractive it becomes. This process also "shields" technologies from new competitors. As a result, what is best or optimal depends on the context that has been created by and for technology. The properties of a technology and its context are acquired in a cumulative coevolutionary process. A technology *becomes* a best solution, even if it was not the best at the time of introduction. Reinforcing effects keep a technology and its context on a particular development path, once its basic features are set. This is, in systems language, a classic case of a positive feedback loop. But technology diffusion can by no means be taken for granted. Sustaining improvements that allow for continued growth, requires continuing, dedicated efforts to sustain a technology's progress along its "learning curve".

Technology growth and diffusion is ultimately limited. The resulting end of growth – saturation – is evident throughout the history of technology. The technology cluster associated with the steam engine and coal saturated in the 1920s, as did the size of inland waterways and canals some 50 years earlier. In some cases saturation is the result of diminishing returns – the value of an additional railway line into the most remote village is much smaller than the value of the initial lines connecting large cities. In other cases, negative externalities become apparent as applications expand. These can be intrinsically functional (e.g., traffic congestion) or environmental (e.g., pollution). The third cause of saturation and ultimate decline is related to new alternatives – better competitors. Steam ships ended the age of sailing ships. In turn, modern aviation replaced trans-Atlantic ocean liner travel.

The persistent S-shaped life cycle model thus begins with slow, initial growth accompanied by continued experimentation. If a technology is one of the fortunate few successful ones, pervasive diffusion may follow, sustained by continuous improvements, adaptation, and new applications, in short, *technological learning*. Ultimately potential improvements are exhausted and saturation sets in, followed by decline and disappearance into historical oblivion. Sailing ships and riding horses have not disappeared altogether, but they have disappeared as important transport technologies. Perhaps the most important lesson is to recognize that change is ubiquitous and pervasive. There is thus absolutely no reason to believe that future technologies will be merely extensions of today's technologies, even if that is how they are usually imagined and how they are usually modeled.

8.3.2. Rates of change

The speed with which new technologies diffuse in space and time depends on numerous factors. Studies have shed light on the importance of comparative advantage (e.g., relative costs), social visibility, complexity, technological interdependence, and others. But there are also more generic forces at work. Sheer size matters – large systems such as transport and energy infrastructures can take up to one century to develop fully. It also matters whether a new technology can integrate easily into a given technological, economic, or social context, for instance, by replacing an existing technology. The automobile initially diffused quickly (in a few decades) because it was replacing another existing individual transport technology – the horse. Further diffusion, however, took about half a century as it depended on infrastructure developments (widespread paved roads and highways), changing settlement patterns (suburbanization), restructured service activities (supermarkets and shopping malls), and of course income growth to pay for all those cars.

Generally, the rate of technological change is closely related to the lifetime of the relevant capital stock and equipment. Replacing black and white television sets with color took less than two decades. Replacing capital intensive, large industrial equipment like power plants or steel plants can take up to half a century. Infrastructures have the longest lifetimes and therefore take the most time to develop fully, or to be replaced. Time constants of up to a century are characteristic. For global environmental transformations such as climate change, the long time-scale of technological change complicates decision making. The time-scale for pervasive technology approximates the time-scale of global environmental change. If climate change proves extremely serious, and we wait to act, it will be too late to implement the massive economic and technological changes required to reduce impacts.[7] To reduce greenhouse gas emissions in sufficient amounts and on the right time-scales would mean starting now. But that conclusion still

---

[7]Alternatively, we could turn to "quick-fix" geoengineering measures such as changing the earth's albedo (cf. e.g., COSEPUP, 1992:447–456), but their environmental side effects remain largely unknown.

leaves a key question: what kind of incentives will induce such large and long-term changes?

While we cannot answer that question satisfactorily at this stage, we can identify some of the factors that will need to be addressed. Foremost among them is the inertia of technological development trajectories. Four principal forces underlie such inertia. First is the cumulative nature – path dependence – of technological change discussed above. Second is the importance of relative factor endowments and associated price structures. These provide the key "signals" that set the direction for productivity growth and the design and adoption of new technologies. Third is the scale of technology investments, particularly in the case of infrastructures. Fourth is technological interrelatedness. Other things being equal, the more individual technologies depend on each other, the more difficult it is to change any one of them independently. The economist Marvin Frankel first analyzed this issue in 1955, but neither the technological nor the economic literature has yet dealt with this definitively.

Environmental objectives will be required to influence the rate and direction of technological change much more in the future than they have done in the past. Incremental improvements will not be enough to significantly reduce potential environmental impacts. What is needed is a technology-led push toward an entirely new "green" technology cluster, characterized especially by dematerialization and decarbonization. Given the trends reviewed in this book, such a push will require major support and take many decades to unfold before there is a complete turnover of capital stock and a fully reconfigured sociotechnical "landscape".

8.4.   Open Issues in Addressing the Technology–Environment Paradox

8.4.1.   Different worldviews

The existence of unambiguous cause-and-effect relationships in the field of global change, and the actions that might be justified by observations of particular environmental problems are matters of serious debate. At one end of the spectrum are those who argue that attention is only required *after* significant environmental impacts have been detected and their causes clearly identified. Such arguments are particularly common in the climate change debate where the long-term nature of impacts and the significant natural "noise" overlying measured temperature increases makes clear determination of causality very difficult, if not impossible.[8] At the other end of

---

[8]E.g., Singer (1996); for a rebuttal see Ehrlich and Ehrlich (1996).

the spectrum are advocates of a strict "precautionary" principle that calls
for avoiding any change on the grounds that it might eventually lead to
unanticipated, adverse environmental impacts.[9]

The empirical evidence put forward in this book suggests that there is no
uniform resolution of the technology–environment paradox. There are cases
where productivity increases due to technology exceed the growth in human
activity and outpace environmental stresses. The reconversion of agricultural fields to forests in North America and Western Europe is one example.
There are cases where the two forces of change roughly balance. Saturating
demands for basic materials in industrialized nations is one example. And
there are cases where the growth in environmental stresses exceeds advances
in productivity, such as the growth in motorized transport.

Embracing a single perspective of technological change – whether Boserupian, Malthusian, technological optimism, or technological pessimism –
yields both bad science and bad policy. But we should not expect that it
will be easy to integrate contrasting perspectives into decision making, or
to easily resolve differences of opinion and different interpretations of fac-
tual evidence. Differences arise partly from different strategies in response
to uncertainties. They arise partly from different cultural biases or opposing "myths of nature" (Timmerman, 1986), ranging from "environmental
conservation" to "no limits to growth". Science and technology cannot resolve such differences in perspective and politics, although they can clarify
such differences. Nor can science and technology resolve possible mismatches
between the rates of change for science and technology and for society. Resolution must come through social and political processes, the complexity of
which would require a separate and even lengthier book than this one.

Science and technology *can* help us learn from past mistakes and replace "inadvertent" environmental experiments (e.g., the wide dispersal of
DDT into varied ecosystems) with nonmaterial simulation experiments (e.g.,
computer models of potential future global warming). It is an increasingly
valuable capability as we have only one planet to live on and experiment
with, and the potential of human activity to transform the planet has already grown to fearsome proportions. An extreme example is that of nuclear
winter, the significant planetary cooling that might result from dust thrown
into the atmosphere in a nuclear war. Less severe but more likely are large
industrial accidents. All are sufficiently consequential to preclude a strategy
of "learning by disaster" and "discovery by accident" (Schelling, 1996). As
the stakes grow, only dematerialized experiments are practical, i.e., using
science and technology.

---

[9]Cf. O'Riordan and Cameron (1994) and for a contrasting perspective Bodansky (1991).

8.4.2. Uncertainty, ignorance, and environmental "surprises"

One recurrent theme in the historical examples presented in this book is that environmental problems almost always arise as genuine "surprises". Either knowledge is unavailable, or it is not communicated. Often the surprise is due to environmental impacts that increase nonlinearly. "Surprise" is thus an issue of technology and organization, psychology and perception, as well as of nonlinear dose–response relationships.

When CFCs were introduced in the 1920s and 1930s, for example, as successful substitutes for risky and environmentally hazardous substances, human knowledge of stratospheric chemistry was simply too limited to even speculate about possible adverse impacts on the atmosphere. Only after technological advances was it possible in the 1970s to measure CFC concentrations as low as parts per trillion. Important advances in the scientific understanding of complex stratospheric chemistry are equally recent. It was only new technology in the form of satellite measurements that finally confirmed theoretical calculations of stratospheric ozone depletion. This is a prime example of a surprise arising out of scientific ignorance.[10] Nuclear weapons are another example. Many effects of atomic bomb explosions, such as electromagnetic impulse and retinal burn, were not anticipated beforehand but discovered only during test explosions (Schelling, 1996). The fact that important nuclear bomb effects were unanticipated during the Manhattan project can hardly be attributed to a lack of scientists or money. A better explanation is more generic. New technologies are always assessed with reference to the characteristics, environmental and otherwise, of existing ones.

A second type of environmental "surprise" stems not from the lack of knowledge per se, but from a failure to communicate critical knowledge to policy-makers and the public at large. Harvey Brooks (1996) refers to this as the "attention management" problem. There is simply a difference between knowledge that may exist somewhere and knowledge that is available in the right form at the right time to the right people. Frequently external events trigger the communication of knowledge from a few individuals (e.g., scientists) to the policy-making process and the public at large. The health impacts of coal smog were intensely studied in Victorian London, but it required the "killer smog" of 1952 to translate that knowledge into concrete policy actions, specifically banning coal use in smokeless zones. The

---

[10]In retrospect, perhaps the only early warning signal was precisely the desirable characteristic of CFCs, i.e., their chemical inertness. Inertness means longevity, and longevity means irreversibility. A similar recent example of irreversibility is the generation of long-lived nuclear isotopes and fission products through military and civilian applications.

greenhouse effect has been known and roughly quantified ever since Svente Arrhenius' paper in 1896, but it had to wait nearly 100 years (and unusually hot summer weather) before attracting widespread attention.

A third type of environmental surprise arises from impacts that are non-linear with respect to the scale of application of particular technologies. The automobile is again a good example. When it was introduced as a replacement for horses, urban environments improved substantially. Automobile emissions were invisible and orders of magnitude below those of horses (per unit service delivered). However, as the automobile population increased new types of impacts emerged nonlinearly, most notably photooxidant smog.

The lesson then is that there will continue to be surprises. We cannot eliminate them. What then, is the best strategy for dealing with uncertainty and surprise? From a technology viewpoint (see also the postscript Chapter 9) the answer is continued experimentation, maintaining technological diversity, and a continued quest for improving technology (via R&D and technological learning). Such an "evolutionary" strategy aims at making use of the best knowledge available but also prepares for changing course as better knowledge becomes available. In other words, technological innovation is the best contingency policy in the face of uncertainty and potential surprises. One unfortunate corollary of this conclusion is that there is no simple answer to the question of which instruments and incentives should be used to follow such an evolutionary technology strategy.

8.4.3. Incentives for change

It is comparatively easy to agree on the defects and environmental drawbacks of existing technologies. It is more difficult to decide on suitable alternatives. It is also inherently difficult to decide on the best incentives and mechanisms to encourage technological change in a particular direction. As the perspectives outlined in this book have made clear, all technological change is *induced*. Change is not exogenous to the economy or society at large. But inducement mechanisms are complex, interwoven, and, at least historically, have also been highly decentralized. It is unclear if singular directed levers are more effective to influence technological change or whether they fare better in choosing particular technological designs rather than decentralized decisions by trial and error. Centralized technology policy has unfortunately frequently bet on wrong horses.[11] At the time they looked like good bets. But with the benefit of hindsight, it seems that more "losers" than "winners" have been picked. From an "evolutionary" technology perspective,

---

[11] Cf. the history of nuclear-powered aircraft, large-scale synthetic fuel plants, or the fast breeder reactor.

this hardly comes as a surprise. After all, the main purpose of technological creativity is to suggest candidates for diffusion, whereas diffusion is decided by economic and social processes that control technology (and not the other way around). Picking any particular technology at the innovation stage thus inevitably entails preempting the economic and social processes that govern technology diffusion. History suggests approaching such a challenge modestly.

Given different worldviews and inevitable surprises as discussed above, what can usefully be said about incentives for change that does not require a whole new book?[12] Let us begin by identifying some principal dilemmas.

Technology and environmental policies face conflicting objectives. The first involves conflicts between long-term and short-term objectives. Short-term strategies of incremental improvements implemented through regulations, taxes, and end-of-the-pipe technologies can improve the environment relatively quickly. But these run the risk of entrenching existing technologies and reducing incentives for more radical long technological changes that could eliminate particular environmental problems entirely. End-of-the-pipe technologies are often favored because they are less risky than major process innovations in that they can be added to existing plants and generally involve fewer learning requirements. In the long run, however, end-of-the-pipe approaches may be inferior – in both their costs and environmental impacts – to "green" technologies designed from scratch. Thus, short-term environmental regulatory policies may work at odds with policies seeking major technological and organizational discontinuities.[13]

The second conflict exists between the short-term environmental benefits of regulatory standards and the possibly counterproductive impact they may have on innovative behavior. Regulation introduces an additional risk to R&D – that a new product cannot meet current or future environmental standards, or both. This can result in less innovation. Moreover, political discrimination in the form of "grandfathering" also makes innovation less attractive. Grandfathering existing facilities and products means that new alternatives are subject to stronger regulations than existing ones. This penalizes innovation. However, there is also the alternative argument that stringent environmental regulation, when introduced with sufficient lead

---

[12]For a good overview of these issues the reader is referred to the "Maastricht Manifesto" (Soete and Arundel, 1993). For a discussion of environmental policy and technical change see also Kemp (1997).

[13]An anonymous reviewer of this manuscript noted that environmental policies are rarely subjected beforehand to gaming approaches, as used for military strategies, to anticipate possible responses. Such a procedural innovation would itself constitute a new kind of policy formulation and analysis "technology" departing from the rigid legal approaches that now dominate environmental policy.

time, can *induce* industrial R&D and innovations. This is California's strategy in setting mandatory future market shares for automobiles with zero emissions.

The third conflict involves the tension between technological diversity and standardization. Most environmental issues and their possible remedies are notoriously poorly understood. This uncertainty argues for maintaining a broad range of alternatives and maximum flexibility to allow adjustments as more is learned. However, there is a constant tendency for technological systems to *reduce* diversity, and for very good reasons – cost reductions through standardization, economies of scale, and shared infrastructures (network externalities).

Act Now or Later?

Timing also influences the costs and impacts of efforts to induce technological change. Consider a recent debate on climate change. Given current uncertainties, should we reduce emissions now or wait? Wigley *et al.* (1996) have argued that for long-term climate change it matters little whether emissions are reduced now or later as long as *cumulative* emissions remain within certain limits. [Currently the science is not well enough developed to quantify those limits so Wigley *et al.* (1996) offer calculations for a range of climate "stabilization" scenarios.] They then argue that technological progress will make mitigation options cheaper with time. Thus reducing emissions later rather than sooner provides equal benefits at lower costs. This logic is correct to the extent that near-term constraints could lead to premature retirement of capital stock. But the bigger issue is the assumed reduction in mitigation costs. This is a standard assumption in all models using the neoclassical paradigm of intertemporal decision making with exogenous technological progress. This book, however, argues strongly against such an assumption. For technological progress to materialize much action is required: concerted R&D, promoting niche markets, experimentation (and even acceptance of failures), and gradually expanding diffusion. None of this happens overnight, and history argues that we should act sooner rather than later if we expect to reap the rewards of technological progress later on. Note that "action" does not necessarily mean massive emission reductions within the next one or two decades. It means enhanced R&D, technology demonstration projects, etc., both nationally and internationally. Those can be induced in a variety of ways, e.g., active efforts to promote innovation, but also credible efforts to cut emissions. However, climate change is a long-term and global issue. Therefore, the real challenge goes beyond that of inducing technological change per se. Given that new technologies will translate into environmental benefits only once they are widely applied, one needs to consider the social and institutional context that could provide for diffusion, or that could slow it down. However, as argued below, knowledge of the social and institutional technologies governing technology diffusion is even more limited than knowledge of "technologies" inducing technical change.

8.4.4. Technological, social, and institutional change

Technologies – both individual and in clusters – cannot evolve without being embedded in an appropriate social and institutional context. The match between technologies and their social and institutional setting is vital in determining how technologies are conceived, what resources society devotes to their development, and whether they are accepted or rejected. However, social structures and institutions change slowly (Linstone, 1996). Because of the cumulative nature of technological change, there is "ever more to change" even if individual technological changes would not proceed faster than in the past. "More to change" basically translates into ever higher rates of technology turnover, that contrasts with the stability and slow rates of change of institutions. It remains an open question whether we as individuals, and collectively as a society, have come to grips with the possible mismatch between the speed at which technology changes and the speed at which our institutions and society can change (see *Box 8.1*).

Institutions also feature when discussing technology diffusion, or lack of diffusion. This becomes especially acute, considering that a majority of the human population on this planet continues to be excluded from the benefits of modern technology. Today institutions are seriously inadequate in their capacity to effectively further social and economic development through better mastery of technology. There are very few successful institutional models of international cooperation promoting technological development and diffusion as a strategy to further development. The most prominent example is the Consultative Group on International Agricultural Research (CGIAR), an international network of agricultural R&D institutions instrumental in exchanging genetic resources and developing and diffusing locally adapted high-yield varieties.[14]

There are very few other successful examples of institutions specifically tailored to issues of *under*development rather than *over*development. Current debates about globalization are more preoccupied with the problems

---

[14]For a discussion of the CGIAR model see Victor (1993:535–538). Notwithstanding its success, CGIAR funding has declined by 20% from a peak of US$ 250 million (constant 1983 prices) in 1989 (Waggoner, 1996:62–63).

Box 8.1: On Change: Technological, Social, and Institutional

We recall voices demanding a halt to invention in the 1920s. The automobile, the radio, the movies, and the telephone, it was argued, were changing life too fast, and we needed time to assimilate the changes. Instead of stopping, invention thundered ahead and has been accelerating to this day. The World Wide Web is only three years old, yet already millions are using it. Contrast that with the centuries after Gutenberg before literacy became general so that his invention could have widespread use.

And what changes in our lives the new inventions and wealth have made! Gone are a restful week crossing the Atlantic, the sympathetic family doctor making house calls, the lifelong association with a spouse. Everything is more efficient now. We travel bunched in a plane with 400 others, we queue at the doctor's office, we have throw-away houses, clothing, spouses.

Paradox and contradiction abound. We are richer than ever at least on average, yet have record numbers of homeless; more labor-saving devices than ever, yet we work no fewer hours. If many of our lives are in disarray it is because we lack institutions that recognize the new technologies. The pill has made marriage seem unnecessary, television overrides what kids learn from parents and teachers, new technology benefits capital more than labor. In tropical Africa mass murder with deadly assault guns replaces tribal hostility played out more ceremonially with drums and spears.

Confusion should end once technology and institutions come into line again. We can attain a humane and equitable social order by adapting our institutions to use the new technologies while avoiding the harm they can do. How to achieve such adaptation as the aim of policy analysis provides a worthy agenda for social science in the 21st century.

Nathan Keyfitz
Harvard University, Cambridge, MA

of industrialized countries as they compete with each other and the rest of the world. However, the real challenge, in the author's view, is to find new institutional and organizational arrangements to smooth the transition away from the "mass production/consumption" technology cluster that is reaching saturation in industrialized countries today, and open new development avenues for those who have so far been largely excluded from the benefits of technology and productivity growth. In this connection, the increasing focus of both industry and governments on short-term results (whether quarterly profits in the private sector or macroeconomic targets in the public sector) is at odds with the long-term strategic goals of R&D, innovation, and institutional reform. There is a basic mismatch between short- and long-term planning, between the time-scales of everyday economic and political processes, and those of large-scale technological and environmental change.

8.4.5. Distributional concerns

Finally, choices about technology, like all social choices, involve important distributional issues. Establishing compatibility between technology and the environment is an intergenerational mission. This, because long lead times separate basic research from widespread diffusion, because both costs and benefits of technological change can extend far into the future, and because of the long lifetimes of technological infrastructures. The intergenerational concern is at the core of the concept of sustainable development.[15] The historical lesson is that perhaps the most important legacy of the current generation will lie in the choices and options made available for future generations due to advances in technology and scientific knowledge. Indeed, the social legacy of wealth – timeless knowledge and useful technological options – may be the only ethical justification for the use of nonrenewable resources by our current generation.

Scientific and technological progress is not free, and those who will benefit are not necessarily those who must pay. The problem of financing scientific and technological progress is thus an important public goods problem. It is compounded by the need to reduce "exclusion" from access to technology and technological skills, and by the reality that benefits arising from knowledge generation and skill formation accrue only over the long term. The accretion of technology confers power to those who can manipulate technology and use it best for their needs and purposes. This creates distributional concerns among social strata within a country, among different countries, and between genders. Abilities to develop, use, and accrue benefits from technology are distributed unequally. As technology changes and old skills become outdated, inequalities shift.

Education is increasingly vital as society relies more heavily on technology. First, education is the necessary prerequisite to advance knowledge and the generation of new technology. As knowledge accumulates, more education is needed. And increasingly the issue is not maximizing the transfer of knowledge and information per se, but building the ability to access, filter, and critically evaluate a torrent of information flowing from the new mass communication technologies. Second, education is needed in order to use technology effectively. Because technological change is more rapid than a

---

[15]Sustainable development may never become an analytically tractable concept, considering the difficulty of anticipating technologies and social preferences far in the future, and the inevitability of surprises. In this book we have preferred the term *environmental compatibility,* with its more modest connotation of trying to address the technology–environment paradox based on current knowledge, rather than trying to resolve it once and for all for future generations.

human lifetime, education becomes a lifetime activity. This is a challenge that social and professional institutions have not yet fully taken up. Nor will it be easy to agree on how it should be financed.

Third and more generically, new skills are required for dealing with technological change, the necessary adjustment costs, and the inevitable uncertainties. We need a new "culture" of dealing with technological change that transcends the two prevalent extremes. At present the tendency is to either blindly accept "progress" or categorically resist change – whether to preserve the environment, jobs, or something else.[16]

Finally, there is the issue of technology and democracy. Controlling technology, limiting its externalities, and ensuring a fair distribution of benefits are all questions of social choice. It is generally agreed that the Industrial Revolution was made possible by far-reaching changes in social and political institutions (cf. Rosenberg and Birdzell, 1990). And given the importance of diversity, uncertainty, and decentralization for technological progress, it is no coincidence that current centers of technological innovation are all in pluralistic societies using the "technology" of democratic institutions to arrive at social choices. Pressures to address environmental problems, both domestic and international, are generally larger in societies with pluralistic structures, democratic institutions, and market-based economies. Free science, free media, the existence of nongovernmental institutions, independent judiciaries, the possibility that prices reflect true costs, and strong public *and* private R&D to promote technological change are all important social "technologies". In the end, it is these *social* technologies that may ultimately be most important in determining the future of the environment and of material technology.

8.5.   A Manifesto

Technology has been an important agent of global change. Technological change has improved local environments and extended human life. At least in the advanced industrialized countries technology has freed human societies from the vagaries of pests, local resource availability, and climate variability. Through technology, humanity has for 300 years increasingly liberated itself from the environment. The job is not yet complete as billions of people continue to be excluded from the benefits of technology. The next immediate task is to assure their inclusion. But what next? We need a utopian vision, a

---

[16]Interestingly enough, although a middle road seems elusive, both attitudes can coexist in one person. Many of those most forcefully opposed to technological change still rush to purchase the latest computer software release despite numerous bugs that would never be accepted in products such as refrigerators, cars, or aircraft.

Box 8.2: The Liberation of the Environment*

Naked, humans are pathetic, vulnerable mammals. Yet we have come to number nearly six billion. Our extraordinary achievement is that we have liberated ourselves from the environment.

Until about 1900 environmental hazards caused about half of all deaths, even in Britain and America. Stagnant, contaminated water was a happy home for cholera, typhoid, and other waterborne diseases. Even more killers came by air, including diphtheria, tuberculosis, and whooping cough, as people crowded into miserably heated and ventilated homes and workplaces. By the middle of the 20th century, water filtration, chlorination, and sewage treatment stopped most of the aquatic killers in the industrialized nations and more recently elsewhere. Refrigeration in homes, shops, and trucks took care of much of the rest.

Now, great longevity, high incomes, and large populations have been achieved in every class of environment on Earth (Ausubel, 1991). Americans manufacture computers in hot, dry Arizona and cool, wet Oregon. Surgeons repair hearts in humid Houston and snowy Moscow. Year round, flowers grow in the Netherlands and vegetables in Belgium. In Berlin and Bangkok we work in climate-controlled office buildings. The Japanese have moved even skiing and sand beaches indoors.

For most of history thick forests and arid deserts, biting insects and snarling animals, ice, waves, and heat slowed or stopped humans. We built up our strength. We burned, cut, dammed, drained, channeled, trampled, paved, and killed. We secured food, water, energy, and shelter. We lost our fear of nature, especially in the aggressive nations of the West. But we also secured a new insecurity. Having liberated ourselves from the traditional challenges to our survival, we now wonder must human ingenuity always slash and burn the environment? Can humans live here *with* others, as part of a part of nature?

My answer is yes. Human culture, utilizing its most powerful tools, science and technology, which have brought us our present paradoxical freedom, can dramatically decouple our goods and services from demands on planetary resources. After long preparation, our science and technology appear ready to do so, to reconcile our economy and the environment. In fact, they are already doing so. Well-established trends, raising the efficiency with which people use energy, land, water, and materials, will cut pollution and leave much more soil unturned.

Yet, the catch for Homo faber, the toolmaker, is that our technology not only spares resources. Technology also expands the human niche. Population growth will require human society in the coming decades to accommodate as many more people as already live on Earth.

A highly efficient hydrogen economy, landless agriculture, industrial ecosystems in which waste virtually disappears: over the coming century these can enable large, prosperous human populations to coexist with the whales and the lions and the eagles and all that underlie them – if we are mentally prepared, which I believe we are. Worldwide, attitudes toward nature are shifting. "Green" is the new religion. Jungles and forests, domains of danger and depravity in popular children's stories until a decade or two ago, are now friendly and romantic. The characterization of animals, from wolves to whales, has changed.

So, our minds as well as our technology seem ready. We have liberated ourselves from the environment. Now it is time to liberate the environment itself.

Jesse H. Ausubel
Rockefeller University, New York

*Based on Ausubel (1996).

new technology manifesto, to set our goals for the 21st century. The author concurs with Jesse Ausubel (see *Box 8.2*), who argues that the task ahead is to progressively liberate the environment from adverse human interference.

For this century-long journey, we will need more technology, not less. As the world's population grows to 10 billion or more, we recognize that nature cannot be shielded perfectly from human intervention any more than we can be shielded perfectly from nature. But technological change can relax our grip and lighten our tread on the natural world. Our choices are constrained by what already exists and the environmental legacy of the past. But over the long term the capacity for social choices to shape technology is endless.

We understand some aspects of technology better than others. Thus, an important precondition for better steering technology's future course is improved knowledge about technology and its interactions with the social and natural environment. As illustrated in this book, good knowledge exists on patterns and characteristic rates of technological change. Improved models exist that enable a better theoretical understanding of the cumulative nature of technological change and of resulting path dependency. Improved models are also becoming available that no longer treat technological change as an externality to the economy and society. Impacts of technological change on productivity, efficiency, and the environment are also empirically well documented. Important knowledge gaps persist in trying to understand the maze of ingenuity and uncertainty surrounding the generation of innovations and the processes by which, ultimately, a few of them begin to diffuse in different social and economic settings. Much remains also to be researched on the role of institutions in promoting or hampering technological innovation and diffusion. Better knowledge is also required on the effectiveness of different instruments to induce technological change in particular directions and how to craft evolutionary technology strategies that prepare us best for a wide range of future contingencies and potential surprises.

Like knowledge, technology is a unique resource. The last word is left to Starr and Rudman (1973:364): "Unlike resources found in nature, technology is a man-made resource whose abundance can be continuously increased, and whose importance in determining the world's future is also increasing".

Chapter 9

Postscript: From Data Muddles to Models

Synopsis

The postscript briefly reviews useful theoretical formulations and empirical data that are available for building improved models of technological change. Elements of a stylized model are outlined, emphasizing uncertainty, mechanisms of continual technological improvement, and their influence on technology diffusion and substitution. Uncertainty introduces stochasticity in model formulations. Technological improvement through R&D and learning by doing introduces nonconvexities due to increasing returns. A number of models with these essential features are presented. The chapter concludes with a simplified model that integrates uncertainty, R&D, and technological learning as sources of technological change. The model demonstrates the feasibility of dealing simultaneously with stochasticity and nonconvexity arising from uncertainty and increasing returns from R&D and learning by doing. The postscript concludes with the optimistic outlook that modeling approaches do exist that can improve the traditional treatment of technological change as an "externality" to the economy and society at large.

9.1. Introduction

9.1.1. Why a postscript?

This book has described the evolution of technology and its relationship to global change largely without recourse to formal models. There are two reasons for this. First, models treating technological change as a process endogenous to the economy and society have been generally disappointing. Second, in the absence of successful models, the first task was to establish a sound empirical base defining essential patterns and impacts of technological change. This we have hopefully done. The task ahead is to begin translating these insights into formal models.

Although there is no single, comprehensive model that treats technological change as an endogenous process and integrates all of the important historical elements presented above, there are approaches that do show promise. Their introduction is the purpose of this postscript. The presentation concentrates on work in progress at IIASA focusing on energy systems.[1]

9.1.2. An initial analysis

Empirical Data and Patterns

Let us start with a summary of the empirical data, characteristic patterns, and modeling insights detailed in the first eight chapters. First, technological change is continuous and pervasive. Change is both incremental and radical (or revolutionary). It is ubiquitous in space and time, across sectors, across societies. It can affect what and how societies produce and consume and how they interact with the environment. Key characteristics of technologies (e.g., costs and performance) change continuously. Such changes are not autonomous, but are induced by forces that include elements of both supply (e.g., new knowledge and human ingenuity) and demand [expressed through, for example, prices (Jørgenson and Fraumeni, 1981) and regulations]. Technologies change the nature and magnitude of human impacts on the environment, in particular when they enable an increase in output, productivity, or both. All of these changes are interrelated.

It is useful to differentiate between potential and actual change, i.e., between technological innovation and diffusion. Innovation refers to new technological combinations arising out of human ingenuity. This has often happened accidentally, but is now more frequently the result of purposeful research and development (R&D). An innovation itself, however, is merely a *potential* source of change. It is only through diffusion that innovations become incorporated into the stock of artifacts (hardware) and practices (software) used by a society. Diffusion often involves the *substitution* of the old by the new, as new technologies interact with existing ones. Historical innovation and diffusion are empirically well documented, and provide a rich data source that can be drawn upon concerning the timing and patterns of diffusion of innovations. In contrast, the history of "failed" innovations that have not diffused remains largely unwritten.

There is always diversity among innovations and technologies in use, and diversity leads to persistent uncertainty. It is inherently difficult to

[1] I wish to thank particularly Yuri Ermoliev, Andrei Gritsevskii, Sabine Messner, and Chihiro Watanabe for the fruitful exchange of ideas and productive scientific collaboration underlying the research described here.

anticipate which innovations will ultimately succeed – "Many are called, but few are chosen" (Matthew 22:14). It is also inherently difficult to antici-pate all possible applications of new technologies (Rosenberg, 1996). But diversity, uncertainty, and continued experimentation are prerequisites for technological evolution. Even when particular technologies dominate their competitors, there is still diversity. The old overlaps with the new (even as their market shares change), and successful technologies are continuously modified as new applications are found and technologies are adapted for di-verse local conditions. Thus it is essential that particular technologies are not analyzed in isolation and that modeling focuses on *change*.

Individuals, firms, and society at large spend substantial resources on innovation, experimentation, and continuous improvements. There are ex-tensive empirical data on the extent and patterns of R&D, and on continuous improvements to existing technologies. Empirical data suggest that both are characterized by increasing returns. Empirical data on so-called learning or experience curves are plentiful.

Technological improvements accrue both from improvements in individ-ual technologies and from a "sequence of replacements" in technological hardware and software. Empirical data indicate that over extended peri-ods such improvements can yield output and productivity gains of up to one or two orders of magnitude. Thus the evidence is against assuming any arbitrary limits to improvements other than those imposed by physical laws of nature. Martino (1983) refers to *a priori* conservativism, or a static view of technology, as a "lack of nerve" on the part of technology modelers. His-tory provides many examples. Recurrent fears since the 1920s of imminent oil scarcity illustrate the pitfalls of ignoring improvements in knowledge and technology – in this case geological knowledge and technologies for explo-ration and production. Such improvements have continuously replenished available crude oil reserves. Conversely, visions of nuclear electricity be-coming "too cheap to meter" and skyrocketing capacity expansion forecasts illustrate the pitfalls of overenthusiasm. Such overenthusiastic forecasts were only gradually brought back to earth by the hard realities of chronic cost overruns and increasing public opposition.

Technological diffusion follows surprisingly regular patterns, which we captured through the technology life-cycle model. Initially, growth and diffu-sion are slow. Experimentation with alternatives is extensive, initial technol-ogy performance is low, and uncertainty is high. As uncertainties are gradu-ally reduced, solutions standardized, technologies improved, and application possibilities widened, growth and diffusion accelerate. Positive feedbacks re-inforce a particular technology choice, as increasing returns through further R&D, technology learning, cost reductions, economies of scale, and widened applications all promote further diffusion. However, all booms eventually bust. The possibilities for improving individual technologies are gradually exhausted, markets become saturated, and negative externalities (most notably environmental and social) become apparent. Diffusion eventually saturates, providing in turn opportunities for renewed experimentation and the possible introduction of new alternatives. The overall pattern of diffusion is S-shaped, and occurs regularly in empirical data ranging from individual artifacts, infrastructures, forms of organization, and even manifestations of social conflict.

Useful Models to Build Upon

Models that do not explicitly treat technology and its dynamics can nonetheless provide a useful starting point. Quantitative growth accounts evaluated with production function models (e.g., Solow, 1957; Denison, 1985) have confirmed the importance of increasing knowledge and technological change in explaining long-run productivity and economic growth, even if these models have treated knowledge and technological change as unexplained "residuals".[2] Microeconomic and sectoral models have greater relevance for modeling technological change. In particular they illustrate three important features governing technological change: (i) uncertainty and expectations as drivers of technological change; (ii) the importance of increasing returns and path-dependency phenomena; and (iii) the importance of R&D and its relationship to technology application and diffusion.

Addressing the first feature, microeconomic (simulation) models have been developed, particularly those within evolutionary economics (e.g., Dosi *et al.*, 1986; Silverberg *et al.*, 1988; Silverberg, 1991), that incorporate uncertainty and diverse expectations. The different strategies that economic agents follow as a result turn out to be the principal drivers of technological change. Similar concepts have also been used in models and empirical econometric analyses of technological diffusion. In these, the *expected* profitability of an innovation explains (among other factors) both its temporal spread and

---

[2]Recently, variants of such macroeconomic models have been used to explore the implications of technological "inertia" for climate change policy. Ha-Duong *et al.* (1997), extend Nordhaus' (1993) DICE model with a single scalar "inertia" coefficient, corresponding to the turnover of capital stock. The dynamic behavior of the model produces more realistic outcomes, consistent with the empirical patterns described here. Ha-Duong *et al.* (1997) correctly argue that with uncertainty and technological inertia, long-term climate stabilization goals call for earlier rather than delayed action. However, the treatment of technological change in the model remains exogenous. Like a sausage, the final product is evidently wholesome but the method of producing tasty results is best left shrouded in mystery.

intersectoral differences in diffusion (e.g., Mansfield, 1968; Mansfield *et al.*, 1971, 1977).

Second, models in the tradition of evolutionary economics also incorporate mechanisms to treat phenomena of increasing returns (e.g., Silverberg *et al.*, 1988) and so-called technological "lock-in" (Arthur, 1983, 1989). The result is *persistence* in the direction of technological change, i.e., path dependency. Cowan (1991) has formulated a number of models in which technological lock-in arises from uncertainty about potential increasing returns (future technological learning) of competing technologies. Because changes are cumulative and build on previous changes, such models (and everyday experience) demonstrate the difficulty of dislodging dominant technologies quickly, even if the conditions that made them attractive initially (e.g., low prices or abundant resources) no longer prevail. Also, as Cowan (1991) demonstrates, strong increasing returns can result in suboptimal technologies being prematurely chosen, i.e., one may "get the hare [rather] than the tortoise" (Cowan, 1991:811).

More neoclassical models offer complimentary perspectives. Theories of induced technical change (e.g., Binswanger and Ruttan, 1978; Ruttan, 1996) and related agricultural production function models (e.g., Hayami and Ruttan, 1985) offer useful formulations for modeling persistence in the direction of technological change. Such models are also powerful tools for explaining differences in the evolution of agricultural productivity over time and between different countries. Similar features are incorporated in engineering and management-type models of technological change that deploy learning-curve formulations (e.g., Dutton and Thomas, 1984; Gulledge and Womer, 1990).

The two central features of technological change, i.e., uncertainty or different expectations, and increasing returns or path dependency, are in stark contrast to the assumptions of perfect information and foresight frequently used in economic models. They are also in contrast to assumptions of quasi instantaneous technological change in response to changing (factor) prices.[3]

Third, useful insights have been provided by models of R&D economics, especially those that depart from the outdated linear model of innovation that assumes a strict temporal and causal sequence between innovation

---

[3]Typically economic models determine technological and economic configurations that optimize the allocation of inputs and outputs in production. When relative factor prices of inputs or outputs change, a new equilibrium is reached quasi instantaneously (subject only to capital turnover constraints). This presupposes that the new and different technological configuration (hardware and software), i.e., a "backstop" technology, is available "off the shelf", and entails no development costs and no initial diffusion lags. It goes without saying that such conditions hardly ever exist.

(R&D) and subsequent diffusion (for a review see OECD, 1992). For instance, input–output models and data have been used to illustrate the intersectoral interdependence between technology generation and usage, whereby a new technology that is developed in one sector enables subsequent changes in another sector (e.g., OECD, 1996). Similar models have also been used to test the hypothesis of "crowding out" phenomena in reallocations of R&D resources due to policy intervention (e.g., Goulder and Schneider, 1996). Models have been developed (particularly for agricultural R&D) that illustrate the importance of "replenishing" the technology knowledge base from which future technological change arises (e.g., Evenson, 1995, 1996). Models have also been developed that treat the critical interdependencies between R&D and actual technology deployment more explicitly and address the inevitable time lags between the two [cf. the discussion of the Watanabe (1995) model below]. Finally, macroeconomic models and corresponding growth accounts have also been extremely useful for estimating the orders of magnitude of the (substantial) economic and social returns to innovative activities as reflected, for example, in R&D expenditures (Mansfield, 1980; Griliches, 1986; Rosenberg, 1990).

9.1.3. Elements of a stylized model

There are many modeling elements upon which we can base a stylized integrated model of technological change consistent with our interpretation of historical evidence. We briefly present these below and then discuss how to translate them into formal models of technological change. A stylized integrated model of technological change should include at least the following elements:

- Uncertainty and experimentation (innovation).
- Continuous change and improvements through both directed activity (R&D) and actual "hands on" experience (learning).
- Diffusion and substitution, i.e., the gradual spread of new technologies in time and space and their interaction with existing technologies.
- Impacts (economic, resource, and environmental) and their possible feedbacks on technological change.

These elements should be linked in the following way. Assuming a competitive economic environment and changing consumer and social preferences, there is an incentive to innovate – to improve both hardware and software. Innovation is both *costly* (it requires resources and time) and *uncertain*. Uncertainty applies to both the actual returns on resources invested, and the ultimate economic and social acceptability of innovations (i.e., their diffusion potentials). But innovation also offers the potential for large rewards. For firms this usually means profits from lowered production costs, for example, increased market shares, or new markets altogether (e.g. for product innovations). For consumers it usually means increased utility provided by new products and services, improved performance, lowered prices, or a combination of all three. For society at large potential rewards come in the form of improved resource allocation, for example, or better environmental performance. In all cases there are powerful incentives for technological change.

Considerable uncertainty exists, however, concerning the possible outcomes of different innovation strategies. It is not simply enough to generate new technological knowledge through research; such knowledge needs to be applied. Applied technological knowledge usually takes the form of artifacts that need to be developed, tested, produced, and marketed. Such artifacts can be new machinery, a new consumer product, or a new piece of computer software. As knowledge is applied, positive feedbacks through learning can improve design characteristics and production economics, and reduce uncertainties. These feedbacks can create powerful incentives to move from innovations to applications in niche markets, to gradual initial diffusion and, ultimately, to pervasive diffusion. Obviously there is also the possibility of negative feedback. New knowledge is expensive to generate but cheap to imitate. Thus benefits may not accrue to the innovator, and this risk is a disincentive for change. The disincentive can be reduced by lessening the risk through, for example, an effective patent system. When negative feedbacks dominate, or there are insufficient returns on investments (e.g., if it proves impossible to improve a technology despite all R&D efforts), then a technology does not diffuse. In this case the effort and money are lost: innovation expenditures *à fonds perdu*. There is only one positive outcome. The knowledge generated is generally not lost and can be used in the next innovation attempt.

Although the model just outlined is conceptually simple, it is still a challenge to translate it into a formal model. First, the formal model needs to incorporate uncertainty (stochasticity) into its decision rules. Second, it needs to incorporate the fact that future outcomes are dependent on intervening actions (increasing returns) and not known beforehand. Assuming some kind of rationality (i.e., attempting to do the best) this inevitably leads to nonconvex optimization problems. Third, to complicate matters even further, uncertainty and increasing returns cannot be viewed separately.

Below we briefly present model formulations that have successfully responded to these challenges. The focus is on energy technologies. Moving from the simple to the more complex, we begin with models incorporating technological uncertainty (stochasticity), continue with models incorporating increasing returns (R&D and learning), and conclude with a simple model integrating both. Characteristic patterns of technology diffusion and productivity growth generated by such models are then presented, and compared with the historical patterns examined throughout this book.

9.2. Modeling Technological Change

9.2.1. Uncertainty

The importance of technological uncertainty has been recognized and explored from the earliest days of global environmental modeling (e.g., Nordhaus, 1973; Starr and Rudman, 1973). Different approaches have been pursued including the use of alternative scenarios (e.g., IIASA–WEC, 1995), model sensitivity analyses (e.g., Nordhaus, 1973, 1979), and sensitivity analyses based on expert polls or Delphi-type methods (e.g., Manne and Richels, 1994). In such sensitivity analyses, the range for varying technological parameters is chosen either by modelers or by the experts whose opinions are polled. Thus, while scenarios and sensitivity analyses show how model outcomes vary as input assumptions change, technological uncertainty is not endogenized into the decision rules represented in the models. That is, although we learn how different future outcomes depend on when, how, and in what direction technological uncertainty is resolved, we remain ignorant of robust or even optimal strategies in the face of uncertainty.

A model that endogenizes technological uncertainty through stochastic optimization has been developed by Golodnikov *et al.* (1995), and Messner *et al.* (1996) and applied to the energy sector.[4] In this model, technological uncertainty translates into both economic risks and opportunities (benefits), and both are directly endogenized into the model's decision rules. A distinctive feature of the model is that subjective definitions of technological uncertainty ranges are replaced by an empirical approach based on detailed statistical analysis (Strubegger and Reitgruber, 1995) of investment costs for current and future energy technologies derived from a large technology inventory, CO2DB, developed at IIASA (Messner and Strubegger, 1991).

The resulting empirically derived uncertainty distributions are incorporated directly into the model, and uncertainty is incorporated into the

---

[4]A stochastic application using the MARKAL model is reported in Fragnière and Haurie (1995), and Ybema *et al.* (1995). The model follows the traditional approach towards uncertainty by exploring alternative model outcomes for scenarios where uncertainty is reduced at various future dates.

model's decision-making (optimization) rule. A risk term, representing additional economic costs if a technology turns out to be more expensive than expected, is added to the objective function. The risk term integrates stochastically drawn data samples (weighted by probabilities) from the uncertainty distribution into the final solution. The model thus computes the optimal technological hedging strategy given the empirically derived uncertainty distributions of current and (estimated) future energy technology costs.

Such a stochastic model thus responds to the weakness inherent in traditional models of assuming perfect foresight. Because uncertainty is endogenized, decision making no longer operates with perfect foresight. The model's behavior thus better approximates real-life decision making, where various economic agents with different expectations and attitudes toward risk show persistent differences in strategies and investments. The result of these differences is technological diversification. Simulations using the stochastic model (cf. Grübler and Messner, 1996) yield more diversified technological configurations than the traditional deterministic models that have perfect foresight. More importantly, simulations with the stochastic model reveal a proinnovation bias and no-risk aversion in investments in technological change. Diversification becomes the optimal strategy in the face of technological uncertainty.

However, the model simulations also show that including uncertainty leads to technological diversification only through incremental innovations. Technology changes only within a "technological neighborhood" (Foray and Grübler, 1990). Radical technological change does not occur. This is because there is no mechanism in the model to reduce the high uncertainties, improve the performance, or lower the initially high costs of technologies that have rarely been tried. Thus, technologies with very high initial costs and uncertainties do not make it to the market. For this to occur in the model – as it does in the real world – the model needs a mechanism whereby uncertainty and costs can be reduced. It needs a way to represent R&D and learning.

9.2.2. Increasing returns (R&D and learning)

Research and development plus actual experimentation (technological learning) are the essential endogenous mechanisms for reducing uncertainty and improving performance and costs. Together, they represent classic examples of increasing returns – the greater the R&D and experimentation (learning), the faster *ceteris paribus* technologies improve and uncertainties diminish. We look first at R&D, then at experimentation, and then at the two together.

R&D

R&D's importance as a source of technological change is self evident.[5] As shown in Chapter 2, research costs are a comparatively small fraction of total research and development costs (and especially in comparison with costs incurred during subsequent technology diffusion). Industry dominates the public sector both as a source of R&D money, and in R&D expenditures, simply because development costs make up the bulk of R&D costs, and industry performs the bulk of technology development work. It is therefore important to always look at total R&D and not just at public- and government-sponsored research alone.

Considering for the moment research, development, and demonstration (RD&D) costs, demonstration costs are usually borne by the users of new technologies (industries and consumers), while R&D costs are borne by suppliers. However, important overlaps exist as close cooperation between suppliers and users is essential for successful technology development and diffusion.

Watanabe (1995, 1997) has developed a conceptual and empirical model of energy technology R&D drawing on the experience of MITI's "sunshine" technology program. The model considers both public and private R&D expenditures and makes use of exceptionally comprehensive time-series data (see *Figure 9.1* and *Figure 9.2*). Watanabe's model also has the added benefit of empirical parametrization obtained through an econometric analysis of the time-series data. In essence it describes a positive feedback loop, or "virtual [virtuous] spin cycle" in the terminology of Watanabe (1995). Public R&D (together with other incentives) stimulates industrial R&D, and both increase the "technology knowledge stock"[6] of a particular technology, which leads to performance and cost improvements. These stimulate demand, increasing the size of niche markets (leading to economies of scale), and thus to increasing learning possibilities (and therefore further cost reductions). All of this feeds back as a further stimulus for industrial R&D and technology improvements.

Watanabe's model clearly identifies the close relationships between R&D and technology demonstration, and between public and private R&D. An interesting result is the length of the time lag between R&D expenditures and their returns in the form of improved technological performance, i.e.,

---

[5]For a discussion see Rosenberg (1990).

[6]This is the sector- or technology-specific equivalent of the knowledge stock in the "new growth theory" production function models (e.g., Romer, 1986, 1990), which also exhibit increasing returns. Interindustry and cross-national R&D spillover effects exist (cf. Mansfield, 1985), including those from purchases of equipment (cf. OECD, 1996), that also increase the technology knowledge stock. These spillover effects are not treated here.

lowered costs. This is estimated at less than three years, indicating a rather effective application of improved technical knowledge gained through systematic R&D.

Technological Learning

Technologies typically increase substantially in performance and productivity as organizations and individuals gain experience with them (cf. Chapter 2). Technological learning phenomena were first described for the aircraft industry by Wright (1936), who reported that unit labor costs in air-frame manufacturing declined significantly with accumulated experience measured by cumulative production. Technological learning has since been analyzed empirically for numerous manufacturing and service activities including aircraft, ships, refined petroleum products, petrochemicals, steam and gas turbines, and even broiler chickens (cf. Clair, 1983). Learning processes have also been documented for a wide variety of human activities, ranging from success rates for new surgical procedures, to productivity in kibbutz farming, and the reliability of nuclear plant operations (Argote and Epple, 1990). In economics, "learning by doing" and "learning by using" have been highlighted since the early 1960s (see e.g., Arrow, 1962; Rosenberg, 1982). A number of detailed studies have tracked the many different sources and mechanisms of technological learning.[7]

Learning phenomena are generally described by "learning" or "experience" curves, where the unit costs of production decrease as experience is gained. Because learning depends on the accumulation of experience and not just on the passage of time, learning curves are generally described in the form of a power function where unit costs decrease exponentially with experience, usually measured as a function of cumulative output (cf. Chapter 2). Frequently, the resulting exponential decay function is plotted with logarithmically scaled axes so it becomes a straight line (see, e.g., *Figure 9.1* which plots the decreasing costs of PV cells in Japan). Such straight line plots should not be misunderstood as "linear" progress that can be maintained indefinitely. Rather, the logarithmic axes mean that each successive doubling takes longer. Cost reductions become smaller over time as each doubling requires more production volume. The potential for cost reductions becomes increasingly exhausted as the technology matures.

Figure 9.1 plots the costs of PV cells per (peak) Watt capacity as a function of total cumulative installed capacity for Japan. Starting from an

---

[7]For a discussion of who learns what, see Cantley and Sahal (1980). For a stylized taxonomy of technological learning processes based on empirical data, see Christiansson (1995).

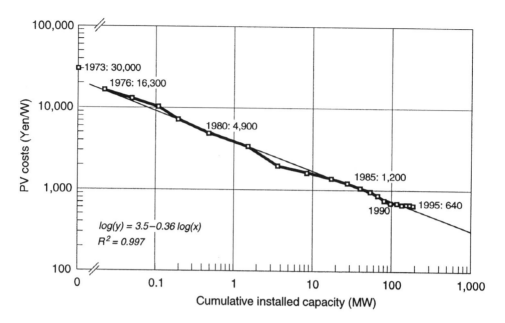

Figure 9.1: Photovoltaic (PV) cell costs (1985 Yen per Watt installed) as a function of cumulative installed capacity (in MW), Japan 1976–1995. Data source: Watanabe (1995, 1997). For the data of this graphic see the Appendix.

extremely high value of 30,000 Yen (in 1985 prices) in the early 1970s, costs fell dramatically: to 16,300 Yen in 1976 and 1,200 Yen in 1985 (a factor of close to 14 in less than 10 years), and then further to 640 Yen in 1995 (another factor of 2 within the next 10 years). The 36% reduction in costs per each doubling of cumulative installed capacity is at the higher end of the range of learning rates observed in the empirical literature (cf. Argote and Epple, 1990; Christiansson, 1995). But it is not unusual in light of the infancy of the technology and the significant role of R&D (see in particular the substantial cost decreases between 1973 and 1976 prior to any installation of demonstration units).

Technological learning is a classic example of "increasing returns", i.e., the more experience is accumulated, the better the performance and the lower the costs of a technology. However, since the experience required for each subsequent reduction in costs takes longer and longer to accumulate and is more and more difficult to achieve, learning itself shows decreasing marginal returns. This is reflected in the increasingly "packed" spacing of observations toward the 1990s in *Figure 9.1*.

Learning phenomena, however, have not been incorporated in many technology models. The most likely reason is the difficulty of dealing algo-rithmically with the associated nonconvexities of the optimization problem. A detailed formulation was first suggested by Nordhaus and van der Hey-den (1983) to assess the potential benefits of enhanced R&D efforts in new energy technologies (the fast breeder reactor in their model). A full-scale op-erational optimization model incorporating systematic technological learning was developed by Messner, 1995 (see also Nakićenović, 1996b, 1997).[8]

Messner used a mixed-integer formulation to introduce learning rates for a number of advanced electricity-generating technologies into a linear programming model of the global energy system. The learning rates were assumed to be known *ex ante*. Future technology costs therefore depended solely on the amount of intervening investments in increased capacity, leading to increased experience and cost reductions (learning).

Messner's results are especially significant for two reasons. First, they indicate that providing for technological learning can lead to radical tech-nological change. Learning enables the diffusion of technologies that are very different in their technological and economic characteristics from those predominantly used today. The resulting technology dynamics in the model yield diffusion patterns that are remarkably consistent with history and the theoretical and empirical findings of the diffusion literature (cf. Grübler, 1991, 1992b). These patterns show slow but early growth in niche markets where initial experience is gained, and subsequent widespread diffusion that ultimately saturates when the technology matures. This is in stark contrast to the typical "flip-flop" behavior of optimization models where technological change (e.g., cost reductions) is introduced exogenously. There, the initial slow growth in niche markets and the necessary up-front investments never appear because the learning that leads to cost reductions comes at no cost, that is, it is treated as an "externality." Second, simulations using Mess-ner's model (Messner, 1995) demonstrate that up-front investments in new technologies stimulate future cost decreases and can therefore be economi-cally optimal, even if at the time of investment the new technology is more expensive and has lower performance than existing alternatives.

There remain two shortcomings at this stage. First, even if the empirical literature and statistical studies (e.g., Christiansson, 1995) report rates and mechanisms of learning in the past, the rates at which a particular tech-nology may improve in the future are uncertain. Thus, instead of treating learning rates as deterministically known *ex ante* one needs to consider un-certainty explicitly. Second, technological change is the result of R&D *and*

---

[8]For details of the methodology, see Messner (1995, 1997).

"hands on" experience (via investments). It is therefore insufficient to consider only investments, even if they do constitute the dominant share of total expenditures on new technologies. We next address the interaction between R&D and learning.

Interaction between R&D and Learning

Let us return to the example of PV cells in Japan. The bottom line in *Figure 9.2* shows PV costs as a function of cumulative R&D expenditures (a proxy for the technology knowledge stock) taking into account the three-year time lag estimated by Watanabe (1995). In the top line of *Figure 9.2* these are added to the cumulative installation investments (a proxy for accumulated experience) calculated from the cumulative installations given in *Figure 9.1*. From 1973 to 1995 a total of 206 billion Yen (in constant 1985 prices)[9] was spent on photovoltaics in Japan. Investments in PV capacity in niche markets and early deployment made up 78% (162 billion Yen) of the total, dominating the 22% (44 billion Yen) spent on R&D. The relationship between PV costs and the combined total of capacity investments and R&D expenditures follows a classic learning curve pattern as shown in *Figure 9.2*. With R&D costs now included on the horizontal axis the learning curve parameter results in a 54% drop in PV costs per doubling of cumulative expenditures (a proxy for accumulated knowledge and experience), compared to a 36% drop per doubling of installed capacity in *Figure 9.1*.

The fact that a classic learning curve pattern emerges when production costs are plotted against both capacity investments and R&D expenditures indicates that the two kinds of investments cannot be treated separately as sources of technological dynamics. A linear model assuming that all R&D precedes actual investments (i.e., demonstration in niche markets, early commercial applications),[10] or that considers the two as independent from each other, is simply not supported by the data.[11] This simplifies our basic model considerably as decreases in production costs in response to both R&D and capacity investments can be modeled by a single learning curve. However, the actual shape of the curve is subject to uncertainty. The next step, therefore, is to integrate uncertainty and learning.

---

[9]This equals approximately US$ 2.5 billion at 1995 prices and exchange rates.

[10]There can be an overlap between demonstrations in niche markets and early commercial applications. Consider the example of PV cells. Their use in remote locations constitutes both an important niche market demonstration, but may also constitute an early commercial investment in many cases.

[11]For further data from other sectors and technologies see also Mori *et al.* (1992) and Baba *et al.* (1995).

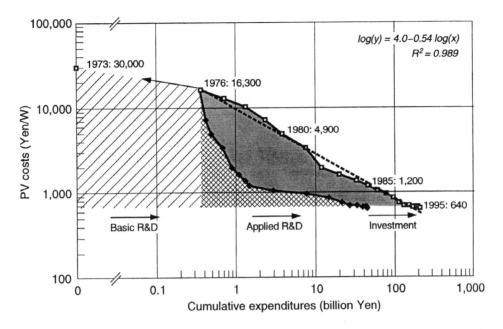

Figure 9.2: Photovoltaic (PV) cell costs (1985 Yen per Watt installed) for Japan (1976–1995) as a function of, first, cumulative R&D expenditures [billion (1985) Yen] (solid boxes) and, second, total expenditures including both R&D and actual investments in capacity expansion (open boxes). R&D expenditures are lagged three years. PV costs follow a classic learning curve pattern with cost reductions of over 50% for each doubling of cumulative expenditures (a proxy for the increasing technology knowledge stock and accumulated experience). Data source: Watanabe (1995, 1997). For the data of this graphic see the Appendix.

9.3. A Model of Uncertain Returns from R&D and Learning

An illustrative model of endogenized technological change through uncertain returns from R&D and learning has been developed by Grübler and Gritsevskii (1997). It combines the above approaches to modeling technological uncertainty (stochastic programming) and technological learning (nonconvex optimization) in an intertemporal optimization framework. Technological change is driven by expected, but uncertain, returns from investments in R&D and niche-market applications. These in turn lead to learning that makes new technologies increasingly competitive, and leads ultimately to their pervasive diffusion.

The model is highly stylized, representing a simplified energy sector with an extremely long simulation horizon of over 100 years. The model includes one category of energy demand, which increases throughout the simulation, and one resource category, whose extraction costs increase with depletion. Demand is satisfied by technologies that convert resource inputs into the final demand. The model selects from three competing technologies that differ in their current costs and in their uncertain potentials for future cost reductions through learning. Technology development costs include both capacity investments and R&D.

The three technologies represent stylized and contrasting generic technological alternatives as follows:

- Existing, mature technology for which no further improvements are possible.
- Incremental technology, which is initially twice as expensive as the existing technology, but has a modest potential for cost reductions through learning (a mean cost reduction of 10% for each doubling of cumulative installed capacity). Uncertainty is comparatively low.
- Revolutionary technology, which is initially 40 times as expensive as the existing technology, but has a high potential for technological learning (a mean cost reduction of 30% for each doubling of cumulative installed capacity). Uncertainty, however, is high.

The environmental characteristics of the three technologies were assumed to vary in a manner similar to costs. Existing technology was characterized by low efficiency and high emissions, versus correspondingly higher efficiencies and lower emissions for the incremental and revolutionary technologies, respectively.

The characteristics of the three technologies and other model parameters were chosen to correspond very roughly to values used in recent well-known energy scenarios. The model has one energy demand category. Demand growth is rapid and increases by a factor of 15 over the next 100 years. This is comparable to electricity growth in high-demand growth scenarios such as the IIASA–WEC (1995) Family "A" scenarios or the IS92e and IS92f scenarios of the IPCC (Pepper *et al.*, 1992). The model parameters for the existing technology resemble conventional coal-fired power stations. The incremental technology resembles advanced coal-fired power stations, with initial costs twice those of the existing technology but with somewhat better environmental performance and a modest rate of technological learning (mean value: 10% per doubled installed capacity). The revolutionary technology resembles photovoltaics, which use a basically free resource (sunlight) and require only a small input of fossil energy for manufacturing. Its initial cost was set quite conservatively at 40 times that of the existing technology, which approximates the comparative costs of PV and conventional coal in the early 1970s. However, the revolutionary technology has, on average, the potential for 30% cost reductions for each doubling of installed capacity.

The model works as follows.[12] The learning rates of the incremental and revolutionary technologies are treated as random values (with means of 10% and 30% for each doubling of installed capacity, respectively). This means that future technology costs are a random function of the intervening total cumulative expenditures on R&D and capacity investments. The probability distributions for the two learning rates are assumed to be lognormally distributed around their respective mean values. Random samples are drawn, and for each draw the resulting nonconvex, nonsmooth optimization problem is solved (cf. Ermoliev and Wets, 1988; Ermoliev, 1995). That is, the model determines for each draw the intertemporal optimum investment profile to achieve future cost reductions. The determined future technology costs are then integrated into an overall objective function that consists of three parts.

The first part corresponds to a deterministic formulation based on the mean (expected) values of the technology learning rates. The second part represents the risk of having overestimated the technological learning rate, i.e., if the learning rate sample drawn from the lognormal distribution is lower than the mean (expected) learning rate. This second term is assumed to be quadratic, i.e., the costs added to the objective function grow quadratically with their deviation from the mean value.[13] The third part of the objective function represents the benefits arising from underestimating the learning rate, i.e., when the sample learning rate drawn from the lognormal distribution is higher than the mean (expected) learning rate. This benefit term is assumed to be linear. The model is then solved for a sufficiently large sample size to determine an overall intertemporal[14] minimum of the overall objective function into which all sample draws[15] are integrated.

The asymmetrical modeling of investment "risks" and "benefits" in the objective function reflects our interpretation of reality.[16] The model conservatively values survival (i.e., avoiding higher costs than one's competitors) more than profitability (i.e., aiming for lower costs than one's competitors).

---

[12] A detailed description and mathematical appendix are given in Grübler and Gritsevskii (1997).

[13] This follows a formulation suggested by Ermoliev (1995) and first applied by Messner *et al.* (1996).

[14] A discount rate of 5% is used in all calculations.

[15] The overall objective function integrates all individual realized objective functions (values from mean learning rate plus/minus the quadratic and linear risk and benefit terms resulting from the learning rate sampled, respectively).

[16] Alternative assumptions are explored in a sensitivity analysis reported in Grübler and Gritsevskii (1997).

The resulting technological uncertainty and asymmetry cannot be expressed simply in terms of the mean and variance of corresponding potential economic gains and losses. Instead, each realization in the uncertainty space represents a diverse individual outcome, with differences amplified due to increasing returns. The final model solution represents the optimal technological investment diversification strategy in light of the uncertain returns from R&D and learning by doing (investments).

Grübler and Gritsevskii (1997) have used the above model to explore a wide spectrum of alternative parameter values and model formulations. Two principal results are summarized here: the diffusion of new technologies, and environmental impacts (emissions).

9.3.1. Technology diffusion

Figure 9.3 compares how the shares of various technologies in new capacity additions change over time. In the figure dotted lines correspond to the existing technology, dashed lines refer to the incremental technology, and solid lines refer to the revolutionary technology. Simulation Runs 0 and 1 represent the more conventional view of technology as either static (Run 0) or exogenously determined (Run 1). In Run 0, 100% of new capacity additions are from the existing technology, represented by the dotted line along the top of the graph. What is significant in this case is what does *not* happen: neither the incremental nor the revolutionary technology ever makes it. This is reflected in the solid line for the revolutionary technology along the bottom of the graph for Run 0 and a similar dashed line for the incremental technology (which cannot be distinguished in the figure from the solid line that overlies it). Run 1 portrays a pattern typical of models that employ exogenous technological change. At some point in the future (2020 in this case) a new technology forcefully enters the picture due to an exogenously prespecified cost reduction.[17] This is represented by the dashed line rising quickly from 0% to 100% starting in 2020. (For clarity we do not show the corresponding fall in the existing technology's share from 100% to 0%. The solid revolutionary technology line lies along the bottom of the graph as for Run 0.) Patterns similar to that of Run 1 occur in models in which an exogenous "backstop" technology suddenly enters the market due to resource

---

[17]In Run 1 the cost of the incremental technology was assumed to fall – for no apparent reason and certainly not because of earlier investments – to the level of the existing technology by 2020. Such a reduction is entirely implausible, but nevertheless represents the current state-of-the-art for representing technological change in both "bottom-up" and "top-down" models of energy/environmental interactions.

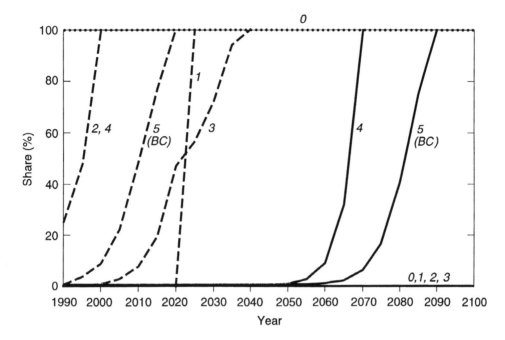

Figure 9.3: Shares of three technologies (in %) of new capacity additions in five different simulations. For all simulations dotted lines represent the share of new capacity made up of the existing technology, dashed lines represent the share made up of incremental technology, and solid lines represent the share made up of revolutionary technology. For clarity only growing shares are shown. Corresponding declining shares are omitted. Run 0 represents static technologies. Run 1 represents exogenous improvements in incremental technology only. Run 2 represents learning in the case of incremental technology only. Run 3 represents uncertain learning for incremental technology only. Run 4 represents learning for both incremental *and* revolutionary technologies. Run 5, which is labeled the base case (BC), represents uncertain learning of both incremental *and* revolutionary technologies. (See text for further discussion.)

depletion or additional exogenous constraints (e.g., environmental limits) that increase the costs of conventional technologies.

More interesting are the results from Runs 2 and 3. There we allow technological learning for the incremental technology (but not for the revolutionary one). In Run 2, the learning rate is certain and amounts to cost reductions of 10% for each doubling of installed capacity. The model's perfect foresight results in a very rapid market introduction. However, when the learning rate is uncertain (Run 3), the result is a more cautious investment strategy in which experimentation is delayed and more gradual. The incremental technology's share of new capacity additions starts at a low level and is only gradually stepped up.

Runs 4 and 5 add learning for the revolutionary technology. Run 4 assumes that the learning rates are known for both the incremental and revolutionary technologies. Run 5 assumes they are uncertain with probability distributions as described above around the mean learning rates of 10% and 30%, respectively. Owing to high initial costs and much greater uncertainty, the market entry of the revolutionary technology is delayed in Run 5 relative to Run 4. But it is important to emphasize that the optimal strategy calculated by the model includes investments in the revolutionary technology from very early on. These small initial investments, which are effectively invisible in *Figure 9.3*, are critical for continued technological learning. Only if they are made will learning occur and the revolutionary technology make it to the market.

Overall, the most significant result from the model simulations is the demonstration of an entirely endogenous mechanism driving technological change. That is, expected returns from R&D and learning that are uncertain but potentially large make gradual technological experimentation and investments the optimal strategy. The decision agent in our model acts entirely rationally (though with uncertainty) by investing up-front in R&D and niche market investments in the expectation of returns in the form of performance improvements and cost reductions (learning).

Figure 9.4 shows the three technologies' market shares of total installed capacity for Run 5. Market share is the usual metric for analyzing technology diffusion phenomena and was used extensively in our earlier chapters. Technologies enter small niche markets slowly; however, with declining costs due to learning, diffuse more widely and rapidly until markets are saturated and technological improvement possibilities (learning potentials) become exhausted. The result is the familiar S-shaped curve. Over time, the result is a pattern of technological evolution characterized by a "sequence of replacements" (Montroll, 1978) of older by newer technologies. This technological structural change is consistent with the diffusion patterns observed historically (cf. Nakićenović, 1997) and formulated by diffusion theory (cf. Rogers, 1983).

9.3.2. Diffusion with additional uncertainties

Obviously technological uncertainty is not the only type of uncertainty that needs to be considered. Other important uncertainties include: (i) the possibility of extreme outcomes; (ii) uncertain demands; and (iii) uncertainty

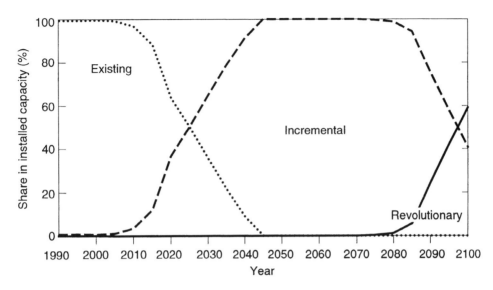

Figure 9.4: Market shares for three technologies as a percentage of total installed capacity under uncertain technological learning (cf. Run 5 in *Figure 9.3*). Note the S-shaped diffusion patterns.

about future environmental constraints. In this section we look at the implications of each one on the diffusion of new technologies.

Extreme Outcomes

In Runs 3 and 5 shown in *Figure 9.3* it was assumed that uncertainty was distributed symmetrically. The chances were the same that technological learning could be higher or lower than the mean value. However, the history of technological change provides many examples of an odd outlier technology suddenly emerging. Diffusion theory (Rogers, 1983) also emphasizes the importance of "outlier" expectations for the early diffusion of new technologies. We therefore look at alternative distribution functions that maintain the same mean and variance (degrees of uncertainty) as those used in the base case (Run 5), but that are highly skewed, such as the Weibull or Gamma distributions. *Figure 9.5* shows the result of adding the slight possibility of an extremely high or low learning rate. The optimal strategy shifts in the direction of earlier and larger up-front investments if positive extremes are possible, and in the direction of delayed or nonexistent diffusion if negative extremes are considered.

Model runs with long-tailed uncertainty distributions also reflect reality, especially for radically new technologies. Empirical distributions of future

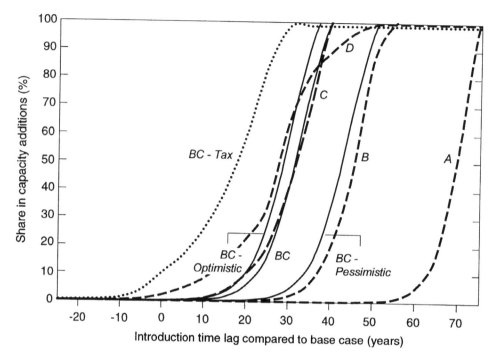

Figure 9.5: The revolutionary technology's share (in %) of new capacity additions for three different types of uncertainty. Solid lines show sensitivity runs for the base case (BC) with positively and negatively skewed distribution functions (although with identical mean and variance) for future learning rates (denoted as BC-Optimistic and BC-Pessimistic, respectively). Dashed lines show four variations in demand growth scenarios denoted as A, B, C, and D. A and B represent different degrees of low demand growth; C and D represent different degrees of high demand growth. The dotted line represents the case of an uncertain environmental constraint (emission tax), denoted as BC-Tax. The time axis shows the diffusion lag in years relative to the base case. Positive values indicate market penetration later than in the base case. Negative values indicate earlier market penetration.

technological "expectations" often show slightly higher frequencies toward the extreme tails, reflecting notorious technological optimism and pessimism. Taken together, the existence of such widely different expectations about future cost improvement (learning) potentials may enhance technological innovation, because innovation is usually carried out by the optimists. In this respect the model described here behaves similarly to simulation models developed within the framework of evolutionary economic theories (cf. the model of Silverberg *et al.*, 1988).

Uncertainty in Demand

Demand growth is the result of complex interacting demographic, economic, and lifestyle forces, and consequently is highly uncertain. It may be even more uncertain than technological parameters. To address these uncertainties, Grübler and Gritsevskii (1997) adopted the following procedure. Instead of sampling from a single uncertainty distribution, the uncertainty distribution was divided into four subsamples, or scenarios. These were not assigned relative probabilities, recognizing that the future may unfold into alternative, mutually exclusive directions for which there are no transitional probabilities. Thus each demand scenario represents a distinctly different set of expectations about future demand, and each was analyzed separately.

Two relatively low demand growth scenarios were analyzed, representing, for example, futures with low fertility and consequently low future population growth. Each had a distinct probability distribution describing uncertainty in demand. Conversely, the two high demand growth scenarios correspond, for example, to high fertility and high population growth. Although the means of the four probability distributions describing demands in the four scenarios were different, the shapes of the distributions about their respective means were assumed to be the same. For each scenario, an optimal strategy was calculated using essentially the same procedure as described above. Each of the four scenarios generated a distinct solution or technology trajectory. The results for the revolutionary technology are shown in *Figure 9.5*. In the face of demand uncertainty the optimal strategies all include investments in R&D and installed capacity (learning). The lower the demand growth, the more cautious the optimal strategy. Investments are lower, but they are still present from the beginning of the scenario even if they are initially impossible to discern in *Figure 9.5*.

9.4.   Environment

Finally, we turn to environmental concerns as possible drivers of technological change. To do this, we illustrate the endogenous technological change viewpoint of the two main recurrent features of environmental issues. The first is the persistence of ignorance, or discovery by "accident", and the resulting high uncertainty surrounding future environmental constraints. The second feature is discontinuity in environmental impacts – the possibility that future environmental problems might be very different from those we anticipate today. (Recall the example of urban horse manure in Chapter 7. In 1900 that was the main environmental concern associated with urban transport. That concern has since evaporated to be replaced by environmental problems from internal combustion engines that nobody anticipated in 1900.)

9.4.1. Uncertainty in environmental limits

Future environmental constraints might emerge, for instance, in the form of emission taxes. The existence, magnitude, and timing of such emission taxes can be treated with the same approach as technological uncertainty. First we assume a cumulative probability distribution of the occurrence of the emission tax over the entire time horizon. Starting near zero in 1990, the cumulative probability distribution rises with time. The illustrative conservative distribution function used by Grübler and Gritsevskii (1997) reflects only a one-third chance that a carbon tax would ever be implemented. The probability of the tax having been introduced rises to one-sixth by 2050 and, by 2100, to nearly one-third. For the magnitude of the tax a distribution with a very small probability of a high tax level was assumed. Formally this was done by constructing a Weibull distribution around the mean value of the tax, which was set arbitrarily at US$50 per ton of carbon, with a 99% probability that it would not exceed US$125 per ton C. Again this represents a conservative assumption.

The results are shown in *Figure 9.5*. The existence of a possible environmental constraint alters the pattern of technological change substantially.[18] Again, investments in R&D and installed capacity that enable subsequent technological learning are shifted earlier in time. This is done in order to prepare for the possibility of a future environmental constraint, even if the probability and absolute magnitude of the constraint do not appear particularly daunting. In this respect, the possibility of an environmental constraint yields patterns similar to those generated by the possibility that future demand will be much higher than expected, or that technological learning rates might be higher than anticipated (i.e., be different from the mean expected value). For the energy sector these are perhaps the three most important unknowns. In all cases earlier R&D and niche market investments are the optimal response strategy. In the face of uncertainty short-term investments into R&D and first applications of new technologies need to be higher in order to stimulate learning, even if these new technologies only penetrate on a massive scale many decades in the future. In essence, the results from an endogenous model of technological change indicate that an optimal strategy *vis*

---

[18] For illustrative purposes conversion efficiencies and carbon emissions per unit output (electricity) representative of conventional and advanced coal systems for the existing and incremental technologies, and of solar PV cells for the revolutionary technology, were assumed in the model.

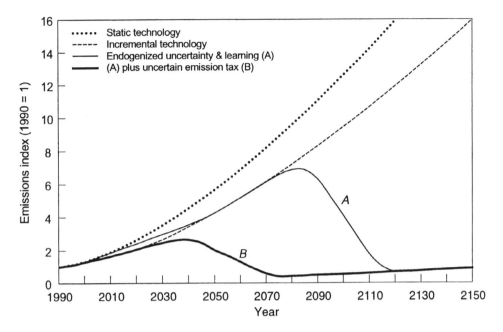

Figure 9.6: Carbon emissions (index, 1990 = 1) of a one-region, three technology model, including an initially expensive but promising zero-carbon option, under alternative assumptions about technological change.

à *vis* future contingencies involves early preparation for potential surprises that may emerge later, as opposed to the "wait and do nothing" strategies that emerge out of conventional approaches (e.g., Wigley *et al.*, 1996).

9.4.2. Environmental discontinuities

We conclude this overview of model runs by noting a final potential "surprise". Models with endogenized technological learning can generate results with pronounced discontinuities in future emission levels. *Figure 9.6* shows an example in which emissions drop substantially, not through an exogenous "shock" such as taxation or emission limits, but through the endogenous dynamics of technological change. Such model behavior is in stark contrast to typical "business as usual" emission trajectories that incorporate either static or incremental technological change. Our model results strongly suggest that future discontinuities might arise not only from having to contend with uncertain limits (e.g., of resource availability), but also from endogenous technological change.[19]

---

[19]Discontinuities from endogenous technological change could also lead to sudden increases in energy emissions (e.g., in energy-intensive hypersonic or space travel).

9.5.   Next Steps

The model of endogenous technological change that we have just outlined
demonstrates the feasibility of such an approach based on the simultaneous
treatment of uncertainty and increasing returns from technological learning.
However, much remains to be done at both the conceptual and modeling
levels. The highly stylized structure of the model must be expanded to
resemble, at least rudimentarily, the complexity of existing technological
systems. Of particular importance is the critical issue of technological *interdependence*. This includes spillover effects and learning externalities, such as
advances in general scientific knowledge or the possibility of "free riding" on
someone else's learning efforts. These are not addressed in the simple, one-
actor model summarized above. Through its stochastic sampling technique,
however, the model does approximate the dynamics of multiactor models
with heterogeneous populations. Currently a multiregion, multiactor model
is being developed at IIASA to study technological trajectories and emissions in the presence of technological interdependence and both positive and
negative spillover effects.

9.6.   Summary

Both conceptual and mathematical approaches are available for modeling
technological change as an endogenous process. The model formulations
and illustrative results presented here capture the most essential features
of technological evolution observed in our historical analysis. These features include innovation amidst substantial uncertainty, enormous advances
in productivity, long lead times in diffusion, and sometimes unanticipated environmental consequences. Nonetheless, the model is only a first step, and
no model can ever resolve deterministically the fundamental uncertainties
inherent in technological creativity and change. There will always be uncertainty and room for speculation about future technological configurations,
their social acceptability, and their environmental implications.

The illustrative model formulations and applications summarized above
suggest promising modeling avenues deserving further exploration. Despite
their limitations and oversimplifications, they justify a number of robust
conclusions. First, operable analytical solutions exist for simultaneously
addressing stochasticity (the uncertainty inherent in technological innovation) and nonconvexity (increasing returns through technological learning).
Second, the S-shaped patterns of technological diffusion observed historically can be replicated in a model with endogenous technological change.
Third, the model reflects path dependency and a wide range of technological

outcomes that can result from small differences in initial conditions when they are combined with uncertain rates of future technological change that accumulate over time. Fourth, diffusion of new technologies can yield pronounced discontinuities in environmental productivity and impacts that should not be ignored either in the formulation of long-term scenarios, or in policy debates. Fifth, and perhaps most importantly, the model illustrates an entirely endogenous mechanism of technological change by which technologies that may appear initially unattractive can diffuse into the market. This does not happen automatically. It happens *only* if required up-front investments in R&D and experimentation (niche-market applications) are made. Indeed, such strategies are optimal in the face of uncertainty about future demand and future environmental limits.

The model simulations also hold important implications for models where technological change is treated as an exogenous variable. For those it is important to perform systematic sensitivity analyses covering a wide range of technology assumptions. Modelers must not demonstrate a "lack of nerve," to recall Martino's phrase. We should never dismiss *a priori* the market potential for new technologies based on their current state. Through learning strategies costs and performance can change drastically – by more than an order of magnitude as has been amply demonstrated.

Our knowledge of future demands – for energy, raw materials, food, or environmental amenities – is extremely uncertain. We are also relatively ignorant of the most basic drivers, such as the world's future population. No model can resolve these fundamental uncertainties, but treating technological change as an endogenous feature of the development of society, the economy, and of energy systems yields insights about how to best prepare for alternative contingencies. The bottom line is *invest*. Invest in R&D, invest in demonstration (niche markets), and invest in gradually expanding commercial markets to foster pervasive diffusion. Such investments all require acting sooner rather than later. Without such investment the technological change needed to face future contingencies will not occur.

Dynamics, both positive and negative, must not be underestimated. Change can be discontinuous and quite rapid due to innovation, but innovation, in order to be ultimately successful, requires much time and resources. For pervasive changes and transitions in large technological systems, time constants of change are well over 50 years. Historical experience suggests that these will be difficult to accelerate. There is much inertia in large sociotechnical systems due to their size, capital intensity, technological interdependence, and slow rates of institutional change. However, long time constants should not be viewed only as undesirable inhibitors of change. They indicate the time required to develop, test, and socially qualify new technological configurations, or to reject them if initial promises do not materialize. They represent solid evidence to help us resist speculating about shortcuts and stay focused on a more cautious, but historically justified, strategy of continued experimentation and learning about the many unanticipated characteristics of new technologies. Ultimately, both in models and reality, there is only one promising strategy: to keep learning, from successes as well as failures.

Chapter 10

Appendix

Synopsis

The Appendix briefly presents data sources and descriptions for representative data sets presented in the preceding chapters that may be useful in coursework and modeling of technological change. After presenting data sources, a description, and formats, instructions are given on how to obtain the data sets in electronic form through internet access.

Data Sets: Sources, Description, and Electronic Access

Overview

This Appendix contains a brief description of the sources, definitions, and comments on a number of data sets presented throughout the book. Their order of presentation follows their chronological order as they appear in the text. For each data set the following information is given: title, figure numbers (as appearing in the preceding chapters), file name, time period covered, unit, and description of data items and sources. For those interested in obtaining more historical data sets, we draw attention to the recently released CD-ROM edition (Carter et al., 1997) of the US Historical Statistics (US DOC, 1975).[1]

Data format

All data sets are stored in two file types: spreadsheet (denoted with the respective file extension .wk1) and in plain, comma-delimited, ascii, UNIX-readable format (denoted with the file extension .csv). Thus, altogether the data sets are stored in 10 files with two formats each, yielding 20 files in the

---

[1] Carter et al., 1997. *Historical Statistics of the United States, Bicentennial Edition on CD-ROM.* ISBN 0521-58541-4, Cambridge University Press, price: US$195. (Unfortunately, neither the publisher nor the editors have ventured to update the data series of the 1975 edition of the *Historical Statistics*. Data series therefore end with the year 1970.)

directory set up for internet access/downloading. The spreadsheet format chosen is Lotus-123, assuring maximum compatibility with higher version releases of this or similar spreadsheet programs (e.g., Excel). The comma-delimited ascii format allows utilization of the data set for any alternative software (and quick preview via any internet browser).

Data are generally formatted in columns, and numbered consecutively, e.g., [1], [2], [3], ... As a rule column [1] refers to year of data. A brief title and legend for each column is given at the top of the file. The column sequence repeats the summary of data series as described below. Data are stored in the spreadsheet format in whatever numerical precision was available from the different data sources (and can therefore change within any particular data series). The numerical precision of the data in the ascii format files is limited by default to two digits after the decimal point for noninteger values. As a rule blank entries refer to unavailable data, and zero entries refer to true zero values, so these can be determined with certainty from available statistics.

Sources and description of data sets

Title: Length of transport infrastructures in the USA
Figure: 2.10
File name: usa-infra
Time period: 1825–1985
Unit: 1,000 miles
Data description:
The length of individual transport infrastructures over the period 1825–1985 is given. *Figure 2.10* shows the growth of these infrastructures between their introduction date and achievement of maximum length (100% saturation level in *Figure 2.10*). The data set extends that shown in the figure, also showing the subsequent stagnation and decline of infrastructure length. Infrastructures are disaggregated as follows.

- Canals: length of canals (excluding navigable natural river waterways) from Grübler (1990a) based on Isard (1942). Comparable data after 1900 are unavailable, but canal length declined further from the 2,000 miles still in existence in 1900.
- Railways: length of railway network from Grübler (1990a), based on US DOC (1975, and consecutive years).
- Surfaced roads: length of surfaced (paved) roads from Grübler (1990a), based on US DOC (1975). Pre-1904 data are model estimates (cf. Grübler, 1990a).

- Oil pipelines: length of crude oil and oil product pipelines (all categories including trunk and gathering lines). Source: API (1971) and US DOC (various volumes). Only pipelines regulated by the Interstate Commerce Commission (and subsequently by the Federal Energy Regulatory Commission) are given, representing 80% of total oil pipeline length in the USA.
- Gas pipelines: length of gas utility mains (including all types of field gathering, transmission and distribution gas pipelines). Source: 1809–1905: US Minerals Yearbook (various volumes); 1933–1968: API (1971); 1968–1985: AGA (1986). Data for the period 1906–1932 are unavailable.
- Telegraphs: miles of wire. Source: US DOC (1975). 1866–1915: Western Union Telegraph only; 1918–1970: all telegraph lines. Comparable data post-1970 are unavailable.

Title: Diffusion of emission controls in US passenger car fleet
Figure: 2.12
File name: usa-cars
Time period: 1965–1984
Unit: million cars in operation
Data description:
The data set updates an earlier analysis of Nakićenović (1986) based on US DOC (1983) and MVMA (1985). With the diffusion of catalytic converter cars almost complete by the mid-1980s, no comparable statistics are available post-1984. *Figure 2.12* aggregates the various first emission controls prior to catalytic converters into a "1st controls" category. A more detailed disaggregation is contained in the data set. *Figure 2.12* shows the relative shares (in total number of cars operated) of three categories of vehicles in logit transform. Vehicle types include:

- None: cars without any emission controls.
- Crankcase: control of unburned fuel emissions (mandated since 1963).
- Exhaust: additional control of hydrocarbon and carbon monoxide emissions (mandated since 1968).
- Evaporation: additional control of fuel evaporation from tanks (since 1971).
- Low-NO_x: additional controls lowering NO_x emissions (but no catalysts), mandated since 1973.
- 1st controls: all emission controls except catalytic converters, i.e., sum of columns [2] to [6].
- Catalyst: cars equipped with catalytic converters (mandated since 1975).

Title: Diffusion of agricultural tractors worldwide and by region
Figure: 5.9
File name: w-tractor
Time period: 1910–1990
Unit: Million tractors
Data description:
Tractors used in agriculture by five major world regions and the global total
are given. Data sources include Woytinsky and Woytinsky (1953), and *FAO
Yearbook: Production* (various volumes). The regional disaggregation data
items (shown as cumulative totals in *Figure 5.9*) include:

- Europe: Western, Central, and Eastern Europe (FAO definition).
- Ex-USSR.
- North America: Canada and USA.
- JANZ: Japan, Australia and New Zealand.
- ROW: rest of world (all other regions not listed separately above, number
  calculated as residual to world total).
- World: World totals as given in original data sources.

Title: World rubber production
Figure: 5.11
File name: w-rubber
Time period: 1900–1990
Unit: 1,000 (metric) tons
Data description:
World production data of rubber (natural, synthetic, and recycled) are given.
As additional information the data set (but not *Figure 5.11*) contains data
on natural rubber production over the 1900–1950 period disaggregated into
rubber collected from forests, and rubber collected in plantations. (This
disaggregation was no longer available for later years, where only total nat-
ural rubber production is reported.) Data were derived from Woytinsky
and Woytinsky (1953), UN *Statistical Yearbook* (various volumes), and *FAO
Yearbook: Production* (various volumes). The data set includes the following
entries:

- Natural rubber production, harvested from forests (wild growing trees),
  1900–1950.
- Natural rubber production, harvested in plantations, 1900–1950.
- Natural rubber production (sum of the above two subcategories), entire
  time period.
- Synthetic (manufactured) rubber, entire time period.
- Recycled rubber, entire time period.

Title: Percent urban population
Figure: 5.19
File name: w-urban
Time period: 1800–1900
Unit: percent
Data description:
The data set gives the percentage of the total population living in urban settlements. These are defined as settlements of more than 2,500 inhabitants (10,000 in the case of Japan) based on Flora (1975) from which data until 1960 have been retained. Since 1960 data are based on the United Nations 1996 revision of percentage urban populations that adopt a similar definition to Flora (1975), without, however, being strictly consistent (in addition, definitions change over time). Data for seven countries are given. As additional information the 1996 UN estimates for developed, developing, and world totals are given for the period 1950–1990. The countries covered include:

- USA.
- England (including Wales), data since 1960 are from the Central Statistical Office (1996, and earlier volumes).
- Japan (note the different definition of urban population given above).
- Germany (FRG for the 1950–1990 period).
- France.
- Ex-USSR (Territory of Tsarist Russia and former USSR).
- Brazil.
- Developed countries (UN definition of "more developed regions", corresponding to the OECD countries in their 1990 definition plus Central and Eastern Europe and the former USSR).
- Developing countries (UN definition of "less developed regions", i.e., all other countries, not classified as "more developed" above).
- World.

Title: World raw steel production by process technology
Figure: 6.4
File name: w-steel
Time period: 1870–1995
Unit: Million tons raw steel
Data description:
The data set updates (based on IISI, various volumes) a set given in Grübler (1987, 1990b), drawing on Roesch (1979), and the statistics of the International Iron and Steel Institute (IISI, various volumes). Prior to 1950 only decadal estimates are available. Since 1950 annual series are given. The total

raw steel tonnage figures reported refer to the sum of output by production technology. IISI (1997) estimates that since 1970 these cover 95% of total global crude steel production (the difference is crude steel production for which no information concerning the production process is available). Figures for open-hearth production since 1990 include some smaller amounts (between 10 and 17 million tons) of Chinese steel production accounted as "other" processes in the IISI statistics. In the absence of further information these have been allocated to open-hearth technology. Production figures for Bessemer technology in 1990 are the author's estimate, thereafter no separate breakdown is available, but global production figures are likely to be negligible. The breakdown by process technology includes the following categories:

- Puddel (production in puddling furnaces and crucible steel).
- Bessemer (production by the Bessemer process).
- Open-hearth (production by Siemens-Martin process).
- Electric arc.
- Basic oxygen (LD, or Linz-Donawitz process).
- Total (total raw steel production; sum of the above categories).

Title: World primary energy use by source and global population
Figures: 6.18, 6.19, 6.29
File name: w-energy
Time period: 1850–1995
Unit: Million tons oil equivalent (energy); million (population)
Data description:
The data set updates an earlier data set developed by Marchetti and Nakićenović (1979) and Nakićenović (1979), drawing on statistics assembled by Schilling and Hildebrandt (1977), based on data of the League of Nations (later to become the United Nations), cf. Darmstadter *et al.* (1971). Unless otherwise specified all data prior to 1965 are based on Marchetti and Nakićenović (1979). Data from 1965 onwards are from *BP Statistical Review of World Energy* (BP, various volumes). This updated data set is identical to the one reported in IPCC (1996a). As a memo item, data on global population since 1850 are also included. To calculate carbon emissions from fossil fuels the following emissions factors are suggested (tons elemental carbon per ton oil equivalent, i.e., tC/toe): wood: 1.25; coal: 1.08; oil: 0.84; natural gas: 0.64. In *Figure 6.29*, the concept of "gross" and "net" carbon intensities are used, where carbon intensity refers to the specific carbon emissions per unit primary energy used (in tC/toe). "Gross" intensities use all energy sources and carbon fluxes in the nominator and denominator, respectively. "Net" intensities exclude (sustainable) fuelwood use and nonenergy feedstock uses

(where carbon is sequestered for extended time periods, e.g., in plastics) from the calculations (for a detailed discussion see Grübler and Nakićenović, 1996). Following a usual accounting convention in the energy industry, non-fossil energy sources (hydropower and nuclear) are accounted for by their "substitution" equivalents, i.e., by the amount of fossil energy that would be required to produce the same electricity output as the nonfossil sources. Following IIASA–WEC (1995), a conversion rate of 38.6% in a modern fossil fuel fired electricity plant was assumed, i.e., one unit hydroelectricity is equivalent to 2.6 units of fossil primary energy. [Note that this accounting convention is different from the one used in the original BP statistics that adopt (inconsistently) different accounting conventions for hydro- and nuclear power.] For the period prior to 1965 the conversion rates as used by the original Schilling and Hildebrandt (1977) reference were retained. Data items include the following:

- Wood: global fuelwood use based on data of Putnam (1954), UN (1952), and FAO (1965 *et passim*). The UN conversion rate of 0.23 toe per m^3 fuelwood was used to convert the FAO statistics into energy equivalents. Note that only part of this fuelwood use refers to commercially traded quantities. Uncertainty margins are correspondingly high. The numbers also do not include other noncommercial renewable energy forms such as agricultural residues, dung, and other traditional renewable energy forms (e.g., animal power). These were estimated at less than 700 Mtoe in 1990, i.e., 8% of global primary energy use (IIASA–WEC, 1995). In all likelihood this value represents a historical high in absolute amounts (however, not in relative/percentage terms).
- Coal: global use of coal (hard and brown coals). Data based on Marchetti and Nakićenović (1979) and BP (various volumes).
- Oil: global use of crude oil (energy use only). Data based on Marchetti and Nakićenović (1979) and BP (various volumes). Nonenergy feedstock uses (cf. next data column) were subtracted from the original data sets of total crude oil use.
- Oil feedstock (nonenergy) use: Data calculated based on UN Energy Statistics (various volumes). Data prior to 1950 are zero-order estimates by the author based on US refinery output structure (from Schurr and Netschert, 1960).
- Natural gas: global use of natural gas (excluding gas flaring). Data based on Marchetti and Nakićenović (1979) and BP (various volumes).
- Hydropower: global primary energy equivalent of electricity generated by hydropower. Data based on Marchetti and Nakićenović (1979) and BP (various volumes). The original BP data were converted to a primary

energy equivalent assuming a 38.6% thermal efficiency of fossil electricity generation. Prior to 1965, the original estimates of Schilling and Hildebrandt (1977) used in the Marchetti and Nakićenović data set were adopted.

- Nuclear: global primary energy equivalent of electricity generated by nuclear reactors. Data based on BP (various volumes) adopting an equivalent accounting convention as for hydropower (38.6% in the original BP data series). Prior to 1965, nuclear electricity generation data were taken from UN Energy Statistics (1973).

- Population (memo item): global population, mid-year estimates. 1850–1949 data were taken from Grübler and Nakićenović (1994), based on data of Durand (1967) and Demeny (1990). 1950–1995 data are from UN (1996).

Title: US population, GNP, and primary energy use by source
Figures: 6.31, 6.32, 6.33
File name: usa-energy
Time period: 1800–1995
Unit: Million tons oil equivalent (energy), million (population), billion US$ in constant 1990 money and prices (GNP)
Data description:
The data set updates an earlier set developed by Nakićenović (1984). Unless otherwise specified, the Nakićenović data set was retained until 1950, and updated by US government and UN statistics (US DOC, various volumes; IMF, 1996; EIA, 1997; FAO, various volumes) for the period thereafter. Despite its unique comprehensiveness, the data set nevertheless excludes important renewable energy sources of the 19th century due to the absence of reliable estimates and statistics such as human energy, animal energy (feed), wind power (mills and water pumps), and illuminants (whale oil and candles). These were important energy sources (perhaps up to 30% of the totals as estimated here) in the pre-1850 period. Thereafter, their omission is likely to result in an underestimation of US primary energy use of less than 15–20%, or less than 5–10% after ca. 1920. By 1950 these traditional energy forms virtually disappeared in the USA, resulting in no underestimation of primary energy use since. Nonfossil energies are accounted with their primary energy equivalents, assuming fossil energies would have provided comparable output (substitution equivalents). For instance, direct water power (mechanical energy from water wheels) is accounted for by the coal needed to produce similar mechanical energy in steam engines. Hydroelectricity is accounted for by the fossil fuel equivalent of generating the same output of electricity from fossil fuels based on the fuel input and conversion efficiencies prevailing

in any particular year. The same substitution equivalent method is also used in contemporary US energy statistics (EIA, 1997). Unless otherwise noted, primary energy use refers to apparent consumption, i.e., production plus net balance from trade and stock changes (cf. EIA, 1997). For the calculation of carbon intensities of different fuels, the following emission factors (IPCC, Working Group II, 1996a) have been retained, i.e., wood: 1.25 tC/toe; coal: 1.08, oil: 0.84, and natural gas: 0.64 tC/toe (tons elemental carbon per ton oil equivalent). For all other energy sources carbon intensities were assumed to be zero. The data set includes the following items:

- Population: Mid-year resident US population (in million) updated US DOC (various volumes) from US DOC (1975).
- Gross National Product (GNP) in billion 1990 US\$. The Nakićenović (1984) data set, based on the estimates of Berry (1978) and US DOC (1975), was retained until 1965. The original 1958 US\$ of the Nakićenović set were converted to 1990 US\$ based on the average GDP deflator of 4.4 (IMF, 1996). After 1965, GNP data are from IMF (1996), expressed in constant 1990 US\$ money and prices.
- Primary energy use, total (sum of the following itemized categories).
- Fuelwood: 1800–1970 from Nakićenović (1984), based on Reynolds and Piersons (1942); Putnam (1954); and US DOC (1975). 1990–1995 data are from EIA (1997). For the period 1970–1990, official EIA statistics under-report fuelwood use. The numbers given are the author's own estimates assuming that fuelwood use has evolved in proportion to 1990 EIA values based on the volumetric fuelwood production statistics as reported in FAO (various volumes).
- Direct waterpower (primary energy equivalent): based on Putnam (1954) and Nakićenović (1984).
- Coal (all categories, i.e., bituminous, sub-bituminous, and anthracite). 1800–1950 data from Nakićenović (1984), based on Putnam (1954), Schurr and Netschert (1960), and US DOC (1975); 1950–1995 data are from EIA (1997).
- Crude oil (and net trade of oil products): 1860–1950 from Nakićenović (1984), based on Putnam (1954) and US DOC (1975); 1950–1995 data are from EIA (1997).
- Natural gas (apparent consumption, excluding gas production flared and repressed into reservoirs): 1850–1950 from Nakićenović (1984), 1950–1995 from EIA (1997). (Data on gas flaring and repressuring are given in Grübler and Nakićenović, 1987.)
- Hyroelectricity (substitution equivalent): 1885–1950 from Nakićenović (1984), 1950–1995 from EIA (1997). The data for hydroelectricity also

include other nonfossil electricity generation such as geothermal energy, and more recently wind and solar energy (for further disaggregation cf. the section on Renewable Energy in EIA, 1997).

- Nuclear energy (substitution equivalent): data as given in EIA (1997).

Title: France: Population, GDP, and mobility by mode
Figure: 7.9
File name: f-transp
Time period: 1800–1994
Unit: Million (population), billion French Francs in constant 1905–1913 prices and money (GDP), million passenger-km traveled per year (mobility)
Data description:
The data set draws on the formidable work of the French quantitative economic history school (cf. Marczewski, 1965), most notably the work of Jean-Claude Toutain (1967, 1987), profiting *inter alia* from the fact that traffic survey records have existed in France since the pre-Revolution period. The data set has been updated (Ann. Stat. Transp., various volumes, 1985–1995) from the one described in more detail in Grübler (1990a), including some revisions (a heroic attempt to provide zero-order estimates for the period during World War I and World War II), as well as amendments (a new estimate of nonmotorized mobility, i.e., walking, cf. the discussion below). The data set includes the following items:

- Population: Data based on Toutain (1987), for the period 1800–1980. Later data are from Ann. Stat. (1996). Population data for the two World War periods (and their aftermath), 1914–1919 and 1939–1948, are zero-order estimates by the author.
- GDP (in constant French Francs of 1905–1913). Source: Toutain (1987) for the period 1800–1980 (with the exceptions noted below). Later data are from Ann. Stat. (1996), calculated using the annual GDP growth rates in constant prices given therein. Data for the period 1914–1919 were estimated given the trend in the GNP estimates of Fontvieille (1976), for the World War I period. Data for the period 1939 were derived in a similar way using Fontvieille (1976). 1940–1948 data are (speculative) zero-order estimates of the author. The GNP data series retain the original unit of Toutain (1987): constant French Francs of 1905–1913. To convert to 1990 money (and to appreciate the extent of inflation since the eve of World War I) a multiplier of 19 will yield approximately a correct order of magnitude.
- Waterways: Data for the period 1800–1913 are from Toutain (1967), referring to all domestic waterborne passenger-km (canals and navigable waterways). Seaborne passenger-km data are unavailable. After 1913

no data are available, but the resulting error in total mobility levels is unlikely to exceed 0.2%, i.e., is insignificant.

- Horses: Passenger traffic data by horses and horse carriages (private and public) are from Toutain (1967), for the period 1800–1940. Subsequent data to 1950 are zero-order estimates by the author based on nonfarm horse population data (Ann. Stat., 1961). Considering the insignificant amounts of mobility by horses thereafter, no attempt was made to extend the data series beyond 1950.

- Railways: Passenger-km by railways (all traffic) for the period 1830–1938 are from Toutain (1967). Later years were obtained from Mitchell (1980), which also included the interwar period estimates, and Ann. Stat. Transp. (1985–1995). Passenger-km provided by TGV trains are subtracted, and are reported as a separate data series.

- Two wheelers (bicycles and motorcycles): 1880–1960: Estimates by the author, based on the number of bicycles and motorcycles registered (data are available since 1880 for bicycles and since 1899 for motorcycles, cf. Ann. Stat., 1961), and average travel distances reported in the traffic surveys given in Sax, 1920 [from where an average of 1,400 (road and passenger) km per year for bicycles and a distance of 3,000 km per year, and a load factor of 1.5, i.e., some 5,000 passenger-km per motorcycle were retained]; the uncertainty margin of this estimate is high: ±50%. Data since 1960 are based on IRF (various volumes, 1970–1995) statistics for motorcycles. No estimate for bicycles was made after 1960, considering that they accounted for less than 3% of total passenger-km in 1960, and a rapidly declining share thereafter.

- Buses: Estimates of Toutain (1967), for the period up to 1965 (missing data for 1941–1944 are the author's estimates). After 1965 the statistics published by IRF (various volumes, 1970–1995) have been used.

- Cars: Estimates of Toutain (1967), for the period up to 1965 (missing data for 1941–1944 are the author's estimates). After 1965 the statistics published by IRF (various volumes, 1970–1995) have been used.

- Air: Domestic and international passenger-km traveled by air by passengers emplaned in France. Data from Ann. Stat. (1961) and Ann. Stat. (various volumes, 1975–1995).

- TGV (superfast trains): Data from Ann. Stat. (various volumes, 1985–1995).

- Walking: Zero-order estimates by the author based on the following simple algorithm: First it is assumed that each citizen spends on average about one hour daily for mobility. Time-budget surveys in the 1960s indicate an average of 58 minutes per day (Grübler, 1990a, based on data from Szalai, 1972) and that figure has remained slightly below

one hour according to the latest French travel surveys (Orfeuil, 1993).
Then for each transport mode an average representative transport speed
was assumed: e.g., 15 km/hr for horse carriages, 30 km/hr for conventional trains, 50 km/hr for cars, 250 km/hr for high speed trains, and
500 km/hr for domestic aircraft trips. Dividing the total daily per capita
passenger-km per mode by the respective average transport speed yields
the travel time spent in each mode. The remaining travel-time budget
(to one hour per day) is then allocated to walking, assuming a mean
speed of 4 km/hr, yielding the passenger-km walked as the final estimate. For simplicity (and in the absence of statistical data) the average
transport speeds by travel mode were kept constant over the entire time
period in the calculations. This might seem surprising considering increasing congestion from denser motorized traffic. However, reduced
speed in dense agglomerations has been compensated by improved infrastructures (more highways) for long distance travel and increasing
suburbanization (André *et al.*, 1993), thus average car speeds in France
have not changed much (Orfeuil, 1996). The resulting estimated modal
split of French passenger-km traveled is in good agreement with the
latest (1982) national transport survey available and the observed significant decline in walking trips that are increasingly substituted by car
travel (Orfeuil, 1993). Nonetheless, the error margin of our estimate
remains substantial: ±25%.

Title: Japan R&D, investments, and costs of photovoltaic (PV) cells
Figures: 9.1, 9.2
File name: j-pvs
Time period: 1973–1995
Unit: 1985 Yen per Watt (costs), kW (installed capacity), billion 1985 Yen
(R&D and investment expenditures)
Data description:
The data set draws on an exceptionally comprehensive empirical analysis of
the history of PV technology development in Japan performed by Watanabe
(1995, 1997). Original as well as derived data underlying *Figures 9.1* and
9.2 are presented, including:

- PV costs: Costs (1985 Yen) of PV cells per Watt (peak) installed.
- Total R&D Expenditures: Annual R&D expenditures, both public
  (through MITI's sunshine technology program) and private (through industry) are included, which makes the data set one of the few available
  accounting for R&D efforts comprehensively (in billion 1985 Yen).
- Cumulative R&D expenditures: Calculated from the above annual expenditures (unit: billion 1985 Yen). This measure is used as a proxy for

the knowledge stock related to PV technology. Watanabe (1995) estimates through econometric analysis that R&D translates into technology improvements (cost reductions) with a time lag of 2.8 years.

- Installed PV capacity (in kW, i.e., 1,000 Watts).
- Investments in PV capacity expansion (annual capacity additions multiplied by the average price of PV cells, as given above). This measure (unit: billion 1985 Yen) is used as a proxy for the cumulative experience gained with PV technology when compared to R&D efforts.
- Total cumulative expenditures in PV technology (in billion 1985 Yen). Retaining Watanabe's (1995) time lag of three years, R&D and investment expenditures (proxy measures for the inputs to the technological learning process) are aggregated so they can be compared to improvements (cost reductions) in PV technology (proxy measure for the output of technological learning).

How to download the data

The easiest way to download the data from the IIASA computer is the following:

1. Access the following address: http://www.iiasa.ac.at/~gruebler.
2. From the index listing (all related to this book) click the relevant file you wish to download (listed in this Appendix).
3. Recall that files with extensions .csv are in plain ascii format and files with extension .wk1 are spreadsheets (for those encountering problems with the spreadsheet download it is recommended to download the ascii formatted text files, and to then open them locally with available spreadsheet software).
4. Download the file selected to your local computer (in most browsers this is done by a right click on the mouse and then clicking the "Save As" – or equivalent – option).
5. For those having no internet access a floppy with the data set can be obtained for a small handling charge from:

Publications Department
International Institute for Applied Systems Analysis
A-2361 Laxenburg, Austria
Phone: +43 2236 807 ext. 342 or 433
Fax: +43 2236 73148 or +43 2236 71313

References

AAMA (American Automobile Manufacturers Association). 1996. *Motor Vehicle Facts and Figures.* AAMA, Detroit, MI, USA.

AGA (American Gas Association). 1986. *Gas Facts 1986: A Statistical Record of the Gas Utility Industry.* AGA, Arlington, VA, USA.

API (American Petroleum Institute). 1971. *Petroleum Facts & Figures.* API, Washington, DC, USA.

Abel, W. 1956. Agrarkonjunktur. In: v. Beckenrath, E., Bente, H., Brinkmann, C., Gutenberg, E., Haberler, G., Jecht, H., Jöhr, W.A., Lütge, E., Predöhl, A., Schaeder, R., Schmidt-Rimpler, W., Weber, W., and v. Wiese, L. (eds.), *Handwörterbuch der Sozialwissenschaften.* G. Fischer, Stuttgart, Germany, pp. 49–59.

Abel, W. 1980. *Agricultural Fluctuations in Europe* (English translation of 3rd edition). Methuen & Co. Ltd., London, UK.

Abernathy, W.J. 1978. *The Productivity Dilemma: Roadblock to Innovation in the Automobile Industry.* The Johns Hopkins University Press, Baltimore, MD, USA.

Abramovitz, M. 1993. The search for the sources of growth: Areas of ignorance, old and new. *Journal of Economic History,* **52**(2):217–243.

Adams, J. 1995. *Risk.* UCL (University College London) Press, London, UK.

Ahmad, S. 1966. On the theory of induced innovation. *Economic Journal,* **76**:344–357.

Akutsu, N. 1996. Personal communication. International Institute for Applied Systems Analysis, Laxenburg, Austria.

Albertson, L.A. 1977. Telecommunication as a travel substitute: Some psychological, organizational and social aspects. *Journal of Communication,* **27**(2):32–43.

Alcamo, J., Shaw, R., and Hordijk, L. 1990. *The RAINS Model of Acidification: Science and Strategies in Europe.* Kluwer Academic Publishers, Dordrecht, Netherlands.

Alcamo, J., Bouwman, A., Edmonds, J., Grübler, A., Morita, T., and Sugandhy, A. 1995. An evaluation of the IPCC IS92 emission scenarios. In: Houghton, J.T. *et al.* (eds.), *Climate Change 1994.* Intergovernmental Panel on Climate Change, Cambridge University Press, Cambridge, UK, pp. 247–304.

Allenby, B.R., and Richards, D.J. 1994. *The Greening of Industrial Ecosystems.* National Academy Press, Washington, DC, USA.

Amann, M., Cofala, J., Dörfner, P., Gyarfas, F., and Schöpp, W. 1995. *Impacts of Energy Systems on Regional Acidification.* Report to the World Energy Council Project 4 on Environment. International Institute for Applied Systems Analysis, Laxenburg, Austria.

Anderberg, S. 1991. Historical land use changes: Sweden. In: Brouwer, F.M., Thomas, A.J., and Chadwick, M.J. (eds.), *Land Use Changes in Europe.* Kluwer Academic Publishers, Dordrecht, Netherlands, pp. 403–426.

Anderberg, S. 1998. Industrial metabolism and the linkages between economics, ethics and the environment. *Ecological Economics* (in press).

Anderberg, S., and Stigliani, W.M. 1994. An integrated approach for identifying sources of pollution: The example of cadmium in the Rhine river basin. *Water Science & Technology,* **23**(3):61–67.

André, M., Hickman, A.J., and Hassel, D. 1993. Usages et conditions de fonctionnement des voitures en Europe. *Recherche-Transport-Sécurité,* **38/39**(June):77–84.

Annuaire Statistique (Ann. Stat.). 1961. *Annuaire Statistique de la France 1961, Volume Rétrospective.* Imprimerie Nationale, Paris, France.

Annuaire Statistique (Ann. Stat.). Various volumes (1975–1996). *Annuaire Statistique de la France.* INSEE, Paris, France.

Annuaire Statistique des Transports (Ann. Stat. Transp.) Various volumes (1985–1995). Ministère des Transports, Département des Statistiques des Transports, Paris, France.

Argawal J.C. 1991. Minerals, energy, and the environment. In: Tester, J.W., Wood, D.O., and Ferrari, N.A. (eds.), *Energy and the Environment in the 21st Century.* MIT Press, Cambridge, MA, USA, pp. 389–395.

Argote, L., and Epple, D. 1990. Learning curves in manufacturing. *Science,* **247**(23 February):920–924.

Armstrong, P. 1984. *Technical Change and Reductions in Life Hours of Work.* The Technical Change Center, London, UK.

Arrhenius, S. 1896. On the influence of carbonic acid in the air upon the temperature of the ground. *Philosophical Magazine and Journal of Science,* Fifth Series (April):237–276.

Arrow, K. 1962. The economic implications of learning by doing. *Review of Economic Studies,* **29**:155–173.

Arthur, W.B. 1983. *On Competing Technologies and Historical Small Events: The Dynamics of Choice Under Increasing Returns.* WP-83-90. International Institute for Applied Systems Analysis, Laxenburg, Austria.

Arthur, W.B. 1988. Competing technologies: An overview. In: Dosi, G., Freeman, C., Nelson, R., Silverberg, G., and Soete, L. (eds.), *Technical Change and Economic Theory.* Pinter, London, UK, pp. 590–607.

Arthur, W.B. 1989. Competing technologies, increasing returns, and lock-in by historical events. *The Economic Journal,* **99**:116–131.

Auer, P.L., Burwell, C.C., and Devine, W.D. 1983. *An Historical Perspective on Electricity and Energy Use.* IEA-83-3, Institute for Energy Analysis, Oak Ridge, TN, USA.

Ausubel, J.H. 1989. Regularities in technological development: An environmental view. In: Ausubel, J.H., and Sladovich, H.E. (eds.), *Technology and Environment.* National Academy Press, Washington, DC, USA, pp. 70–91.

Ausubel, J.H. 1990. Hydrogen and the green wave. *The Bridge,* **20**(1):17–22.

Ausubel, J.H. 1991. Does climate still matter? *Nature,* **350**:649–652.

Ausubel, J.H. 1996. The liberation of the environment. *Dædalus*, **125**(3):1–17.

Ausubel, J.H., and Grübler, A. 1995. Working less and living longer: Long-term trends in working time and time budgets. *Technological Forecasting & Social Change*, **50**(3):195–213.

Ausubel, J.H., and Marchetti, C. 1996. *Elektron*: Electrical systems in retrospect and prospect. *Dædalus*, **125**(3):139–169.

Ausubel, J.H., and Sladovich, H.E. (eds.). 1989. *Technology and Environment*. National Academy Press, Washington, DC, USA.

Ausubel, J.H., Grübler, A., and Nakićenović, N. 1988. Carbon dioxide emissions in a methane economy. *Climatic Change*, **12**:245–263.

Ayres, R.U. 1987. *Industry Technology Life Cycles: An Integrating Meta-Model?* RR-87-3. International Institute for Applied Systems Analysis, Laxenburg, Austria.

Ayres, R.U. 1988. Complexity, reliability, and design: Manufacturing implications. *Manufacturing Review*, **1**(1):26–35.

Ayres, R.U. 1989a. *Technological Transformations and Long Waves*. RR-89-1. International Institute for Applied Systems Analysis, Laxenburg, Austria.

Ayres, R.U. 1989b. *Energy Inefficiency in the US Economy: A New Case for Conservation*. RR-89-12. International Institute for Applied Systems Analysis, Laxenburg,

Ayres, R.U. 1989c. Industrial metabolism. In: Ausubel, J.H., and Sladovich, H.E. (eds.), *Technology and Environment*. National Academy Press, Washington, DC, USA, pp. 23–49.

Ayres, R.U., and Ayres, L.W. 1996. *Industrial Ecology: Towards Closing the Materials Cycle*. Edward Elgar, Cheltenham, UK.

Ayres, R.U., and Ezekoye, I. 1991. Competition and complementarity in diffusion: The case of octane. In: Nakićenović, N., and Grübler, A. (eds.), *Diffusion of Technologies and Social Behavior*. Springer-Verlag, Berlin, Germany, pp. 433–450.

Ayres, R.U., Schlesinger, W.H., and Socolow, R.H. 1994. Human impacts on the carbon and nitrogen cycles. In: Socolow, R. *et al.* (eds.), *Industrial Ecology and Global Change*. Cambridge University Press, Cambridge, UK, pp. 121–155.

BP (British Petroleum). Various volumes (1967–1997). *BP Statistical Review of World Energy*. BP, London, UK.

Baba, Y., Kikuchi, J., and Mori, S. 1995. Japan's R&D strategy reconsidered: Departure from the managable risks. *Technovation*, **15**(2):65–78.

Bairoch, P. 1982. International industrialization levels from 1750 to 1980. *Journal of European Economic History*, **11**:269–333.

Bandura, A. 1977. *Social Learning Theory*. Prentice Hall, Englewood Cliffs, NJ, USA.

Barnett H.J., and Morse, C. 1967. *Scarcity and Growth: The Economics of Natural Resource Availability*. The Johns Hopkins University Press, Baltimore, MD, USA.

Basalla, G. 1988. *The Evolution of Technology*. Cambridge University Press, Cambridge, UK.

Benzoni, L. 1992. Le rythme de l'innovation: l'anomalie de l'industrie des circuits intégrés. *Communications and Strategies*, **3**(5):13–45.

Berry, B.J.L. 1990. Urbanization. In: Turner II, B.L. *et al.* (eds.), *The Earth As Transformed by Human Action*. Cambridge University Press, Cambridge, UK, pp. 103–119.

Berry, T. 1978. *Revised Annual Estimates of American Gross National Product, Preliminary Annual Estimates of Four Major Components of Demand, 1789–1889*. Bostwick Paper No. 3, University of Richmond, The Bostwick Press, Richmond, VA, USA.

Binswanger, H.P., and Ruttan, V.W. (eds.). 1978. *Induced Innovation: Technology, Institutions and Development*. The Johns Hopkins University Press, Baltimore, MD, USA.

BioPharm. 1994. Genzyme Transgenics Corp. develops transgenic proprietary technology. *BioPharm*, **14**(December):14.

Bodanksy, D. 1991. Scientific uncertainty and the precautionary principle. *Environment*, **33**(7):4–5, 43.

Boden T.A., Kaiser, D.P., Sepanski, R.J., and Stoss, F.W. 1994. *Trends '93: A Compendium of Data on Global Change*. ORNL/CDIAC-65, Carbon Dioxide Information Analysis Center, Oak Ridge, TN, USA.

Bolin, B. 1995. L'eventualité d'un changement de climat. Keynote address to the 16th World Energy Congress, 9 October, Tokyo. CSC 1, World Energy Council, London, UK.

Bonny, S. 1993. Is agriculture using more and more energy? A French case study. *Agricultural Systems*, **43**:51–66.

Borderon, D. 1990. Aircraft and rail development: Europe 2010. Internal report. Group Planning, Shell International, London, UK.

Boserup, E. 1981. *Population and Technological Change: A Study of Long-term Trends*. University of Chicago Press, Chicago, IL, USA.

Boyer, R. 1988a. Technical change and the theory of *régulation*. In: Dosi, G., Freeman, C., Nelson, R., Silverberg, G., and Soete, L. (eds.), *Technical Change and Economic Theory*. Pinter, London, UK, pp. 67–94.

Boyer, R. 1988b. Formalizing growth regimes. In: Dosi, G., Freeman, C., Nelson, R., Silverberg, G., and Soete, L. (eds.), *Technical Change and Economic Theory*. Pinter, London, UK, pp. 608–630.

Brimblecombe, P. 1987. *The Big Smoke: A History of Air Pollution in London Since Medieval Times*. Methuen, London, UK.

Broadberry, S.N., and Crafts, N.F.R. 1990. European productivity in the twentieth century: Introduction. *Oxford Bulletin of Economics and Statistics*, **52**(4):331–341.

Brooks, H. 1986. The typology of surprises in technology, institutions, and development. In: Clark, W.C., and Munn, T. (eds.), *Sustainable Development of the Biosphere*. Cambridge University Press, Cambridge, UK, pp. 325–350.

Brooks, H. 1996. The problem of attention management in innovation for sustainability. *Technological Forecasting & Social Change*, **53**(1):21–26.

Brown, L., Kane, H., and Roodman, D.M. 1994. *Vital Signs 1994: The Trends that are Shaping our Future*. Earthscan, London, UK.

Buol, S.W. 1994. Soils. In: Meyer W.B., and Turner II, B.L. (eds.), *Changes in Land Use and Land Cover: A Global Perspective*. Cambridge University Press, Cambridge, UK, pp. 211–229.

Buringh, P., and Dudal, R. 1987. Agricultural land use in space and time. In: Wolman, M.G., and Fournier, F.G.A. (eds.), *Land Transformation in Agriculture*. SCOPE 32, John Wiley & Sons, Chichester, UK, pp. 9–43.

COSEPUP (Committee on Science, Engineering and Public Policy, US Academy of Engineering). 1992. *Policy Implications of Global Warming*. National Academy Press, Washington, DC, USA.

CSST (Committee on Separation Science and Technology, US National Research Council). 1987. *Separation and Purification: Critical Needs and Opportunities*. National Academy Press, Washington, DC, USA.

Cameron, R. 1989. *A Concise Economic History of the World: From Paleolithic Times to the Present*. Oxford University Press, Oxford, UK.

Cantley, M.F. and Sahal, D. 1980. *Who Learns What? A Conceptual Description of Capability and Learning in Technological Systems*. RR-80-42. International Institute for Applied Systems Analysis, Laxenburg, Austria.

Carson, R. 1962. *Silent Spring*. Riverside Press, Cambridge, MA, USA.

Carter, S., Gertner, S., Haines, M., Olmstead, A., Sutch, R., and Wright, G. 1997. *Historical Statistics of the United States, Bicentennial Edition on CD-ROM*. Cambridge University Press, Cambridge, UK.

Central Statistical Office (UK). Various volumes (1975–1996). *Annual Abstract of Statistics*. HMSO, London, UK.

Christiansson, L. 1995. *Diffusion and Learning Curves of Renewable Energy Technologies*. WP-95-126. International Institute for Applied Systems Analysis, Laxenburg, Austria.

Cipolla, C.M. 1981. *Before the Industrial Revolution: European Society and Economy, 1000–1700*. 2nd edition. Methuen & Co. Ltd., London, UK.

Clair, D.R. 1983. *The Perils of Hanging On*. European Petrochemical Association 17th Annual Meeting, Monte Carlo, Monaco.

Clark, C. 1940. *The Conditions of Economic Progress*. Revised 2nd edition, 1951. Macmillan Press, London, UK.

Clark, W.C. and Munn, T. (eds.). 1986. *Sustainable Development of the Biosphere*. Cambridge University Press, Cambridge, UK.

Coase, R.H. 1937. The nature of the firm. *Economica*, 4:386–405.

Cohen, J.E. 1995. *How Many People Can the Earth Support?* W.W. Norton & Company, Inc., New York, NY, USA.

Cook, E. (ed.). 1996. *Ozone Protection in the United States*. World Resources Institute, Washington, DC, USA.

Cowan, R., 1991. Tortoises and hares: Choice among technologies of unknown merit. *The Economic Journal* **101**:801–814.

Crutzen, P.J., and Graedel, T.E. 1986. The role of atmospheric chemistry in environment-development interactions. In: Clark, W.C., and Munn, T. (eds.), *Sustainable Development of the Biosphere*. Cambridge University Press, Cambridge, UK, pp. 213–250.

Crutzen, P.J., Aselmann, I., and Seiler, W. 1986. Methane production by domestic animals, wild ruminants, other herbivorous fauna, and humans. *Tellus*, **38B**:271–284.

Darby, H.C. 1956. The clearing of the woodland in Europe. In: Thomas, W.L. (ed.), *Man's Role in Changing the Face of the Earth* (2 volumes). University of Chicago Press, Chicago, IL, USA, Vol. I, pp. 183–216.

Darmstadter, J., Teitelbaum, P.D., and Polach, J.G. 1971. *Energy in the World Economy: A Statistical Review of Trends in Output, Trade, and Consumption Since 1925*. The Johns Hopkins University Press, Baltimore, MD, USA.

David, P.A. 1975. *Technical Choice, Innovation and Economic Growth: Essays on American and British Experience in the Nineteenth Century*. Cambridge University Press, Cambridge, UK.

David, P.A. 1985. CLIO and the economics of QWERTY. *American Economic Review*, **75**(2):332–337.

David, P.A. 1990. The dynamo and the computer: A historical perspective on the modern productivity paradox. *American Economic Review*, **80**(2):355–361.

David, P.A., and Wright, G., 1996. *The Origins of American Resource Abundance*. WP-96-15. International Institute for Applied Systems Analysis, Laxenburg, Austria.

Davis, S.C., and McFarlin, D.N. 1996. *Transportation Energy Data Book: Edition 16*. Oak Ridge National Laboratory ORNL-6898, Oak Ridge, TN, USA.

Debecker, A., and Modis, T. 1994. Determination of the uncertainties in S-curve logistic fits. *Technological Forecasting & Social Change*, **46**(2):153–173.

Demeny, P. 1990. Population. In: Turner II, B.L. *et al.* (eds.), *The Earth As Transformed by Human Action*. Cambridge University Press, Cambridge, UK, pp. 41–54.

Denison, E.F. 1962. *The Sources of Economic Growth in the United States and the Alternatives Before Us*. Supplementary Paper No. 13. Committee for Economic Development, New York, NY, USA.

Denison, E.F. 1985. *Trends in American Economic Growth, 1929–1982*. The Brookings Institution, Washington, DC, USA.

Desmond, K. 1987. *Harwin Chronology of Inventions, Innovations, Discoveries*. Constable, London, UK.

Devine, W.D. 1982. *An Historical Perspective on the Value of Electricity in American Manufacturing*. ORAU/IEA-82-8(M), Institute for Energy Analysis, Oak Ridge, TN, USA.

Devine W.D. 1983. From shafts to wires: Historical perspective on electrification. *Journal of Economic History*, **43**(2):347–372.

Donkin, R.A. 1978. *The Cistercians: Studies in the Geography of Medieval England and Wales*. Pontifical Institute for Mediaeval Studies, Toronto, Ontario, Canada.

Dosi, G., Orsenigo, L., and Silverberg, G. 1986. *Innovation, Diversity and Diffusion: A Self-Organization Model*. Science Policy Research Unit (SPRU), University of Sussex, Brighton, UK.

Dosi, G., Pavitt, K., and Soete, L. 1990. *The Economics of Technical Change and International Trade*. Harvester Wheatsheaf, New York, NY, USA.

Douglas, I. 1990. Sediment transfer and siltation. In: Turner II, B.L. *et al.* (eds.), *The Earth As Transformed by Human Action.* Cambridge University Press, Cambridge, UK, pp. 215–234.

Douglas, P.H. 1934. *A Theory of Wages.* Macmillan, New York, NY, USA.

van Duijn, J.J. 1983. *The Long Wave in Economic Life.* Allen and Unwin, London, UK.

Dumazedier, J. 1989. France: Leisure sociology in the 1980s. In: Olszewska, A., and Roberts, K. (eds.), *Leisure and Life-style.* Sage Studies in International Sociology, Volume 38. Sage Publications Ltd, London, UK, pp. 143–161.

Duncan, O.D. (ed.). 1964. *William Fielding Ogburn on Culture and Social Change.* University of Chicago Press, Chicago, IL, USA.

Durand, J.D. 1967. The modern expansion of world population. *Proceedings of the American Philosophical Society,* Vol. 111, No. 3(June):136–159.

Dutton, J.M., and Thomas, A. 1984. Treating progress functions as a managerial opportunity. *Academy of Management Review,* 9(2):235–247.

Dwyer, J.L. 1984. Scaling up bio-product separation with high performance liquid chromatography. *Bio/Technology,* November:957–964.

EIA (US Energy Information Administration, Department of Energy). 1997. *Annual Energy Review 1996.* www.eia.doe.gov (go to menu "Other Energy Sources" and then to "Historic Data").

EMEP (Cooperative Programme for Monitoring and Evaluation of the Long-Range Transmission of Air Pollutants in Europe). 1996. Anthropogenic emissions of sulfur (1980–2010) in the ECE region. Personal communication, E. Berge, EMEP, Oslo, Norway.

EPA (Environmental Protection Agency), with Lashof, D.A., and Tirpak, D.A. (eds.). 1990. *Policy Options for Stabilizing Global Climate.* EPA, Washington, DC, USA.

EPA (Environmental Protection Agency). 1992. *1990 Toxics Release Inventory: Public Data Release.* Office of Toxic Substances, EPA, Washington, DC, USA.

EPA (Environmental Protection Agency). 1995. *National Air Pollutant Emission Trends 1900–1994.* EPA-454/R-95-11, EPA, Washington, DC, USA.

Economist. 1990. *Vital World Statistics: A Complete Guide to the World in Figures.* The Economist Books Limited, London, UK.

Ehrlich, P.A., and Ehrlich, A.H. 1996. *The Betrayal of Science and Reason.* Island Press/Shearwater Books, Washington, DC, USA.

Elliott J.F. 1991. Energy, the environment, and iron and steel technology. In: Tester, J.W., Wood, D.O., and Ferrari, N.A. (eds.), *Energy and the Environment in the 21st Century.* MIT Press, Cambridge, MA, USA, pp. 373–382.

Engel, E. 1857. Die Produktions- und Consumptionsverhältnisse des Königreichs Sachsen. *Zeitschrift des Statistischen Bureaus des königlich-sächsischen Ministeriums des Inneren,* Band 3, Dresden, Germany.

Engel, J. (ed.). 1979. *Großer historischer Weltatlas, 2. Teil: Mittelalter.* Bayrischer Schulbuchverlag, München, Germany.

Engels, F. 1845. *Zur Lage der arbeitenden Klasse in England.* English translation, 1958. *The Condition of the Working Class in England.* Blackwell, Oxford, UK.

Ermoliev, Y. 1995. Optimization algorithms incorporating risk functions. Internal communication. International Institute for Applied Systems Analysis, Laxenburg, Austria.

Ermoliev, Y., and Wets, R. (eds.). 1988. *Numerical Techniques for Stochastic Optimization.* Springer-Verlag, Berlin, Germany.

Evenson, R.E. 1995. *The Valuation of Crop Genetic Resource Preservation, Conservation and Use.* Economic Growth Center, Yale University, New Haven, CT, USA.

Evenson, R.E. 1996. The economic principles of research resource allocation. In: Evenson, R.E., Herdt, R.W., and Hossain, M. (eds.), *Rice Research in Asia: Progress and Priorities.* CAB (Commonwealth Agricultural Bureau) International, Wallingford, UK.

FAO (Food and Agriculture Organization). Various volumes (1965–1996). *FAO Yearbook: Production.* FAO, Rome, Italy.

Falkenmark, M., and Widstrand, C. 1992. Population and water resources: A delicate balance. *Population Bulletin,* **47**:3–17, Population Reference Bureau, Washington, DC, USA.

Fettweis, G.B. 1979. *World Coal Resources: Methods of Assessment and Results.* Elsevier Science Publishers B.V., Amsterdam, Netherlands.

Fischer, G., and Rosenzweig, C. 1996. *The Impacts of Climate Change, CO_2, and SO_2 on Agricultural Supply and Trade.* WP-96-5. International Institute for Applied Systems Analysis, Laxenburg, Austria.

Fischer, G., Frohberg, K., Keyzer, M.A., and Parikh, K.S. 1988. *Linked National Models: A Tool for International Policy Analysis.* Kluwer Academic Publishers, Dordrecht, Netherlands.

Fischer, G., Frohberg, K., Parry, M.L., and Rosenzweig, C. 1994. Climate change and world food supply, demand and trade. Who benefits, who loses? *Global Environmental Change,* 4(1):7–23.

Fisher, J.C. 1974. *Energy Crises in Perspective.* John Wiley & Sons, New York, NY, USA.

Fisher, J.C., and Pry, R.H. 1971. A simple substitution model of technological change. *Technological Forecasting & Social Change,* **3**:75–88.

Fishlow, A. 1965. *American Railroads and the Transformation of the Antebellum Economy.* Harvard University Press, Cambridge, MA, USA.

Fleck, J. 1988. *Innofusion or Diffusation? The Nature of Technological Development in Robotics.* ESRC Programme on Information and Communication Technologies Working Paper series, University of Edinburgh, Edinburgh, Scotland.

Flora, P. 1975. *Indikatoren der Modernisierung.* Westdeutscher Verlag, Opladen, Germany.

Flora, P., Kraus, F., and Pfennig, W. 1987. *State, Economy, and Society in Western Europe, 1815–1975.* Volumes I and II. Campus Verlag, Frankfurt, Germany.

Fogel, R.W. 1970. *Railroads and American Economic Growth: Essays in Economic History.* The Johns Hopkins University Press, Baltimore, MD, USA.

Fontvieille, L. 1976. *Evolution et Croissance de l'Etat Français: 1815–1969.* Cahiers de l'ISMEA, Serie AF, No. 13. Institut de Sciences Mathématiques et Economiques Appliquées. Paris, France.

Foray, D., and Grübler, A. 1990. Morphological analysis, diffusion and lock-out of technologies: Ferrous casting in France and the FRG. *Research Policy,* **19**(6):535–550.

Forrester, J.W. 1971. *World Dynamics.* Wright-Allen Press, Cambridge, MA, USA.

Fourastié, J. 1949. *Le Grand Espoir du XXe Siècle: Progrès Technique, Progrès Economique, Progrès Social.* Presses Universitaires de France, Paris, France.

Fragnière, E., and Haurie, A. 1995. A stochastic programming model for energy/ environment choices under uncertainty. *International Journal of Environment and Pollution,* **6**(4–6):587–603.

Frankel, M. 1955. Obsolescence and technological change in a maturing economy. *American Economic Review,* **45**:296–319.

Freeman, C. (ed.). 1983. *Long Waves in the World Economy.* Butterworths, London, UK.

Freeman, C. 1989. *The Third Kondratieff Wave: Age of Steel, Electrification and Imperialism.* Research Memorandum 89-032, MERIT, Maastricht, Netherlands.

Freeman, C. 1994. The economics of technical change. *Cambridge Journal of Economics,* **18**:463–514.

Freeman, C. (ed.). 1996. *Long Wave Theory.* Elgar Reference, Cheltenham, UK.

Freeman, C., and Perez, C. 1988. Structural crises of adjustment, business cycles and investment behavior. In: Dosi, G., Freeman, C., Nelson, R., Silverberg, G., and Soete, L. (eds.), *Technical Change and Economic Theory.* Pinter, London, UK, pp. 38–66.

Freeman, C., Clark, J., and Soete, L. 1982. *Unemployment and Technological Innovation: A Study of Long Waves and Economic Development.* Frances Pinter, London, UK.

Fremdling, R. 1975. *Eisenbahnen und deutsches Wirtschaftswachstum 1840–1879.* Gesellschaft für Westfälische Wirtschaftsgeschichte E.V., Dortmund, Germany.

Friedlander, S.K. 1989. Environmental issues: Implications for engineering design and education. In: Ausubel, J.H., and Sladovich, H.E. (eds.), *Technology and Environment.* National Academy Press, Washington, DC, USA, pp. 167–181.

Frosch, R.A. 1984. Improving American innovation: The role of industry in innovation. In: Coles, J.S. (ed.), *Technological Innovation in the 1980s.* Prentice Hall, Englewood Cliffs, NJ, USA, pp. 56–81.

Frosch, R.A. 1992. Industrial ecology: A philosophical introduction. *Proceedings of the National Academy of Sciences of the United States of America,* **89**(3):800–803.

Frosch, R.A. 1996. The customer for R&D is always wrong! *Research Technology Management,* November–December:22–27.

Frosch, R.A., and Gallopoulos, N.E. 1989. Strategies for manufacturing. *Scientific American,* **261**(3):144–152.

Fussel, G.E. 1958. Agriculture: Techniques of farming. In: Singer, C., Holmyard, E.J., Hall, A.R., and Williams T.I. (eds.), *A History of Technology, Vol. IV:*

The Industrial Revolution c. 1750–c. 1850. Clarendon Press, Oxford, UK, pp. 13–43.

Ganzel, B. 1984. *Dust Bowl Descent.* University of Nebraska Press, Lincoln, NE, USA.

Gaspari, C., and Millendorfer, J. 1976. Non-economic and economic factors in societal development: The general production function. In: Bruckmann, G. (ed.), *Latin American World Model: Proceedings of the Second International Institute for Applied Systems Analysis Symposium on Global Modelling.* CP-76-8. International Institute for Applied Systems Analysis, Laxenburg, Austria, pp. 175–187.

General Motors. 1983. *The First 75 Years of Transportation Products.* General Motors, Detroit, MI, USA.

Gerschenkron, A. 1962. *Economic Backwardness in Historical Perspective.* Belknap Press, Cambridge, MA, USA.

Gershuny, J. 1978. *After Industrial Society: The Emerging Self-Service Economy.* Macmillan Press, London, UK.

Gershuny, J. 1984. Growth, social innovation and time use. In: Boulding, K.E. (ed.), *The Economics of Human Betterment.* Macmillan Press, London, UK, pp. 36–57.

Gershuny, J. 1989. Technical change and the work/leisure balance: A new system of socio-economic accounts. In: Silberston, A. (ed.), *Technology and Economic Progress.* Macmillan Press, London, UK, pp. 181–215.

Gershuny, J. 1991. Time budget research. Paper presented at the workshop on Social Behavior, Lifestyles and Energy Use, 24–26 June, International Institute for Applied Systems Analaysis, Laxenburg, Austria.

Gershuny, J.I. 1983. *Social Innovation and the Division of Labour.* Oxford University Press, Oxford, UK.

Gershuny, J.I. 1992. La répartition du temps dans les societés post-industrielles. *Futuribles,* **165–166**:215–226.

Gershuny, J.I., and Robinson, J. 1989. Multinational comparisons of change in the household division of labor. Paper presented at the International Workshop on the Changing Use of Time, Brussels, Belgium, 17–18 April, European Foundation for the Improvement of Living and Working Conditions, Dublin, Ireland.

Gilfillan, S.C. 1935. *The Sociology of Invention.* Follett Publishing Co., Chicago, IL, USA.

Gilfillan, S.C. 1937. The prediction of innovations; and, Social effects of inventions. In: Subcommittee on Technology, National Resources Committee, *Technological Trends and National Policy.* 75th Congress, 1st Session, House Document No. 360, US Government Printing Office, Washington, DC, USA, pp. 15–23, and 24–38.

Gille, B. 1978. *Histoire des Techniques.* La Pléiade, Paris, France.

Glas, J.P. 1989. Protecting the ozone layer: A perspective from industry. In: Ausubel, J.H., and Sladovich, H.E. (eds.), *Technology and Environment.* National Academy Press, Washington, DC, USA, pp. 137–155.

Glaser, P. 1975. The satellite solar power station option. In: Veziroglu, T.U. (ed.), *Remote Sensing: Energy Related Studies*. John Wiley & Sons, New York, NY, USA, pp. 367–394.

Glass, D. (ed.). 1953. *Introduction to Malthus*. Watts, London, UK.

Glaziev, S. 1991. Some general regularities of techno-economic evolution. In: Nakićenović, N., and Grübler, A. (eds.), *Diffusion of Technologies and Social Behavior*. Springer-Verlag, Berlin, Germany, pp. 295–315.

Gleick, P.H. (ed.). 1993. *Water in Crisis: A Guide to the World's Freshwater Resources*. Oxford University Press, Oxford, UK.

Godlund, S. 1952. *Ein Innovationsverlauf in Europa, dargestellt in einer vorläufigen Untersuchung über die Ausbreitung der Eisenbahninnovation*. Lund Studies in Geography, Ser.B. Human Geography No. 6, C.W.K Gleerup Publishers, Lund, Sweden.

Goeller, H.E., and Weinberg, A.M. 1976. *The Age of Substitutability, or What Do We Do When the Mercury Runs Out?* Report 76-1. Institute for Energy Analysis, Oak Ridge, TN, USA.

Goldemberg, J. 1991. "Leap-frogging": A new energy policy for developing countries. *WEC Journal*, December:27–30.

Golodnikov, A., Gritsevskii, A., and Messner, S. 1995. *A Stochastic Version of the Dynamic Linear Programming Model MESSAGE III*. WP-95-94. International Institute for Applied Systems Analysis, Laxenburg, Austria.

Goulder, L.H., and Schneider, S.H. 1996. *Induced Technical Change, Crowding Out, and the Attractiveness of CO_2 Emissions Abatement*. Mimeo, Stanford University, Stanford, CA, USA.

Gray, P.E. 1989. The paradox of technological development. In: Ausubel, J.H., and Sladovich, H.E. (eds.), *Technology and Environment*. National Academy Press, Washington, DC, USA, pp. 192–204.

Grenon, M. 1979. The WELMM (water, energy, land, materials, manpower) approach to coal mining. In: Grenon, M. (ed.), *Future Coal Supply for the World Energy Balance*. Pergamon Press, Oxford, UK, pp. 379–396.

Grenon, M., and Lapillonne, B. 1976. *The WELMM Approach to Energy Strategies and Options*. RR-76-19. International Institute for Applied Systems Analysis, Laxenburg, Austria.

Grigg, D.B. 1980. *Population Growth and Agrarian Change: An Historical Perspective*. Cambridge University Press, Cambridge, UK.

Grigg, D.B. 1982. *The Dynamics of Agricultural Change: The Historical Experience*. Hutchinson & Co. Ltd., London, UK.

Grigg, D.B. 1987. The industrial revolution and land transformation. In: Wolman, M.G., and Fournier, F.G.A. (eds.), *Land Transformation in Agriculture, SCOPE 32*, John Wiley & Sons, Chichester, UK, pp. 79–109.

Griliches, Z. 1957. Hybrid corn: An exploration in the economics of technical change. *Econometrica,* **25**:501–522.

Griliches, Z. 1986. Productivity, R&D and basic research at the firm level in the 1970s. *American Economic Review,* **76**:141–154.

Griliches, Z. 1996. The discovery of the residual: A historical note. *Journal of Economic Literature,* **34**(September):1324–1330.

Grossman, G., and Helpman, E. 1991. *Innovation and Growth in the Global Economy*. MIT Press, Cambridge, MA, USA.

Grubb, M. 1996. Technologies, energy systems, and the timing of CO_2 emissions abatement: An overview of economic issues. In: Nakićenović, N., Nordhaus, W.D., Richels, R., and Toth, F.L. (eds.), *Climate Change: Integrating Science, Economics, and Policy*. CP-96-1. International Institute for Applied Systems Analysis, Laxenburg, Austria, pp. 249–270.

Grübler, A. 1987. Technology diffusion in a long wave context: The case of the steel and coal industries. *Proceedings of the International Workshop on Life Cycles and Long Waves*, Montpellier, France, 8–10 July, International Institute for Applied Systems Analysis, Laxenburg, Austria (also published in Grübler, 1990b).

Grübler, A. 1990a. *The Rise and Fall of Infrastructures, Dynamics of Evolution and Technological Change in Transport*. Physica-Verlag, Heidelberg, Germany.

Grübler, A. 1990b. Technology diffusion in a long wave context: The case of the steel and coal industries. In: Vasko, T., Ayres, R., and Fontvieille, L. (eds.), *Life Cycles and Long Waves* (Lecture Notes in Economics and Mathematical Systems 340). Springer-Verlag, Berlin, Germany, pp. 117–146.

Grübler, A. 1991. Diffusion: Long-term patterns and discontinuities. *Technological Forecasting & Social Change*, **39**(1–2):159–180.

Grübler, A. 1992a. *Technology and Global Change: Land-use, Past and Present*. WP-92-2. International Institute for Applied Systems Analysis, Laxenburg, Austria.

Grübler, A. 1992b. Introduction to Diffusion Theory. In: Ayres, R.U., Haywood, W., and Tchijov, I. (eds.), *Computer Integrated Manufacturing*, Vol. II. Chapman & Hall, London, UK, pp. 3–52.

Grübler, A. 1993a. Energy and environment: Post-UNCED. In: Ghosh, P., and Jaitly, A. (eds.), *The Road from Rio: Environment and Development Policy Issues in Asia*. Tata Energy Research Institute, New Delhi, India, pp. 104–120.

Grübler, A. 1993b. The transportation sector: Growing demand and emissions. *Pacific and Asian Journal of Energy*, **3**(2):179–199.

Grübler, A. 1994a. Technology. In: Meyer W.B., and Turner II, B.L. (eds.), *Changes in Land Use and Land Cover: A Global Perspective*. Cambridge University Press, Cambridge, UK, pp. 287–328.

Grübler, A. 1994b. Industrialization as a historical phenomenon. In: Socolow, R. *et al.* (eds.), *Industrial Ecology and Global Change*. Cambridge University Press, Cambridge, UK, pp. 43–68.

Grübler, A. 1995a. *Industrialization as a Historical Phenomenon*. WP-95-29. International Institute for Applied Systems Analysis, Laxenburg, Austria.

Grübler, A. 1995b. *Time for a Change: Rates of Diffusion of Ideas, Technologies, and Social Behaviors*. WP-95-82. International Institute for Applied Systems Analysis, Laxenburg, Austria.

Grübler, A., and Gritsevskii, A. 1997. A model of endogenous technological change through uncertain returns on learning (R&D and investments). Paper presented at the International Workshop on Induced Technical Change and the

Environment, 26–27 June, International Institute for Applied Systems Analysis, Laxenburg, Austria.

Grübler, A., and Messner, S. 1996. Technological uncertainty. In: Nakićenović, N., Nordhaus, W.D., Richels, R., and Toth, F.L. (eds.), *Climate Change: Integrating Science, Economics, and Policy.* CP-96-1. International Institute for Applied Systems Analysis, Laxenburg, Austria, pp. 295–314.

Grübler, A., and Nakićenović, N. 1987. *The Dynamic Evolution of Methane Technologies.* WP-87-002. International Institute for Applied Systems Analysis, Laxenburg, Austria.

Grübler, A., and Nakićenović, N. 1990. *Economic Map of Europe: Transport, Communication and Energy Infrastructures in a Wider Europe.* Mimeo, International Institute for Applied Systems Analysis, Laxenburg, Austria.

Grübler, A., and Nakićenović, N. 1991. *Evolution of Transport Systems: Past and Future.* RR-91-8. International Institute for Applied Systems Analysis, Laxenburg, Austria.

Grübler, A., and Nakićenović, N. 1994. *International Burden Sharing in Greenhouse Gas Reduction.* RR-94-9. International Institute for Applied Systems Analysis, Laxenburg, Austria.

Grübler, A. and Nakićenović, N. 1996. Decarbonizing the global energy system. *Technological Forecasting & Social Change,* **53**(1):97–110.

Grübler, A., and Nowotny, H. 1990. Towards the fifth Kondratiev upswing: Elements of an emerging new growth phase and possible development trajectories. *International Journal of Technology Management,* **5**(4):431–471.

Grübler, A., Nakićenović, N., and Schäfer, A. 1992. Dynamics of transport and energy systems: History of development and a scenario for the future. In: *Proceedings of the 15th World Energy Congress Madrid,* 20–25 September, Division 3, Technical Session 3.3, World Energy Council, London, UK, pp. 219–240 (cf. also Grübler *et al.,* 1993b).

Grübler, A., Messner, S., Schrattenholzer, L., and Schäfer, A. 1993a. Emission reduction at the global level. *Energy,* **18**(5):539–581.

Grübler, A., Nakićenović, N., and Schäfer, A. 1993b. *Dynamics of Transport and Energy Systems: History of Development and a Scenario for the Future.* RR-93-19. International Institute for Applied Systems Analysis, Laxenburg, Austria.

Gulledge, T.R., and Womer, N.K. 1990. Learning curves and production functions: An integration. *Engineering Costs and Production Economics,* **20**:3–12.

HABITAT (United Nations Centre for Human Settlements). 1987. *Global Report on Human Settlements 1986.* Oxford University Press, Oxford, UK.

HABITAT (United Nations Centre for Human Settlements). 1996. *An Urbanizing World: Global Report on Human Settlements 1996.* Oxford University Press, Oxford, UK.

HDP (Human Dimensions of Global Environmental Change Programme). 1990. *A Framework for Research on the Human Dimensions of Global Environmental Change.* HDP Report No. 1, HDP Secretariat, Barcelona, Spain.

Habakkuk, H.J. 1962. *American and British Technology in the Nineteenth Century.* Cambridge University Press, Cambridge, UK.

Häfele, W., Barnert, H., Messner, S., Strubegger, M., Anderer, J. 1986. Novel integrated energy systems: The case of zero emissions. In: Clark, W.C., and Munn, R.E. (eds.), *Sustainable Development of the Biosphere.* Cambridge University Press, Cambridge, UK, pp. 171–193.

Hägerstrand, T. 1967. *Innovation Diffusion as a Spatial Process.* University of Chicago Press, Chicago, IL, USA.

Hareven, T.K. 1982. *Family Time and Industrial Time: The Relationship Between the Family and Work in a New England Industrial Community.* Cambridge University Press, Cambridge, UK.

Harris, M. 1981. *Why Nothing Works: The Anthropology of Daily Life.* Simon & Schuster, New York, NY, USA.

Hart, J.F. 1991. *The Land that Feeds Us.* W.W. Norton & Company, Inc., New York, NY, USA.

Haustein, H.-D., and Neuwirth, E. 1982. *Long Waves in World Industrial Production, Energy Consumption, Innovations, Inventions, and Patents and Their Identification by Spectral Analysis.* WP-82-9. International Institute for Applied Systems Analysis, Laxenburg, Austria.

Hayami, Y., and Ruttan, V.W. 1971. *Agricultural Development: An International Perspective.* 1st edition. The Johns Hopkins University Press, Baltimore, MD, USA.

Hayami, Y., and Ruttan, V.W. 1985. *Agricultural Development: An International Perspective.* 2nd edition. The Johns Hopkins University Press, Baltimore, MD, USA.

Hays, S.P. 1959. *Conservation and the Gospel of Efficiency.* Harvard University Press, Cambridge, MA, USA.

Heilig, G. 1995. *Lifestyles and Global Land Use Change: Data and Theses.* WP-95-91. International Institute for Applied Systems Analysis, Laxenburg, Austria.

Heilig, G. 1996. *Anthropogenic Driving Forces of Land Use Change in China.* WP-96-11. International Institute for Applied Systems Analysis, Laxenburg, Austria.

von Hippel, E. 1988. *The Sources of Innovation.* Oxford University Press, Oxford, UK.

Hobsbawn, E.J., and Rudé, G. 1968. *Captain Swing.* Pantheon Books, New York, NY, USA.

Hoffmann, W.G. 1955. *British Industry 1700–1950.* Basil Blackwell, Oxford, UK.

Hoffmann, W.G. 1958. *The Growth of Industrial Economies.* Manchester University Press, Manchester, UK (revised English translation).

Holbrook, W.S., Hoskins, H., Wood, W.T., Stephen, R.A., and Lizarralde, D. 1996. Methane hydrate and free gas on the Blake Ridge from vertical seismic profiling. *Science,* **273**(27 September):1840–1842.

Holland, H.D., and Petersen, U. 1995. *Living Dangerously: The Earth, Its Resources, and the Environment.* Princeton University Press, Princeton, NJ, USA.

Holtfrerich, C.L. 1973. *Quantitative Wirtschaftsgeschichte des Ruhrkohlenbergbaus im 19. Jahrhundert: Eine Führungssektoranalyse.* Gesellschaft für Westfälische Wirtschaftsgeschichte E.V., Dortmund, Germany.

Houghton, R.A., and Skole, D.L. 1990. Carbon. In: Turner II, B.L. *et al.* (eds.), *The Earth as Transformed by Human Action*. Cambridge University Press, Cambridge, UK, pp. 393–408.

Houghton, J.T., Meira Filho, L.G., Bruce, J., Lee, H., Callander, B.A., Haites, E., Harris, N., and Maskell, K. (eds.). 1995. *Climate Change 1994, Reports of Working Groups I and III of the IPCC*. Cambridge University Press, Cambridge, UK.

Hughes, T.P. 1983. *Networks of Power: Electrification in Western Society 1880–1930*. The Johns Hopkins University Press, Baltimore, MD, USA.

Hughes, T.P. 1988. The seamless web: Technology, science, et cetera, et cetera. In: Elliott, B. (ed.), *Technology and Social Process*. Edinburgh University Press, Edinburgh, Scotland, pp. 9–19.

Hugill, P.J. 1993. *World Trade Since 1431: Geography, Technology, and Capitalism*. The Johns Hopkins University Press, Baltimore, MD, USA.

IEA (International Energy Agency). 1991. *Energy Statistics of OECD Countries 1960–1979*. Volumes I and II. IEA, Paris, France.

IEA (International Energy Agency). 1993. *Energy Statistics and Balances of Non-OECD Countries 1971–1991*. IEA, Paris, France.

IEA (International Energy Agency). 1994. *Energy in Developing Countries: A Sectorial Analysis*. IEA, Paris, France.

IEM (Institut d'Estudis Metropolitans de Barcelona). 1986. *Cities: Statistical, Administrative and Geographical Information on the Major Urban Areas of the World* (5 volumes). IEM, Barcelona, Spain.

IIASA–WEC (International Institute for Applied Systems Analysis and World Energy Council). 1995. (Grübler, A., Jefferson, M., McDonald, A., Messner, S., Nakićenović, N., Rogner, H-H., and Schrattenholzer, L.). *Global Energy Perspectives to 2050 and Beyond*, WEC, London, UK.

IIED (International Institute for Environment and Development). 1996. *Towards a Sustainable Paper Cycle*. IIED, London, UK.

IISI (International Iron and Steel Institute). Various volumes (1975–1996). *World Steel Statistics*. IISI, Brussels, Belgium.

IISI (International Iron and Steel Institute). 1997. Phillip Norris, personal communication, Statistics Department, IISI, Brussels, Belgium.

ILO (International Labor Office). Various volumes (1985–1995). *Yearbook of Labour Statistics*. ILO, Geneva, Switzerland.

IMF (International Monetary Fund). Various volumes (1975–1997). *International Financial Statistics: Yearbook*. IMF, Washington, DC, USA.

IPCC (Intergovernmental Panel on Climate Change). 1990. *Climate Change: The IPCC Scientific Assessment*. Cambridge University Press, Cambridge, UK.

IPCC (Intergovernmental Panel on Climate Change). 1994. *Radiative Forcing of Climate Change: Summary for Policy Makers*. IPCC, Geneva, Switzerland.

IPCC (Intergovernmental Panel on Climate Change). 1995. *Climate Change 1994. Radiative Forcing of Climate Change and an Evaluation of the IPCC IS92 Emission Scenarios, Reports of Working Groups I and III*, Cambridge University Press, Cambridge, UK.

IPCC (Intergovernmental Panel on Climate Change). 1996a. *Climate Change 1995. Contributions of Working Groups I, II, and III to the Second Assessment Report of the IPCC* (3 volumes). Cambridge University Press, Cambridge, UK.

IPCC (Intergovernmental Panel on Climate Change). 1996b. *Climate Change 1995: The Science of Climate Change. Summary for Policy Makers.* IPCC, Geneva, Switzerland.

IRF (International Road Federation). Various volumes (1970–1995). *World Road Statistics.* IRF, Washington, DC, USA, and Geneva, Switzerland.

Imhoff, A.E. 1981. *Die gewonnenen Jahre: Von der Zunahme unserer Lebensspanne seit dreihundert Jahren oder von der Notwendigkeit einer neuen Einstellung zu Leben und Sterben.* C.H. Beck, München, Germany.

Isard, W. 1942. A neglected cycle: The transport building cycle. *Review of Economic Statistics* **XXIV**(2):348–364.

Jackson, R.S. 1994. *Wine Science: Principles and Applications.* Academic Press, Orlando, FL, USA.

Jahoda, M. 1988. Time: A social psychological perspective. In: Young, M., and Schuller, T. (eds.), *The Rhythms of Society.* Routledge, London, UK, pp. 154–172.

Janauschek, P.L. 1877. *Originum Cisterciensium.* Tomus I. A. Hoeler, Vienna, Austria.

Japanese Statistics Bureau. Various volumes (1975–1995). *Japan Statistical Yearbook.* Management and Coordination Agency, Tokyo, Japan.

Jevons, W.S. 1865. *The Coal Question: An Enquiry Concerning the Progress of the Nation and the Probable Exhaustion of Our Coal-mines.* Macmillan Press, London, UK.

Jewkes, J., Sawers, D., and Stillermann, R. 1969. *The Sources of Innovation.* Macmillan Press, London, UK (revised edition).

Johnson, C.L., and Smith, M. 1985. *More than my Share of it All.* Smithsonian Institution Press, Washington, DC, USA.

Jones, D.W. 1991. How urbanization affects energy-use in developing countries. *Energy Policy,* **19**(7):621–630.

Jorgenson, D.W., and Fraumeni, B.M. 1981. Relative prices and technical change. In: Bendt, E., and Fields, B. (eds.), *Modeling and Measuring Natural Resource Substitution.* MIT Press, Cambridge, MA, USA.

Kamann, D.J., and Nijkamp, P. 1991. Technogenesis: Origins and diffusion in a turbulent environment. In: Nakićenović, N., and Grübler, A. (eds.), *Diffusion of Technologies and Social Behavior.* Springer-Verlag, Berlin, Germany, pp. 93–124.

Kates, R.W. 1996. Population, technology, and the human environment: A thread through time. *Dædalus,* **125**(3):51.

Kemp, R. 1997. *Environmental Policy and Technical Change.* Edward Elgar, Cheltenham, UK.

Kennedy, C. 1964. Induced bias in innovation and the theory of distribution. *Economic Journal,* **74**:541–547.

Kennedy, P. 1987. *The Rise and Fall of the Great Powers: Economic Change and Military Conflict from 1500–2000.* Random House, New York, NY, USA.

Kennedy, P. 1993. *Preparing for the 21st Century*. Random House, New York, NY, USA.

Keyfitz, N. 1991. The middle class lifestyle. Paper presented at the Workshop on Social Behavior, Lifestyles and Energy Use, 24–26 June, International Institute for Applied Systems Analysis, Laxenburg, Austria.

Kirsch, D. 1996. The electric car and the burden of history: Studies in automotive systems rivalry in America 1890–1996. PhD Thesis, Department of History, Stanford University, Stanford, CA, USA.

Kleinknecht, A. 1987. *Innovation Patterns in Crisis and Prosperity, Schumpeter's Long Cycle Reconsidered*. Macmillan Press, London, UK.

Kline, S.J. 1985. What is technology? *Bulletin of Science, Technology and Society*, 5(3):215–219.

Kolmhofer, D. 1987. Weltproduktion von Dünge- und technischem Stickstoff. Internal paper, Chemie Linz AG, Linz, Austria.

Kondratiev, N.D. 1926. Die langen Wellen in der Konjunktur. *Archiv für Sozialwissenschaft und Sozialpolitik*, 56:573–609.

Kuznets, S.S. 1958. *Six Lectures on Economic Growth*. Free Press, New York, NY, USA.

Kuznets, S.S. 1971. *Economic Growth of Nations: Total Output and Production Structure*. Belknap Press, Cambridge, MA, USA.

Landes, D.S. 1969. *The Unbound Prometheus: Technological Change and Industrial Development in Western Europe From 1750 to the Present*. Cambridge University Press, Cambridge, UK.

Langdon, J. 1986. *Horses, Oxen, and Technological Innovation*. Cambridge University Press, Cambridge, UK.

Lay, G.M. 1992. *Ways of the World: A History of the World's Roads and the Vehicles That Used Them*. Rutgers University Press, New Brunswick, NJ, USA.

Lazarus, M.L., Greber, J., Hall, C., Bartels, S., Bernow, S., Hansen, E., Raskin, P., and von Hippel, D. 1993. *Towards a Fossil Free Energy Future: The Next Transition*. A technical analysis for Greenpeace International, Stockholm Environmental Institute, Boston Center, Boston, MA, USA.

Leach, G. 1995. *Global Land and Food in the 21st Century: Trends and Issues for Sustainability*. Stockholm Environment Institute (SEI), POLESTAR Series Report No. 5, SEI, Stockholm, Sweden.

Lebergott, S. 1993. *Pursuing Happiness: American Consumers in the Twentieth Century*. Princeton University Press, Princeton, NJ, USA.

Lebergott, S. 1996. *Consumer Expenditures: New Measures and Old Motives*. Princeton University Press, Princeton, NJ, USA.

Leontieff, W. 1978. Worksharing, unemployment, and economic growth. In: *National Commission for Manpower Policy, Work Time and Employment: A Conference Report*. Special Report No. 28, U.S. Government Printing Office No. 052-003-00686-3, Washington, DC, USA, pp. 129–135.

Liebowitz, S.J., and Margolis, S.E. 1990. The fable of the keys. *Journal of Law and Economics*, 33:1–25.

Liebowitz, S.J., and Margolis, S.E. 1995. Path dependence, lock-in, and history. *Journal of Law, Economics, and Organization,* **11**(1):205–226.

van Lier, H.N. 1991. Historical land use changes: The Netherlands. In: Brouwer, F.M. (ed.), *Land Use Changes in Europe.* Kluwer Academic Publishers, Dordrecht, Netherlands, pp. 379–401.

Liesner, T. (ed.). 1985. *Economic Statistics 1900–1983.* The Economist Publications Limited, London, UK.

Linstone, H.A. 1996. Technological slowdown or societal speedup: The price of complexity? *Technological Forecasting & Social Change,* **51**(2):195–205.

Lovelock, J.E. 1971. Atmospheric fluorine compounds as indicators of air movements. *Nature,* **230**(9 April):379.

Lovelock, J.E. 1979. *GAIA: A New Look at Life on Earth.* Oxford University Press, Oxford, UK.

L'vovich, M., and White, G.F. 1990. Use and transformation of terrestrial water systems. In: Turner II, B.L. *et al.* (eds.), *The Earth as Transformed by Human Action.* Cambridge University Press, Cambridge, UK, pp. 235–252.

MVMA (Motor Vehicle Manufacturers Association of the United States, Inc.). 1985. *MVMA Motor Vehicle Facts and Figures.* MVMA, Detroit, MI, USA.

MVMA (Motor Vehicle Manufacturers Association of the United States, Inc.). 1991. *World Motor Vehicle Data.* MVMA, Detroit, MI, USA.

MacDonald, G.J. 1990. The future of methane as an energy resource. *Annual Review of Energy,* **15**:53–83.

Mackenzie, D. 1991. *Economic and Sociological Explanation of Technical Change.* Mimeo, Department of Sociology, University of Edinburgh, Edinburgh, Scotland.

Maddison, A. 1991. *Dynamic Forces in Capitalist Development: A Long-run Comparative View.* Oxford University Press, Oxford, UK.

Maddison, A. 1995. *Monitoring the World Economy 1820–1992.* Development Centre Studies, Organisation for Economic Co-operation and Development (OECD), Paris, France.

Malone, T.F. 1995. Reflections on the human prospect. *Annual Review of Energy and the Environment,* **20**:1–29.

Malthus, T. 1798. *An Essay on the Principle of Population.* 1986 Reprint: Penguin Paperback, London, UK (see also Glass, 1953).

Manne, A., and Richels, R. 1992. *Buying Greenhouse Insurance: The Economic Costs of CO_2 Emission Limits.* MIT Press, Cambridge, MA, USA.

Manne, A., and Richels, R. 1994. The costs of stabilizing global CO_2 emissions: A probabilistic analysis based on expert judgements. *The Energy Journal,* **15**(1):31–56.

Mansfield, E., 1968. *The Economics of Technological Change.* W.W. Norton & Co., New York, NY, USA.

Mansfield, E. 1980. Basic research and productivity increase in manufacturing. *American Economic Review,* **70**:863–873.

Mansfield, E. 1985. How rapidly does new industrial technology leak out? *Journal of Industrial Economics,* **34**(2):217–223.

Mansfield, E., Rapoport, J., Schnee, J., Wagner, S., and Hamburger, M. 1971. *Research and Development in the Modern Corporation.* W.W. Norton & Company, Inc., New York, NY, USA.

Mansfield, E., Rapoport, J., Romeo, A., Wagner, S., and Beardsley, G. 1977. *The Production and Application of New Industrial Technology.* W.W. Norton & Company, Inc., New York, NY, USA.

Marchand, O. 1992. Une comparaison internationale de temps de travail. *Futuribles,* **165–166**(5–6):29–39.

Marchetti, C. 1978. *On 10^{12}: A Check on the Earth Carrying Capacity for Man.* RR-78-7. International Institute for Applied Systems Analysis, Laxenburg, Austria.

Marchetti, C. 1980. Society as a learning system: Discovery, invention and innovation cycles revisited. *Technological Forecasting & Social Change,* 18:267–282.

Marchetti, C., and Nakićenović, N. 1979. *The Dynamics of Energy Systems and the Logistic Substitution Model.* RR-79-13. International Institute for Applied Systems Analysis, Laxenburg, Austria.

Marczewski, J. 1965. *Introduction à l'Histoire Quantitative.* Librairie Droz, Geneva, Switzerland.

Marland, G. 1989. The role of forests in addressing the CO_2 greenhouse. In: White, J.C., and Wagner, W. (eds.), *Global Climate Change Linkages: Acid Rain, Air Quality and Stratospheric Ozone.* Elsevier Publishing Co. Inc., New York, NY, USA, pp. 199–212.

Marland, G., and Pippin, A. 1990. United States emissions of carbon dioxide to the earth's atmosphere by economic activity. *Energy Systems and Policy,* **14**:319–336.

Marland, G., and Weinberg, A. 1988. Longevity of infrastructure. In: Ausubel, J.H., and Herman, R. (eds.), *Cities and Their Vital Systems: Infrastructure Past, Present and Future.* National Academy Press, Washington, DC, USA, pp. 312–332.

Marsh, G.P. 1864. *Man and Nature: Or, Physical Geography as Modified by Human Action.* 1965 Reprint: Lowenthal, D. (ed.), Harvard University Press, Cambridge, MA, USA.

Marshall, C. 1938. *A History of British Railways Down to the Year 1830.* Oxford University Press, Oxford, UK.

Martin, H.P., and Schumann, H. 1996. *Die Globalisierungsfalle: Der Angriff auf Demokratie und Wohlstand.* Rowohlt Verlag, Hamburg, Germany.

Martino, J.P. 1983. *Technological Forecasting for Decision Making.* 2nd edition. North Holland, New York, NY, USA.

Marx, L., and Mazlish, B. (eds.). 1996. *Progress: Fact of Illusion?* University of Michigan Press, Ann Arbor, MI, USA.

Maslow, A.H. 1954. *Motivation and Personality.* Harper & Row, New York, NY, USA.

McEvedy, C., and Jones, R. 1978. *Atlas of World Population History.* Penguin Books, London, UK.

Meadows, D., Meadows, D., Randers, J., and Behrens, W.W. 1972. *The Limits to Growth. A Report for the Club of Rome's Project on the Predicament of Mankind.* Signet, New York, NY, USA.

Mensch, G. 1979. *Stalemate in Technology.* Ballinger, Cambridge, MA, USA.

Messner, S. 1995. *Endogenized Technological Learning in an Energy Systems Model.* WP-95-114. International Institute for Applied Systems Analysis, Laxenburg, Austria.

Messner, S. 1997. Endogenized technological learning in an energy systems model. *Journal of Evolutionary Economics,* 7(3):291–313.

Messner, S., and Strubegger, M. 1991. *User's Guide to CO2DB: The IIASA CO_2 Technology Data Bank – Version 1.0.* WP-91-31. International Institute for Applied Systems Analysis, Laxenburg, Austria.

Messner, S., Golodnikov, A., and Gritsevskii, A. 1996. A stochastic version of the dynamic linear programming model MESSAGE III. *Energy,* 21(9):775–784.

Metallstatistik (Metallgesellschaft AG). 1993. *Metallstatistik 1991–1992.* Metallgesellschaft AG, Frankfurt a.M., Germany.

Metcalfe, S. 1987. Technical change. In: Eatwell, J., Milgate, M., and Newman, P. (eds.), *The New Palgrave, A Dictionary of Economics,* Vol. 4. Macmillan Press, London, UK, pp. 617–620.

Meyer, W.B. 1996. *Human Impact on the Earth.* Cambridge University Press, Cambridge, UK.

Meyer-Abich, K.M. 1996. Humans and nature: Toward a physiocentric philosophy. *Dædalus,* 125(3):213–234.

Michaelis, L. (convening lead author). 1996. Mitigation options in the transportation sector. In: Watson, R.T., Zinyowera, M.C., and Moss, R.H. (eds.), *Climate Change 1995.* Intergovernmental Panel on Climate Change (IPCC) and Cambridge University Press, Cambridge, pp. 679–712.

Miklin, P.P. 1996. Introductory remarks on the Aral issue. In: Miklin, P.P., and Williams, W.D. (eds.), *The Aral Sea Basin,* NATO ASI Series Environment Volume 12. Springer-Verlag, Berlin, Germany, pp. 3–8.

Minge-Kalman, W. 1980. Does labor time decrease with industrialization: A survey of time allocation studies. *Current Anthropology,* 21:279–287.

Mitchell, B.R. 1980. *European Historical Statistics: 1750–1975.* MacMillan Press, London, UK.

Mitchell, B.R. 1982. *International Historical Statistics: Africa and Asia.* MacMillan Press, London, UK.

Mitchell, B.R. 1983. *International Historical Statistics: The Americas and Australia.* MacMillan Press, London, UK.

Mokyr, J. 1990. *The Lever of Riches: Technological Creativity and Economic Progress.* Oxford University Press, Oxford, UK.

Montroll, E.W. 1978. Social dynamics and the quantifying of social forces. *Proceedings of the National Academy of Sciences USA (October 1978 Applied Mathematical Sciences),* 75(10):4633–4637.

Montroll, E.W. 1981. On the entropy function in sociotechnical systems. *Proceedings of the National Academy of Sciences USA,* 78(12):7839–7843.

Montroll, E.W., and Badger, W.E. 1974. *Introduction to Quantitative Aspects of Social Phenomena*. Gordon and Breach, New York, NY, USA.

Mori, S., Kikuchi, J., Baba, Y., and Morino, Y. 1992. Dynamics of industrial research and development. *Journal of Scientific & Industrial Research*, **51** (August–September):658–668.

Morill, R.L. 1968. Waves of spatial diffusion. *Journal of Regional Science*, 8:1–18.

Mothes, F. 1950. Das Wachstum der Eisenbahnen. *Zeitschrift für Ökonometrie*, **1**:85–104.

Mowery, D.C., and Rosenberg, N. 1979. The influence of market demand upon innovation: A critical review of some recent empirical studies. *Research Policy*, 8:102–153.

Mumford, L. 1934. *Technics and Civilization*. Harcourt, New York, NY, USA.

Mumford, L. 1966. Technics and the nature of man. In: Smithson, J. (ed.), *Knowledge among Men*. Simon & Schuster, New York, NY, USA. (Reprinted in: Mitcham, C., and Mackey, R. (eds.). 1972. *Philosophy and Technology*. The Free Press, New York, NY, USA, pp. 77–85.)

Mylona, S. 1993. Trends of sulfur dioxide emissions, air concentrations and depositions of sulfur in Europe since 1880. EMEP/MSC-W Report 2/93. Norwegian Meteorological Institute, Oslo, Norway.

Mylona, S. 1996. Sulphur dioxide emissions in Europe 1880–1991 and their effect on sulphur concentrations and depositions. *Tellus*, **48B**:662–689.

NAE (National Academy of Engineering). 1987. *Technology and Global Industry: Companies and Nations in the World Economy*. Guile, B., and Brooks, H. (eds.), National Academy Press, Washington, DC, USA.

NPTS (Nationwide Personal Transportation Survey). 1992. Summary of Travel Trends (Report FHWA-PL-92-027) and Public Use Data Tapes. Federal Highway Administration, Washington, DC, USA.

NRC (National Research Council). 1994. *Information Technology in the Service Society: A Twenty-First Century Lever*. NRC Computer Science and Telecommunications Board, National Academy Press, Washington, DC, USA.

Nagayama, S. and Funk, J. 1985. *A Market Analysis of Japanese and U.S. Corporate and National Strategies*. Yokogawa Hokushin Electric Co., Tokyo, Japan.

Nakićenović, N. 1979. *Software Package for the Logistic Substitution Model*. RR-79-12. International Institute for Applied Systems Analysis, Laxenburg, Austria.

Nakićenović, N. 1984. Growth to limits, long waves and the dynamics of technology. Dissertation, Sozial- und Wirtschaftswissenschaftliche Fakultät, Universität Wien, Vienna, Austria (and International Institute for Applied Systems Analysis, Laxenburg, Austria).

Nakićenović, N. 1986. The automobile road to technological change. *Technological Forecasting & Social Change*, **29**:309–340.

Nakićenović, N. 1990. Dynamics of change and long waves. In: Vasko, T., Ayres, R., and Fontvieille, L. (eds.), *Life Cycles and Long Waves*. Lecture Notes in Economics and Mathematical Systems, Springer-Verlag, Berlin, Germany, pp. 147–192.

Nakićenović, N. 1996a. Freeing energy from carbon. *Dædalus*, **125**(3):95–112.

Nakićenović, N. 1996b. Technological change and learning. In: Nakićenović, N.,
    Nordhaus, W.D., Richels, R., and Toth, F.L. (eds.), *Climate Change: Inte-
    grating Science, Economics, and Policy.* CP-96-1. International Institute for
    Applied Systems Analysis, Laxenburg, Austria, pp. 271–294.

Nakićenović, N. 1997. Technological change as a learning process. Paper presented
    at the International Workshop on Induced Technical Change and the Envi-
    ronment, 26–27 June, International Institute for Applied Systems Analysis,
    Laxenburg, Austria.

Nakićenović, N., and Rogner, H.H. 1996. Financing global energy perspectives to
    2050. *OPEC Review,* **XX**(1):1–23.

Nakićenović, N., Grübler, A., Bodda, L., and Gilli, P.-V. 1990. *Technological
    Progress, Structural Change and Efficient Energy Use: Trends Worldwide and
    in Austria.* International part of a study supported by the Österreichische Elek-
    trizitätswirtschaft AG. International Institute for Applied Systems Analysis,
    Laxenburg, Austria.

Nakićenović, N., Grübler, A., Ishitani, H., Johansson, T., Marland, G., Moreira,
    J.R., and Rogner, H.H. 1996. Energy primer. In *Climate Change 1995.* In-
    tergovernmental Panel on Climate Change, Cambridge University Press, Cam-
    bridge, UK, pp. 77–92.

Nakićenović, N., Nordhaus, W.D., Richels, R., and Toth, F.L. (eds.). 1996. *Climate
    Change: Integrating Science, Economics, and Policy.* CP-96-1. International
    Institute for Applied Systems Analysis, Laxenburg, Austria.

Nasbeth, L., and Ray, G.F. (eds.). 1974. *The Diffusion of New Industrial Processes:
    An International Study.* Cambridge University Press, Cambridge, UK.

National Resources Committee (Subcommittee on Technology), US Congress. 1937.
    Technological Trends and National Policy. 75th Congress, 1st Session, House
    Document No. 360, US Government Printing Office, Washington, DC, USA.

National Science Board. 1993. *Science and Engineering Indicators 1993.* US Gov-
    ernment Printing Office, Washington, DC, USA.

Nelson, R.R., and Winter, S. 1977. In search of a useful theory of innovations.
    Research Policy, **6**:36–77.

Nelson, R.R. and Winter, S. 1982. *An Evolutionary Theory of Economic Change.*
    Harvard University Press, Cambridge, MA, USA.

Newman, L.F. (ed.). 1990. *Hunger in History.* Basil Blackwell, Cambridge, MA,
    USA.

Nilsson, S. 1994. Market trends in Europe for wastepaper: Environmental impacts
    and other peculiarities of increased recycling. In: *Recycling and Incineration
    of Waste.* Energy Research Foundation, Ås, Norway.

Nordhaus, W.D. 1973a. The allocation of energy resources. *Brookings Papers on
    Economic Activity,* **3**:529–576.

Nordhaus, W.D. 1973b. Some skeptical thoughts on the theory of induced innova-
    tion. *Quarterly Journal of Economics,* **87**:208–219.

Nordhaus, W.D. 1979. *The Efficient Use of Energy Resources.* Cowles Foundation
    Monograph 26, Yale University Press, New Haven, CT, USA.

Nordhaus, W.D. 1997. Traditional productivity measures are asleep at the (tech-
    nological) switch. *The Economic Journal,* **107**:1548–1559.

Nordhaus, W.D., and van der Heyden, L. 1983. Induced technical change: A programming approach. In: Schurr, S.H., Sonenblum, S., Wood, D.O. (eds.), *Energy, Productivity, and Economic Growth*. Oelgeschlager, Gunn & Hain, Cambridge, MA, USA, pp. 379–404.

Nordhaus, W.D., and Tobin, J. 1972. *Is Growth Obsolete?* National Bureau of Economic Research, General Series 96, Volume 5. Columbia University Press, New York, NY, USA.

Nowotny, H. 1989. *Eigenzeit: Entstehung und Strukturierung eines Zeitgefühls*. Suhrkamp, Frankfurt, Germany.

Nriagu, J.O. 1996. A history of global metal pollution. *Science*, **272**(12 April):223–224.

OECD (Organisation for Economic Co-operation and Development). 1991. The resistance to agricultural reform. *OECD Observer*, **171**(August/September): 4–8.

OECD (Organisation for Economic Co-operation and Development). 1992. *Technology and the Economy: The Key Relationships*. OECD, Paris, France.

OECD (Organisation for Economic Co-operation and Development). 1993. *OECD Environmental Data. Compendium 1993*. OECD, Paris, France.

OECD (Organisation for Economic Co-operation and Development). 1994. *Global Change of Planet Earth. Report from the OECD Megascience Forum*. OECD, Paris, France.

OECD (Organisation for Economic Co-operation and Development). 1996. *Technology and Industrial Performance: Technology Diffusion, Productivity, Employment and Skills, International Competitiveness*. OECD, Paris, France.

ÖIR (Österreichisches Institut für Raumplanung). 1972. *Simulationsmodell "Polis" - Wien*. Arb. Nr. 301.1, ÖIR, Vienna, Austria.

O'Brien, P. (ed.). 1983. *Railways and the Economic Development of Western Europe 1830–1914*. Macmillan Press, London, UK.

Ogburn, W.F. 1950. *Social Change with Respect to Culture and Original Nature*. Viking Press, New York, NY, USA (revised edition).

Orfeuil, J.P. 1993. France: A centralized country in between regional and European development. In: Salomon, I., Bovy, P., and Orfeuil, J.P. (eds.), *A Billion Trips a Day, Tradition and Transition in European Travel Patterns*. Kluwer Academic Publishers, Dordrecht, Netherlands, pp. 241–256.

Orfeuil, J.P. 1996. Personal communication. Economics of Space and Mobility Division, INRETS, Arcueil, France.

O'Riordan, T., and Cameron, J. (eds.). 1994. *Interpreting the Precautionary Principle*. Earthscan, London, UK.

Owen, J.D. 1978. Hours of work in the long run: Trends, explanations, scenarios, and implications. In: *National Commission for Manpower Policy, Work Time and Employment: A Conference Report*. Special Report No. 28, U.S. Government Printing Office No. 052-003-00686-3, Washington, DC, USA, pp. 331–364.

Owen, J.D. 1979. *Working Hours: An Economic Analysis*. D.C. Heath, Lexington, MA, USA.

Pacey, A. 1976. *The Maze of Ingenuity: Ideas and Idealism in the Development of Technology*. MIT Press, Cambridge, MA, USA.

Pavitt, K. 1984. Sectorial patterns of technical change: Towards a taxonomy and a theory. *Research Policy,* **13**:343–373.

Payson, S. 1994. *Quality Measurement in Economics: New Perspectives on the Evolution of Goods and Services.* Edward Elgar, Aldershot, UK.

Pemberton, E. 1936. The curve of culture diffusion rate. *American Sociological Review,* **1**(4):547–556.

Pepper, W., Legett, J., Swart, R., Wasson, J., Edmonds, J., and Mintzer, I. 1992. *Emission Scenarios for the IPCC. An Update: Assumptions, Methodology and Results.* Intergovernmental Panel on Climate Change (IPCC), Geneva, Switzerland.

Perez, C. 1983. Structural change and the assimilation of new technologies in the economic and social system. *Futures,* **15**(4):357–375.

Perkins, D.H. 1969. *Agricultural Development in China 1368–1968.* Edinburgh University Press, Edinburgh, UK.

Perkins, D.H., and Yusuf, S. 1984. *Rural Development in China.* The Johns Hopkins University Press, Baltimore, MD, USA.

Perrow, C. 1984. *Normal Accidents: Living with High Risk Technologies.* Basic Books, New York, NY, USA.

Peters, T.J. 1986. The mythology of innovation: A skunkworks tale. Part 1 and Part 2. *Chemtech,* May:270–276, and August:472–477.

Phelps Brown, E.H. 1973. Levels and movements of industrial productivity and real wages internationally compared, 1860–1970. *The Economic Journal,* **83**(329):58–71.

Pinch, T.J., and Bijker, W.E. 1987. The social construction of facts and artefacts: Or how the sociology of science and the sociology of technology might benefit each other. In: Bijker, W.E., Hughes, T.P., and Pinch, T. (eds.), *The Social Construction of Technological Systems.* MIT Press, Cambridge, MA, USA, pp. 17–50.

Ponting, C. 1991. *A Green History of the World: The Environment and the Collapse of Great Civilizations.* Sinclair-Stevenson Ltd., London, UK (also published in 1993 as a Penguin paperback).

Porter, M.E. 1983. The technological dimension of competitive strategy. In: Rosenblom, R.S. (ed.), *Research on Technological Innovation, Management and Policy.* JAI Press, Greenwich, UK, pp. 1–34.

Porter, M.E. 1990. *The Competitive Advantage of Nations.* Macmillan, Basingstoke, UK.

Putnam, P.C. 1954. *Energy in the Future.* Macmillan Press, London, UK.

Prieler, S., and Anderberg, S. 1996. *Assessment of Long-term Impacts of Cadmium and Lead Load to Agricultural Soils in the Upper Elbe and Oder River Basins.* WP-96-113. International Institute for Applied Systems Analysis, Laxenburg, Austria.

Putzger, F.W. 1965. *Historischer Weltatlas.* 44th edition. Österreichischer Bundesverlag, Vienna, Austria.

Quinn, J.B. 1987. The impacts of technology in the services sector. In: B. Guile and H. Brooks (eds.), *Technology and Global Industry: Companies and Nations in the World Economy.* National Academy Press, Washington, DC, USA, pp. 119–159.

Raskin, P., Hansen, E., and Margolis, R. 1995. *Water and Sustainability: A Global Outlook.* Stockholm Environment Institute (SEI), POLESTAR Series Report No. 4, SEI, Stockholm, Sweden.

Ray, G.F. 1989. Full circle: The diffusion of technology. *Research Policy,* **18**:1–18.

Reddy, A.K.N., and Goldemberg, J. 1990. Energy for the developing world. *Scientific American,* **263**(3):110–118.

Rennie, J. 1997. 13 Vehicles that went nowhere. *Scientific American,* Special Issue: The Future of Transportation, October:40–43.

Rescher, N. 1996. *Priceless Knowledge? Natural Science in Economic Perspective.* Rowman & Littlefield Publishers, Lanham, MD, USA.

Reynolds, R.V., and Pearson, A.H. 1942. *Fuel Wood Used in the United States 1630–1930.* US Department of Agriculture, Circular No. 641, February. US Department of Agriculture, Washington, DC, USA.

Richards, J.F. 1990. Land transformations. In: Turner II, B.L. *et al.* (eds.), *The Earth As Transformed by Human Action.* Cambridge University Press, Cambridge, UK, pp. 163–178.

Robinson, J.P., and Converse, P.E. 1967. *66 Basic Tables of Time Budget Research Data for the U.S.* Survey Research Center, University of Michigan, Ann Arbor, MI, USA.

Roesch, K. 1979. *3500 Jahre Stahl: Abhandlungen und Berichte, Vol. 2.* Deutsches Museum, Munich, Germany.

Rogers, E. 1962. *Diffusion of Innovations.* 1st edition. The Free Press, New York, NY, USA.

Rogers, E. 1983. *Diffusion of Innovations.* 3rd edition. The Free Press, New York, NY, USA.

Rogers, P. 1994. Hydrology and water quality. In: Meyer W.B., and Turner II, B.L. (eds.), *Changes in Land Use and Land Cover: A Global Perspective.* Cambridge University Press, Cambridge, UK, pp. 231–257.

Rogner, H.H. 1996. *An Assessment of World Hydrocarbon Resources.* WP-96-56. International Institute for Applied Systems Analysis, Laxenburg, Austria.

Romer, P.M. 1986. Increasing returns and long-run growth. *Journal of Political Economy,* **94**(5):1002–1037.

Romer, P.M. 1990. Endogenous technological change. *Journal of Political Economy,* **98**(5):S71–S102.

Rosegger, G. 1996. *The Economics of Production and Innovation: An Industrial Perspective.* 3rd edition. Butterworth Heinemann Ltd., Oxford, UK.

Rosegger, G., and Baird, R.N. 1987. Entry and exit of makes in the automobile industry, 1895–1960: An international comparison. *Omega – The International Journal of Management Science,* **15**(2):93–102.

Rosenberg, N. 1982. *Inside the Black Box: Technology and Economics.* Cambridge University Press, Cambridge, UK.

Rosenberg, N. 1990. Why do firms do basic research (with their own money)? *Research Policy,* **XIX**(2):165–174.

Rosenberg, N. 1991. *The Role of Science in Industrial Development: Some Historical Perspectives.* Mimeo, Economics Department, Stanford University, Stanford, CA, USA.

Rosenberg, N. 1996. Uncertainty and technological change. In: Landau, R., Taylor, T., and Wright, G. (eds.), *The Mosaic of Economic Growth.* Stanford University Press, Stanford, CA, USA, pp. 334–353.

Rosenberg, N., and Birdzell, L.E. 1986. *How the West Grew Rich: The Economic Transformation of the Industrial World.* I.B. Tauris & Co., London, UK.

Rosenberg, N., and Birdzell, L.E. 1990. Science, technology and the Western miracle. *Scientific American,* **263**(5):18–25.

Ross, M., Goodwin, R., Watkins, R., Wang, M.Q., and Wenzel, T. 1995. *Real-World Emissions from Model Year 1993, 2000, and 2010 Passenger Cars.* Report LBL-37977 UC-400, Lawrence Berkeley National Laboratory, University of California, Berkeley, CA, USA.

Rostow, W.W. 1960. *The Stages of Economic Growth.* Cambridge University Press, Cambridge, UK.

Rostow, W.W. 1978. *The World Economy: History and Prospect.* University of Texas Press, Austin, TX, USA.

Rozanov, B., Targulian, V., and Orlov, D. 1990. Soils. In: Turner II, B.L. *et al.* (eds.), *The Earth As Transformed by Human Action.* Cambridge University Press, Cambridge, UK, pp. 203–214.

Ruskin, J. 1884. *The Storm Cloud of the 19th Century.* Reprint: Wheeler, M. (ed.), 1995. Manchester University Press, New York, NY, USA.

Ruttan, V.W. 1996. Induced innovation and path dependence: A reassessment with respect to agricultural development and the environment. *Technological Forecasting & Social Change,* **53**(1):41–59.

Sarma, J.S., and Gandhi, V.P. 1990. *Production and Consumption of Foodgrains in India: Implications for Accelerated Economic Growth and Poverty Alleviation.* International Food Policy Research Institute, Washington, DC, USA.

Sax, E. 1920. *Die Verkehrsmittel in Volks- und Staatswirtschaft.* Vols. 1 and 2, Second revised edition, Springer-Verlag, Berlin, Germany.

Schelling, T.C. 1996. Research by accident. *Technological Forecasting & Social Change,* **53**(1):15–20.

Schilling, H.D., and Hildebrandt, R. 1977. *Die Entwicklung des Verbrauchs an Primärenergieträgern und an elektrischer Energie in der Welt, den U.S.A. und in Deutschland seit 1860 oder 1925.* Glückauf Verlag, Essen, Germany.

Schimel, D., Enting, I.G., Heiman, M., Wigley, T.M.L., Raynaud, D., Alves, D., and Siegenthaler, U. 1995. CO_2 and the carbon cycle. In: *Climate Change 1994.* Intergovernmental Panel on Climate Change, Cambridge University Press, Cambridge, UK, pp. 39–71.

Schipper, L., Bartlett, S., Hawk, D., and Wine, E. 1989. Linking life-styles and energy use: A matter of time? *Annual Review of Energy,* **14**:273–320.

Schor, J. 1991. *The Overworked American.* Basic Books, New York, NY, USA.

Schumpeter, J.A. 1911. *Theorie der wirtschaftlichen Entwicklung.* Duncker & Humblot, Leipzig, Germany.

Schumpeter, J.A. 1934. *The Theory of Economic Development: An Inquiry into Profits, Capital, Credit, Interest, and the Business Cycle.* Harvard University Press, Cambridge, MA, USA (English translation of the second [1926] edition of *Theorie der wirtschaftlichen Entwicklung*).

Schumpeter, J.A. 1935. The analysis of economic change. *Review of Economic Statistics,* **17**:2–10.

Schumpeter, J.A. 1939. *Business Cycles: A Theoretical, Historical and Statistical Analysis of the Capitalist Process.* Volumes I and II. McGraw Hill, New York, NY, USA.

Schumpeter, J.A. 1942. *Capitalism, Socialism and Democracy.* Harper & Brothers, New York, NY, USA.

Schurr, S.H., and Netschert, B.C. 1960. *Energy in the American Economy, 1850–1975.* The Johns Hopkins University Press, Baltimore, MD, USA.

Sentance, A. 1996. Innovation, imitation and growth in a changing world economy. *Economic Outlook,* May. London Business School, London, UK.

Sheldon, R.P. 1982. Phosphate rock. *Scientific American,* **246**(6):45–51.

Shiklomanov, I. 1993. World freshwater resources. In: Gleick, P.H. (ed.), *Water in Crisis: A Guide to the World's Fresh Water Resources.* Oxford University Press, New York, NY, USA, pp. 13–24.

Silverberg, G. 1991. Adoption and diffusion of technology as a collective evolutionary process. In: Nakićenović, N., and Grübler, A. (eds.), *Diffusion of Technologies and Social Behavior.* Springer-Verlag, Berlin, Germany, pp. 209–229.

Silverberg, G., and Verspagen, B. 1994. Learning, innovation and economic growth: A long-run model of industrial dynamics. *Industrial and Corporate Change,* **3**:199–223.

Silverberg, G., Dosi, G., and Orsenigo, L. 1988. Innovation, diversity and diffusion: A self-organisation model. *The Economic Journal,* **98**(December):1032–1054.

Simon, H.A. 1957. *Models of Man.* John Wiley & Sons, New York, NY, USA.

Simon, H.A. 1982. *Models of Bounded Rationality.* 2 volumes. MIT Press, Cambridge, MA, USA.

Simon, H.A. 1988. Prediction and prescription in system modeling. Paper presented at the *IIASA 1988 Conference.* International Institute for Applied Systems Analysis, Laxenburg, Austria.

Simon, H.A. (with M. Egidi, R. Marris, and R. Viale). 1992. *Economics, Bounded Rationality and the Cognitive Revolution.* Edward Elgar, Aldershot, UK.

Simon, J.L., and Khan, H. (eds.). 1984. *The Resourceful Earth: A Response to Global 2000.* Blackwell, Oxford, UK.

Singer, S.F. 1996. A preliminary critique of the IPCC's Second Assessment of Climate Change. In: Emsley, J. (ed.), *The Global Warming Debate.* European Science and Environment Forum, London, UK, pp. 146–157.

Singer, C., Holmyard, E.J., and Hall, A.R. 1954–1979. *A History of Technology.* Volumes I to VII. Oxford University Press, Oxford, UK.

Slicher van Bath, B.H. 1963a. Yield ratios 810–1820. *Afdeling Agrarische Geschiedenis: A.A.G. Bijdragen 10.* Landbouwhogeschool, Wageningen, Netherlands.

Slicher van Bath, B.H. 1963b. *The Agrarian History of Western Europe A.D. 500–1850.* Edward Arnold Ltd., London, UK.

Smil, V. 1990. Nitrogen and phosphorus. In: Turner II, B.L. *et al.* (eds.), *The Earth As Transformed by Human Action.* Cambridge University Press, Cambridge, UK, pp. 423–436.

Smil, V. 1994. *Energy in World History.* Westview Press, Boulder, CO, USA.

Smil, V. 1997. *Cycles of Life: Civilization and the Biosphere.* Scientific American Library, W.H. Freeman and Co., New York, NY, USA.

Smith, A.D. (ed.). 1986. *Technological Trends and Employment V: Commercial Service Industries.* Gower, Aldershot, UK.

Smith, K. 1995. Interactions in knowledge systems: Foundations, policy implications and empirical methods. *STI Review* No. 16:69–102.

Smith, K.R. 1988. Air pollution: Assessing total exposure in developing countries. *Environment,* **30**(10):16–35.

Smith, K.R. 1993. Fuel combustion, air pollution exposure, and health: The situation in developing countries. *Annual Review of Energy and the Environment,* **18**:529–566.

Smith, M.R., and Marx, L. 1994. *Does Technology Drive History?* MIT Press, Cambridge, MA, USA.

Socolow, R., Andrews, C., Berkhout, F., and Thomas, V. (eds.). 1994. *Industrial Ecology and Global Change.* Cambridge University Press, Cambridge, UK.

Soete, L., and Arundel, A. 1993. *An Integrated Approach to European Innovation and Technology Diffusion Policy: A Maastricht Memorandum.* Commission of the European Communities, SPRINT Programme, Luxembourg.

Solow, R.M. 1957. Technical change and the aggregate production function. *Review of Economics and Statistics,* **39**:312–320.

de Solla-Price, D.J. 1963. *Little Science, Big Science.* Columbia University Press, New York, NY, USA.

Stahel, W.R. 1994. The utilization-focused service economy: Resource efficiency and product-life extension. In: Allenby, B.R., and Richards, D.J. (eds.), *The Greening of Industrial Ecosystems.* National Academy Press, Washington, DC, USA. pp. 178–190.

Starr, C. 1990. Economic growth, social change and energy. In: Sharma, H.D. (ed.), *Energy Alternatives: Benefits and Risks.* University of Waterloo Press, Waterloo, Ontario, Canada.

Starr, C. 1996. Sustaining the human environment: The next 200 years. *Dædalus,* **125**(3):235–253.

Starr, C., and Rudman, R. 1973. Parameters of technological growth. *Science,* **182**(26 October):358–364.

Steinhart, C., and Steinhart, J. 1974. *Energy: Sources, Use, and Role in Human Affairs.* Duxbury Press, North Scituate, MA, USA.

Stern, B.J. 1937. Resistance to the adoption of technological innovations. In: Subcommittee on Technology, National Resources Committee, *Technological Trends and National Policy.* 75th Congress, 1st Session, House Document No. 360, US Government Printing Office, Washington, DC, USA, pp. 39–66.

Stigliani, W.M., Jaffé, P., and Anderberg, S. 1994. Metal loading in the environment: Cadmium in the Rhine basin. In: Socolow, R. *et al.* (eds.), *Industrial Ecology and Global Change.* Cambridge University Press, Cambridge, UK, pp. 287–296.

Strasser, S. 1982. *Never Done: A History of American Housework.* Pantheon, New York, NY, USA.

Strubegger, M., and Reitgruber, I. 1995. *Statistical Analysis of Investment Costs for Power Generation Technologies.* WP-95-109. International Institute for Applied Systems Analysis, Laxenburg, Austria.

Swedberg, R. 1991. *Joseph A. Schumpeter: His Life and Work.* Polity Press, Cambridge, UK.

Szalai, A. 1972. *The Use of Time: Daily Activities of Urban and Suburban Populations in Twelve Countries.* Mouton, The Hague, Netherlands.

TERI (Tata Energy Research Institute). 1993. *Impact of Road Transport Systems on Energy and the Environment: An Analysis of Metropolitan Cities of India.* TERI, New Delhi, India.

Tarde, G., 1895. *Les Lois de l'Imitation: Etude Sociologique.* 2nd edition. Germer Baillière, Paris, France.

Tarr, J.A., and Dupuy, G. (eds.). 1988. *Technology and the Rise of the Networked City in Europe and America.* Temple University Press, Philadelphia, PA, USA.

Thomas, W.L. (ed.). 1956. *Man's Role in Changing the Face of the Earth.* University of Chicago Press, Chicago, IL, USA.

Thompson, E.P. 1967. Time, work-discipline, and industrial capitalism. *Past and Present,* **38**:56–97.

Thompson, F.M.L. 1976. Nineteenth-century horse sense. *Economic History Review,* **29**(1):60–81.

Tikovsky, H. 1996. Personal communication. Entsorgungsbetriebe Simmering, Vienna, Austria.

Timmerman, P. 1986. Mythology and surprise in the sustainable development of the biosphere. In: Clark, W.C., and Munn, R.E. (eds.), *Sustainable Development of the Biosphere.* Cambridge University Press, Cambridge, UK, pp. 435–453.

Tinbergen, J. 1942. Zur Theorie der langfristigen Wirtschaftsentwicklung. *Weltwirtschaftliches Archiv,* **1**:511–549.

Toutain, J.-C. 1967. *Les Transports en France de 1830 à 1965.* Cahiers de l'ISEA, Serie AF, No. 9. Institut de Sciences Economiques Appliquées. Paris, France.

Toutain, J.-C. 1987. *Le Produit Interérieur Brut de la France de 1789 à 1982.* Cahiers de l'ISMEA, Serie AF, No. 15. Institut de Sciences Mathématiques et Economiques Appliquées. Paris, France.

Toynbee, A. 1896. *Lectures on the Industrial Revolution in the 18th Century in England.* 3rd edition, Longman, Greens & Co., London, UK. Reprint 1956: Beacon, Boston, MA, USA.

von Tunzelmann, G.N. 1982. Structural change and leading sectors in British manufacturing 1907–1968. In: Kindleberger, C.P., and di Tella, G. (eds.), *Economics in the Long View, Vol. 3, Part 2: Applications and Cases.* New York University Press, New York, NY, USA, pp. 1–49.

von Tunzelmann, G.N. 1995. *Technology and Industrial Progress: The Foundations of Economic Growth.* Edward Elgar, Aldershot, UK.

Turner II, B.L., Clark, W.C., Kates, R.W., Richards, J.F., Mathews, J.T., and Meyer, W.B. (eds.). 1990. *The Earth as Transformed by Human Action.* Cambridge University Press, Cambridge, UK.

UN (United Nations). 1952. *World Energy Supplies in Selected Years 1929–1950.* Statistical Office of the United States, Department of Economic Affairs, Statistical Papers Series J No. 1. UN, New York, NY, USA.

UN (United Nations). Various volumes (1950–1995). *Statistical Yearbook.* UN, New York, NY, USA.

UN (United Nations) Energy Statistics. Various volumes (1973–1995). *Energy Statistics Yearbook.* UN, New York, USA.

UN (United Nations). 1980. *Patterns of Urban and Rural Population Growth.* UN, New York, NY, USA.

UN (United Nations). 1996. *Annual Populations 1950–2050: The 1996 Revision.* UN Population Division, United Nations, New York, NY, USA (data on diskettes).

UNDP (United Nations Development Programme). 1993. *Human Development Report 1993.* Oxford University Press, Oxford, UK.

UNEP (United Nations Environmental Programme). 1991. *Freshwater Pollution.* UNEP, Nairobi, Kenya.

UNEP (United Nations Environmental Programme). 1993. *Environmental Data Report 1993–1994.* Blackwell Publishers, Oxford, UK.

US DOC (United States Department of Commerce). 1975. *Historical Statistics of the United States: Colonial Times to 1970* (2 volumes). US DOC, Bureau of the Census, Washington, DC, USA.

US DOC (United States Department of Commerce). Various volumes (1970–1996). *Statistical Abstract of the United States.* US DOC Bureau of the Census, Washington, DC, USA.

US NRC (National Research Council). 1983. *Risk Assessment and the Federal Government: Managing the Process.* National Academy Press, Washington, DC, USA.

US Minerals Yearbook (United States Bureau of Mines, Department of the Interior). Various volumes (1905–1995). *Minerals Yearbook.* US Bureau of Mines, Washington, DC, USA.

Vanek, J. 1974. Time spent in housework. *Scientific American,* November:116–120.

Vasko, T. (ed.). 1987. *The Long-Wave Debate.* Springer-Verlag, Berlin, Germany.

Veblen, T. 1904. *The Theory of Business Enterprise.* Scribner, New York, NY, USA.

Veblen, T. 1921. *The Engineers and the Price System.* Huebsch, New York, NY, USA.

Veblen, T. 1953. *The Theory of the Leisure Class: An Economic Study of Institutions.* Revised edition. New American Library, New York, NY, USA.

Victor, D. 1993. Overt diffusion as technology transfer. *Energy,* **18**(5):535–538.

Victor, D., and Salt, J. 1995. *Keeping the Climate Treaty Relevant: An Elaboration.* Mimeo, International Institute for Applied Systems Analysis, Laxenburg, Austria.

Virtanen, Y., and Nilsson, S. 1993. *Environmental Impacts of Waste Paper Recycling.* Earthscan, London, UK.

WEC (World Energy Council). 1993a. *Energy for Tomorrow's World.* WEC Commission global report, Kogan Page, London, UK.

WEC (World Energy Council). 1993b. *Renewable Energy Resources: Opportunities and Constraints 1990–2020.* WEC, London, UK.

WHO (World Health Organization) and UNEP (United Nations Environment Programme). 1993. *Urban Air Pollution in Megacities of the World.* 2nd edition. Blackwell Publishers, Oxford, UK.

WRI (World Resources Institute). 1990. *World Resources 1990–91.* WRI, Washington, DC, USA.

WRI (World Resources Institute). 1992. *World Resources 1992–93.* WRI, Washington, DC, USA.

WRI (World Resources Institute). 1996. *World Resources 1996–97.* WRI, Washington, DC, USA.

WRI (World Resources Institute). 1997a. *Resource Flows: The Material Basis of Industrial Economies.* WRI, Washington, DC, USA.

WRI (World Resources Institute). 1997b. *Worldwatch Data Base Disk.* WRI, Washington, DC, USA.

Waggoner, P.E. 1996a. How much land can ten billion people spare for nature? In: Ausubel, J.H., and Langford, H.D. (eds.), *Technological Trajectories and the Human Environment.* National Academy Press, Washington, DC, USA, pp. 56–73.

Waggoner, P.E. 1996b. How much land can be spared for nature? *Dædalus,* **125**(3): 73–93.

Walsh, V. 1984. Invention and innovation in the chemical industry: Demand pull or discovery push? *Research Policy,* **13**:211–234.

Ward, W.H. 1967. The sailing ship effect. *Bulletin of the Institute of Physics and the Physical Society,* **18**:169.

Watanabe, C. 1995. Identification of the role of renewable energy. *Renewable Energy,* **6**(3):237–274.

Watanabe, C. 1997. Personal communication. Tokyo Institute of Technology, based on MITI statistics.

Weinberg, A. 1985. "Immortal" energy systems and intergenerational justice. *Energy Policy,* **13**(1):51–59.

Wernick, I.K. 1996. Consuming materials: The American way. *Technological Forecasting & Social Change,* **53**(1):111–122.

Wernick, I.K. 1997. Personal communication. Rockefeller University, New York, NY, USA.

Wernick, I.K., and Ausubel, J.H. 1995. National materials flows and the environment. *Annual Review of Energy and the Environment,* **20**:463–492.

Wernick, I.K., Herman, R., Govind, S., and Ausubel, J.H. 1996. Materialization and dematerialization: Measures and trends. *Dædalus,* **125**(3):171–198.

White, L. 1967. The historical roots of our ecological crisis. *Science,* **CLV**(10 March):1203–1207.

Wigley, T., Richels, R., and Edmonds, J. 1996. Economic and environmental choices in the stabilization of atmospheric CO_2 concentrations. *Nature,* **379**:240–243.

Williams, R.H., Larson, E.D., and Ross, M.H. 1987. Materials, affluence, and industrial energy use. *Annual Review of Energy,* **12**:99–144.

Williamson, O.E., and Masten, S.E. 1995. *Transaction Cost Economics.* Volume I. Edward Elgar, Aldershot, UK.

Wittfogel, K.A. 1957. *Oriental Despotism: A Comparative Study of Total Power.* Yale University Press, New Haven, CT, USA.

Woodwell, G.M. (ed.). 1990. *The Earth in Transition: Patterns and Processes of Biotic Impoverishment.* Cambridge University Press, Cambridge, UK.

World Bank. 1992. *World Development Report 1992: Development and the Environment.* Oxford University Press, Oxford, UK.

World Bank. Various volumes (1990–1996). *World Tables.* The Johns Hopkins University Press, Baltimore, MD, USA.

Woytinsky, W.L. 1926. *Die Welt in Zahlen, Vol. 3: Die Landwirtschaft.* Rudolf Mosse Verlag, Berlin, Germany.

Woytinsky, W.L. 1927. *Die Welt in Zahlen, Vol. 5: Handel und Verkehr.* Rudolf Mosse Verlag, Berlin, Germany.

Woytinsky, W.L., and Woytinsky, E.S. 1953. *World Population and Production: Trends and Outlook.* The Twentieth Century Fund, New York, NY, USA.

Wright, T.P. 1936. Factors affecting the costs of airplanes. *Journal of the Aeronautical Sciences,* **3**(February):122–128.

Yates, R.D.S. 1990. War, food shortages, and relief measures in early China. In: Newman, L.F. (ed.), *Hunger in History.* Basil Blackwell, Cambridge, MA, USA, pp. 147–177.

Ybema, J.R., Boonekamp, P.G.M., and Smit, J.T.J., 1995. *Including Climate Change in Energy Investment Decisions.* ECN-C-95-073. Netherlands Energy Foundation, Petten, Netherlands.

Young, M. 1988. *The Metronomic Society: Natural Rhythms and Human Timetables.* Thames and Hudson, London, UK.

Young, M., and Schuller, T. 1991. *Life After Work.* Harper Collins, Glasgow, UK.

Zimmermann, E.W. 1951. *World Resources and Industries.* Harper, New York, NY, USA.

Index

NOTE:
b = box
f = figure
n = note
t = table

Made in the USA
Middletown, DE
28 September 2016